Secure Localization and Time Synchronization for Wireless Sensor and Ad Hoc Networks

Advances in Information Security

Sushil Jajodia

Consulting Editor
Center for Secure Information Systems
George Mason University
Fairfax, VA 22030-4444
email: jajodia@gmu.edu

The goals of the Springer International Series on ADVANCES IN INFORMATION SECURITY are, one, to establish the state of the art of, and set the course for future research in information security and, two, to serve as a central reference source for advanced and timely topics in information security research and development. The scope of this series includes all aspects of computer and network security and related areas such as fault tolerance and software assurance.

ADVANCES IN INFORMATION SECURITY aims to publish thorough and cohesive overviews of specific topics in information security, as well as works that are larger in scope or that contain more detailed background information than can be accommodated in shorter survey articles. The series also serves as a forum for topics that may not have reached a level of maturity to warrant a comprehensive textbook treatment.

Researchers, as well as developers, are encouraged to contact Professor Sushil Jajodia with ideas for books under this series.

Additional titles in the series:

Additional information about this series can be obtained from
http://www.springer.com

Secure Localization and Time Synchronization for Wireless Sensor and Ad Hoc Networks

Edited by

Radha Poovendran
University of Washington, USA

Cliff Wang
Army Research Office, USA

Sumit Roy
University of Washington, USA

Radha Poovendran
University of Washington
Dept. Computer Science & Engineering
P.O.Box 352350
Seattle WA 98195
radha@ee.washington.edu

Cliff Wang
Computing and Information Science Div.
U.S. Army Research Office
P.O. Box 12211
Research Triangle Park, NC 27709-2211
cliff.wang@us.army.mil

Sumit Roy
University of Washington
Dept. Computer Science & Engineering
P.O.Box 352350
Seattle WA 98195
roy@ee.washington.edu

Secure Localization and Time Synchronization for Wireless Sensor and
Ad Hoc Networks edited by Radha Poovendran, Cliff Wang, Roy Sumit

ISBN-13: 978-1-4419-4096-4

e-ISBN-13: 978-0-387-46276-9
e-ISBN-10: 0-387-46276-7

Printed on acid-free paper.

9 8 7 6 5 4 3 2 1

springer.com

Foreword

During the past three decades, every major advance in computing introduced new and largely unanticipated security challenges. Wireless sensor networks are only the latest technology that confirms this observation. These networks, which represent a basic tenet of what we call ubiquitous computing, are now or will soon be deployed in physical environments that are vulnerable not only to the vicissitudes of nature but also to acts that could be easily viewed as hostile attacks by potent adversaries. Indeed, unattended operation of sensor-network nodes in hostile environments requires that we rethink the definition of our adversary, its capabilities and modes of attack.

There are few problems of wireless sensor network design and analysis that are as challenging as localization and time synchronization. Yet both are fundamental building blocks not just for new applications and but also security services themselves. Localization complexity is, to a significant degree, the result of deployment and operation in environments that lack of unobstructed line-of-site connectivity, reference points, and communications. Further, time synchronization gains added complexity due to the limited computing resources sensor nodes possess. As a consequence, the natural interplay between space and time measurements and bounds, which are basic to both localization and time synchronization, produces a largely uncharted research territory. And, of course, the new capabilities and attack modes of the new adversary complicates the landscape in unanticipated ways.

This book represents a snapshot of our understanding in solving problems of robust, resilient and secure localization and time synchronization at the inception of the sensor network technology development. It offers a clear view of the essential challenges posed by localization and time synchronization in sensor networks, subtleties of potential solutions, and extensive discussion of specific protocols and mechanisms required by these solutions. In short, the book is an indispensable reference to both researchers and developers, and an invaluable aid to students.

I am pleased and honored to have been asked to write the foreword for this book. The authors, all active researchers in the area of sensor network security, should be congratulated for providing this valuable reference book for the research community.

September 2006, College Park, Maryland   *Virgil D. Gligor*

Preface

This book is an outcome of a special workshop on Localization in Wireless Sensor Networks, held between June 13-14 of 2005, at the University of Washington, Seattle.

During several technical discussions, Dr. Radha Poovendran of University of Washington and ARO Information Assurance (IA) program director Dr. Cliff Wang felt that robust and resilient localization for wireless sensor networks is an important research area and a special workshop was needed to address the research challenges and to promote innovative ideas for solutions. Dr. Sumit Roy from the University of Washington later joined the organizing committee. The workshop was organized and held successfully. Over 30 researchers participated in the workshop and a total of 18 presentations were made, covering various aspects of the localization problem.

This book is a direct outcome of this special workshop. We have also expanded the scope of this book to include secure time synchronization since the techniques used for localization distance bounding protocols are dependent on correct time synchronization of wireless sensor networks. A total of sixteen contributed papers are received from both workshop participants and researchers active in wireless sensor network research. The collection of these high quality papers makes this edited volume a valuable resource for both researchers and engineers in related fields. We believe that this book will serve as a reference as well as the starting point of research in the exciting areas of secure location estimation, secure time synchronization, verification of sensor security protocols, and location privacy.

The book is organized into three parts. The chapters in Part I present approaches for sensor location estimation under a benign environment and technical discussions focus on the quality of location estimation. The chapters in the Part II of the book contain the latest work on resilient sensor location estimation in the presence of an adversary that may inject Byzantine errors into the localization process. Also in Part II of the book, there is one chapter dedicated to distance bounding protocol verification and there is another chapter that focuses specifically on privacy protection against location tracking. The Part III of the book contains chapters addressing the problem of secure time synchronization in wireless sensor networks.

We would like to express our thanks to Professor Sushil Jajodia for including this book in his series. We thank Susan Lagerstrom-Fife and Sharon Palleschi of Springer, and Krishna Sampigethaya of University of Washington for working closely with us during the production of this book. We also thank Krishna Sampigethaya, Loukas Lazos, Mingyan Li, Patrick Tague, and Javier Salido for their help and support during the workshop.

<table>
<tr><td>September 2006, University of Washington</td><td>Radha Poovendran</td></tr>
<tr><td>September 2006, ARO/NCSU</td><td>Cliff Wang</td></tr>
<tr><td>September 2006, University of Washington</td><td>Sumit Roy</td></tr>
</table>

Contents

List of Contributors

Prathima Agrawal
200, Broun Hall
Auburn University
Auburn, Alabama 36830
agrawpr@auburn.edu

Farooq Anjum
Telcordia Technologies
One Telcordia Drive
Piscataway, NJ 08854
fanjum@telcordia.com

Nirupama Bulusu
Department of Computer Science
Portland State University
Portland, OR 97207-0751
nbulusu@cs.pdx.edu

Srdjan Čapkun
Informatics and Mathematical Modelling Department,
Technical University of Denmark
DK-2800 Lyngby, Denmark
sca@imm.dtu.dk

LiWu Chang
U.S. Naval Research Laboratory,
Code 5543
Washington, DC 20375
lchang@itd.nrl.navy.mil

Jose A. Costa
Center for the Mathematics of
Information
California Institute of Technology
1200 E. California Blvd.
Pasadena, CA 91106
jcosta@caltech.edu

Thanh X Dang
Department of Computer Science
Portland State University
Portland, OR 97207-0751
dangtx@cs.pdx.edu

Wenliang Du
Department of Electrical Engineering
and Computer Science
Syracuse University
3-114 Sci-Tech Building
Syracuse, NY 13244
wedu@ecs.syr.edu

Lei Fang
Department of Electrical Engineering
and Computer Science
Syracuse University
3-114 Sci-Tech Building
Syracuse, NY 13244
lefang@ecs.syr.edu

Saurabh Ganeriwal
Networked and Embedded Systems Lab
University of California
Los Angeles, CA 90095-1594
saurabh@ee.ucla.edu

Simon Han
Networked and Embedded Systems Lab
University of California
Los Angeles, CA 90095-1594
simonhan@ee.ucla.edu

Tian He
Department of Computer Science and
Engineering
University of Minnesota
200 Union Street SE
Minneapolis, MN 55455
tianhe@cs.umn.edu

Alfred O. Hero III
Department of Electrical Engineering
and Computer Science
University of Michigan
1301 Beal Avenue
Ann Arbor, MI 48109-2122
hero@umich.edu

Leping Huang
Nokia Research Center/University of
Tokyo
1-8-1, Shimomeguro, Meguro-ku
Tokyo, Japan
leping.huang@nokia.com

Sanjay Jha
School of Computer Science and
Engineering,
University of New South Wales
Sydney 2052, Australia
sjha@cse.unsw.edu.au

Manali Joglekar
WINLAB
Rutgers University
671 Route 1 South
North Brunswick, N.J. 08902-3390
manali@winlab.rutgers.edu

Bhaskar Krishnamachari
Department of Electrical Engineering-
Systems
University of Southern California
3740 McClintock Avenue
Los Angeles, CA 90089
bkrishna@usc.edu

Loukas Lazos
Network Security Lab
Department of Electrical Engineering
Box 352500
University of Washington
Seattle, WA 98195-2500
llazos@u.washington.edu

Zang Li
WINLAB
Rutgers University
671 Route 1 South
North Brunswick, N.J. 08902-3390
zang@winlab.rutgers.edu

Donggang Liu
Department of Computer Science and
Engineering
University of Texas at Arlington
330 Nedderman Hall
Arlington, Texas 76019-0015
dliu@cse.uta.edu

Dimitrios Lymberopoulos
Department of Electrical Engineering
Yale University
51 Prospect St.
New Haven, CT 06511
dimitrios.lymberopoulos
@yale.edu

Michael Manzo
Department of Electrical Engineering
and Computer Sciences
University of California at Berkeley
333 Cory Hall
Berkeley, CA 94720
mike@manzo.org

Kanta Matsuura
University of Tokyo
4-6-1 Komaba, Meguro-ku
Tokyo, Japan
kanta@iis.u-tokyo.ac.jp

Catherine Meadows
U.S. Naval Research Laboratory,
Code 5543
Washington, DC 20375
meadows@itd.nrl.navy.mil

Badri Nath
Computer Science Department
Rutgers University
110 Frelinghuysen Road
Piscataway, NJ 08854
badri@cs.rutgers.edu

Peng Ning
Department of Computer Science
North Carolina State University
890 Oval Dr.
Raleigh, NC 27695-8206
pning@ncsu.edu

Santosh Pandey
200, Broun Hall
Auburn University
Auburn, Alabama 36830
pandesg@auburn.edu

Pubudu N Pathirana
School of Engineering and Technology
Deakin University
Geelong 3217, Australia
pubudu@deakin.edu.au

Neal Patwari
Department of Electrical Engineering
and Computer Science
University of Michigan
1301 Beal Avenue
Ann Arbor, MI 48109-2122
npatwari@umich.edu

Dusko Pavlovic
Kestrel Institute,
3260 Hillview Avenue
Palo Alto, CA 94304
dusko@kestrel.edu

Radha Poovendran
Network Security Lab
Department of Electrical Engineering
Box 352500
University of Washington
Seattle, WA 98195-2500
rp3@u.washington.edu

Tanya Roosta
Department of Electrical Engineering
and Computer Sciences
University of California at Berkeley
333 Cory Hall
Berkeley, CA 94720
roosta@eecs.berkeley.edu

Shankar Sastry
Department of Electrical Engineering
and Computer Sciences
University of California at Berkeley
514 Cory Hall
Berkeley, CA 94720
sastry@eecs.berkeley.edu

Andrey V Savkin
School of Electrical Engineering and
Telecommunications
University of New South Wales
Sydney 2052, Australia
a.savkin@unsw.edu.au

Andreas Savvides
Department of Electrical Engineering
Yale University
51 Prospect St.
New Haven, CT 06511
andreas.savvides@yale.edu

Kaoru Sezaki
University of Tokyo
4-6-1 Komaba, Meguro-ku
Tokyo, Japan
sezaki@iis.u-tokyo.ac.jp

Mani Srivastava
Networked and Embedded Systems Lab
University of California
Los Angeles, CA 90095-1594
mbs@ee.ucla.edu

John A. Stankovic
Department of Computer Science
University of Virginia
151 Engineer's Way, P.O. Box 400740
Charlottesville, VA 22904-4740
stankovic@cs.virginia.edu

Radu Stoleru
Department of Computer Science
University of Virginia
151 Engineer's Way, P.O. Box 400740
Charlottesville, VA 22904-4740
stoleru@cs.virginia.edu

Kun Sun
Department of Computer Science
North Carolina State University
890 Oval Drive
Raleigh, NC 27695-8206
ksun3@ncsu.edu

Paul Syverson
U.S. Naval Research Laboratory,
Code 5543
Washington, DC 20375
syverson@itd.nrl.navy.mil

Wade Trappe
WINLAB
Rutgers University
671 Route 1 South
North Brunswick, N.J. 08902-3390
trappe@winlab.rutgers.edu

Cliff Wang
Army Research Office
4300 S Miami Blvd.
RTP, NC 27709
cliff.wang@us.army.mil

Hiroshi Yamane
University of Tokyo
4-6-1 Komaba, Meguro-ku
Tokyo, Japan
yamane@mcl.iis.u-
tokyo.ac.jp

Kiran Yedavalli
Department of Electrical Engineering-
Systems
University of Southern California
3740 McClintock Avenue
Los Angeles, CA 90089
kyedaval@usc.edu

Yanyong Zhang
WINLAB
Rutgers University
671 Route 1 South
North Brunswick, N.J. 08902-3390
yyzhang@winlab.rutgers.edu

Part I

Localization Techniques

Range-Free Localization

Radu Stoleru[1], Tian He[2] and John A. Stankovic[3]

[1] Department of Computer Science, University of Virginia,
stoleru@cs.virginia.edu
[2] Department of Computer Science and Engineering, University of Minnesota
tianhe@cs.umn.edu
[3] Department of Computer Science, University of Virginia,
stankovic@cs.virginia.edu

1 Introduction

Advances in micro-electro-mechanical systems have triggered an enormous interest in wireless sensor networks (WSN). WSN are formed by large numbers of densely deployed nodes enabled with sensing and actuating capabilities. These nodes have very limited processing and memory capabilities, limited energy resources and it is envisioned that they will be mass produced, to reduce costs.

Several challenging problems exist in wireless sensor networks. Among these is how to obtain location information for sensor nodes and events present in the network. From this perspective, we categorize the localization problem as: node localization, target localization and location service. Node localization is the process of determining the coordinates of the sensor nodes in the WSN. Target localization is the process of obtaining the coordinates of an event or a target present in the sensor network. The location of a target can be obtained either passively (the nodes sense the target) or actively, when the target cooperates and communicates with the sensor network. A location service acts as a repository that can be used to answer questions like "where is entity X?". In the remaining part of this chapter we focus on the node localization problem in WSN.

Node localization is a complicated and important problem for wireless sensor networks (WSN). The aspects of this problem that have challenged the research community can be summarized as follows:

- **Assumptions** - The node localization problem remains a difficult challenge to be solved practically. To make the problem practically tractable, its complexity had to be reduced, by making simplifying assumptions. As a result, many localization schemes proposed solutions that are based on assumptions that do not always hold or are not practical. Examples of such assumptions are: circular radio range, symmetric radio connectivity, additional hardware (e.g., ultrasonic), lack of obstructions, lack of line-of-sight, no multipath and flat terrain.

- **Localization Protocol Design** - The problem of localization in WSN is further complicated by the large number of parameters that need to be considered when designing a localization system for a particular WSN deployment. Among these parameters are: the deployment method for the sensor network; the existence of a line-of-sight between sensor nodes and a remote, central point; the time required by the localization scheme; the presence of reference points (anchors) in the network, and the density; the cost for localization, represented by additional hardware (form factor) and energy expenditure (messages exchanged or time necessary for localization).

- **Cost/Accuracy trade-off** - Due to the mostly static nature of many WSN, obtaining the location information by each sensor node is often a one time or rare event. Adding hardware to each sensor node, to assist in the localization, is a costly solution, and, so far, has been ruled out from real system deployments. For example, GPS is a typical high-end solution, which requires sophisticated hardware to achieve high resolution time synchronization with satellites. The constraints on power and cost for tiny sensor nodes and the need for a line of sight from a sensor node to four or more satellites preclude this as a viable solution. Other solutions require per node devices that can perform ranging among neighboring nodes. The difficulties of these approaches are two-fold. First, under constraints of form factor and power supply, the effective ranges of such devices are very limited. For example the effective range of an ultrasonic transducer is on the order of a few meters, when the sender and receiver are not facing each other. Second, since most sensor nodes are static, i.e., the location is not expected to change, it is not cost-effective to equip these sensors with special circuitry just for a one-time localization.

- **Performance Evaluation** - The problem of localization in wireless sensor networks has been studied and evaluated predominantly in simulators. Due to the severe hardware constraints imposed on wireless sensor nodes, real system implementations of the proposed simulated solutions have not produced encouraging results. Solutions that use the most tempting means of evaluating relative distances between sensor nodes - RF signal strength, have largely failed in practice, due to the unreliable nature and irregular pattern of the radio communication. Localization schemes that are based on the receive signal strength indicator (RSSI) have been, however, intensively studied in simulators.

- **Security** - Since localization is a critical factor in WSN, attacks on it can render the sensor network ineffective. To date, very little work has been done on creating robust and secure localization schemes. A few notable exceptions are [15] [14] [17] [16] [5].

For wireless sensor networks ranging is a difficult option. The hardware cost (hardware used only for localization), the energy expenditure, the form factor, the small range, all are difficult compromises, and it is hard to envision cheap, unreliable and resource-constraint devices make use of range-based localization solutions. Their high accuracy in localization is very desirable, however.

To overcome the limitations of the range-based localization schemes, many range-free solutions have been proposed. These solutions estimate the location of sensor nodes by, either, exploiting the radio connectivity information among neighboring nodes, or exploiting the sensing capabilities that each sensor node possesses. Due to the distinct characteristics of these two approaches, we categorize the range-free localization schemes into: anchor-based schemes (which assume the presence of sensor nodes in the network that have knowledge about their location) and anchor-free schemes, which require no special sensor nodes for localization. The range-free localization schemes eliminate the need of high-cost specialized hardware on each sensor node. The fact that the radio propagation characteristics vary over time and are environment dependent, imposes higher calibration costs for the anchor-based localization schemes.

In this chapter we review a *representative* set of range-free localization schemes, from the perspective of the above proposed taxonomy: anchor-based and anchor-free solutions. We point out that hybrid solutions exist and, sometimes, one solution does not neatly fit in either one of the categories. Also, in addition to the localization schemes described below, many more have been proposed. To name a few: the ELA [32], Thunder [35], Hop-TERRAIN [26], KPS [7], RIPS [18], Resilient LSS [13], Robust Quadrilaterals [19] and MAL [23] . In the remaining part of this chapter, we use R to denote the radio range of a sensor node.

2 Anchor-Based Solutions

The location of a sensor node has to be expressed in a coordinate system. In a 2D space, three anchor nodes (three fixed points in the space) uniquely determine a coordinate system. In a 3D space, four anchor nodes are required. In this section, to demonstrate a wide range of possible solutions, we present several range-free localization schemes that use radio connectivity to infer proximity to a set of anchor nodes.

2.1 Centroid

The Centroid scheme was proposed by Bulusu et al. in [2]. This localization scheme assumes that a set of anchor nodes (A_i, $1 \leq i \leq n$), with overlapping regions of coverage, exist in the deployment area of the WSN. The main idea is to treat the anchor nodes, located at (X_i, Y_i), as point masses m_i and to find the center of gravity (centroid) of all these masses. In the most general form, the coordinates of the centroid of n point masses m_i are given by:

$$(X_G, Y_G) = \left(\frac{\sum_{i=1}^{n} m_i X_i}{\sum_{i=1}^{n} m_i}, \frac{\sum_{i=1}^{n} m_i Y_i}{\sum_{i=1}^{n} m_i} \right)$$

which, for equal masses m_i simplifies to:

$$(X_G, Y_G) = \left(\frac{\sum_{i=1}^{n} X_i}{n}, \frac{\sum_{i=1}^{n} Y_i}{n} \right)$$

An example of how the Centroid scheme works is shown in Figure 1, where a sensor node N_k is within communication range to four anchor nodes, $A_1...A_4$. The node N_k localizes itself to the centroid of the quadrilateral $A_1 A_2 A_3 A_4$ (for the case of a quadrilateral, the centroid is at the point of intersection of the bimedians - the lines connecting the middle points of opposite sides).

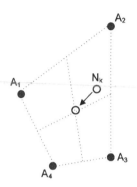

Fig. 1. Centroid localization - node N_k localizes to the centroid of the $A_1 A_2 A_3 A_4$ quadrilateral.

The steps of the localization scheme are the following:

- Each anchor node A_i broadcasts its position.
- Each sensor node N_k listens for beacons from anchors and computes a connectivity metric, for each anchor node A_i it has received beacons from. This metric is defined as follows:

$$CM_{k,A_i} = \frac{N_{recv}(A_i, t)}{N_{sent}(A_i, t)}$$

where $N_{recv}(A_i, t)$ and $N_{sent}(A_i, t)$ are the numbers of beacons received from anchor A_i and sent by anchor A_i, respectively, in the time interval t.

- A node N_k computes its location as the average of all the anchor nodes A_i it has heard from with a connectivity higher than a threshold, e.g., $CM_{k,A_i} 90\%$, as follows:

$$(X_k, Y_k) = \left(\frac{X_{A_{i1}} + ... + X_{A_{ij}}}{j}, \frac{Y_{A_{i1}} + ... + Y_{A_{ij}}}{j} \right)$$

where j is the number of anchors with a higher connectivity than the threshold.

In subsequent work [3], the authors explored adaptive mechanisms for placing additional anchor nodes in a WSN, in order to reduce the average localization error.

2.2 APIT

APIT [8] is an area-based range-free localization scheme. It assumes that a small number of nodes, called anchors, are equipped with high-powered transmitters and know their location, obtained via GPS or some other mechanism. Using beacons from these anchors, APIT employs a novel area-based approach to perform location estimation by isolating the environment into triangular regions between anchor nodes as shown in Figure 2. A node's presence inside or outside of these triangular regions allows a node to narrow down the area in which it can potentially reside. By utilizing different combinations of anchors, the size of the estimated area in which a node resides can be reduced, to provide a good location estimate.

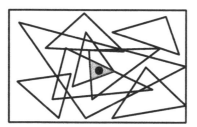

Fig. 2. Area-based APIT Algorithm Overview

The theoretical method used to narrow down the possible area in which a target node resides is called the *Point-In-Triangulation Test* (PIT). For three given anchors: $A(a_x, a_y), B(b_x, b_y), C(c_x, c_y)$, the Point-In-Triangulation test determines whether a point M with an unknown position is inside triangle $\triangle ABC$ or not. APIT repeats this PIT test with different anchor combinations until all combinations are exhausted or the required accuracy is achieved. At this point, APIT calculates the center of gravity (COG) of the intersection of all of the triangles in which a node resides to determine its estimated position. These steps are shown in Algorithm 1.

Algorithm 1 APIT

1: Receive location beacons (X_i, Y_i) from N anchors;
2: $InsideSet = \emptyset$;
3: **for** each triangle $T_i \in \binom{N}{3}$ triangles **do**
4:     **if** Point-In-Triangle-Test (T_i) == TRUE **then**
5:         $InsideSet = InsideSet \bigcup T_i$;
6:     **end if**
7: **end for**
8: Estimated Position = CenterOfGravity($\bigcap T_i \in InsideSet$);

In [8], the authors provide a perfect, albeit theoretical, solution for perfect Point-In-Triangulation test as follows:

Perfect P.I.T. Test Theory: If there exists a direction such that a point adjacent to M is further/closer to points A, B, and C simultaneously, then M is outside of △ABC. Otherwise, M is inside △ABC (Figure 3).

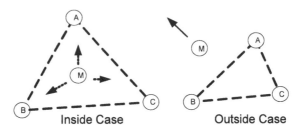

Fig. 3. Cases for Point-In-Triangulation Test

The perfect P.I.T. test can correctly decide whether a point M is inside triangle △ABC (the formal proofs can be found in [8]). However, there are two major issues to apply this theory practically in wireless sensor networks: First, how does a node recognize directions of departure from an anchor without moving? Second, how to exhaustively test all possible directions in which node M might depart/approach vertexes A, B, C simultaneously? The answer to the first question is to use RSSI comparisons. Through experiments, the authors confirm that in a narrow direction, the further away a node is from the anchor, the weaker the received signal strength (RSSI) will be. Through signal strength comparisons, a node can determine whether the direction towards a neighboring node is closer to a given anchor or not. To address the second issue, the authors propose an approximation (APIT) for the perfect PIT test, which uses neighbor information, through RSSI comparisons, to emulate the node movement in the Perfect PIT test. With a finite number of neighbors, APIT can only evaluate a limited number of directions. Consequently, APIT could make an incorrect decision. Fortunately, experiments indicate that the percentage of APIT tests exhibiting such an error is relatively small (14% in the worst case). Figure 4 demonstrates this error percentage as a function of node density. When node density increases, APIT can evaluate more directions, considerably reducing false positive, i.e. APIT returns true, while a node is outside of triangle (OutToInError). On the other hand, false negative (InToOutError) will slightly increase due to the increased chance of edge effects.

Once the individual APIT tests finish, APIT aggregates the results (inside/outside decisions among which some may be incorrect) through a grid SCAN algorithm (Figure 5). In this algorithm, a grid array is used to represent the maximum area in which a node likely resides. In the experiments, the length of a grid side is set to 0.1R, to guarantee that estimation accuracy is not noticeably compromised.

For each APIT inside decision (a decision where the APIT test determines the node is inside a particular region (Figure 5) the values of the grid regions over which the corresponding triangle resides are incremented. For an outside decision, the grid area is similarly decremented. Once all triangular regions are computed, the resulting

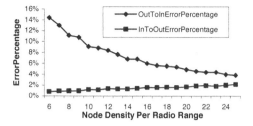

Fig. 4. APIT Error under Varying Node Densities

0	0	0	0	0	0	1	0	0	0
0	0	1	0	1	1	1	0	0	0
0	0	1	1	1	1	1	0	0	0
0	1	2	2	1	1	0	1	0	0
1	1	2	2	1	1	0	-1	-1	0
0	0	2	2	2	1	0	-1	-1	-1
0	0	1	1	1	0	0	-1	-1	-1

Fig. 5. APIT Error under Varying Node Densities

information is used to find the maximum overlapping area (e.g., the grid area with value 2 in Figure 5). Since the majority (more than 85% in the worst case shown in Figure 4) of APIT tests are correct, the correct decisions build up on the grid and the small number of errors only serves as a slight disturbance to the final estimation. Evaluation in [8] indicates APIT works better than other range-free solutions under irregular radio patterns and random node placement. However, it should be pointed out that APIT has a more demanding requirement on the number of anchors used in localization.

2.3 SeRLoc

SeRLoc [15] is another area-based range-free localization. The authors assume two types of nodes: normal nodes and locators (i.e., anchors). Normal nodes are equipped with omnidirectional antennas, while locators are equipped with directional sectored antennas and their locations of locators are known a priori. In SeRLoc, a sensor estimates it location based on the information transmitted by the locators. Figure 6 shows the main idea, with node N_k within radio range to locators A_1, A_2 and A_3:

SeRLoc localizes the sensor nodes in four steps. First, a locator transmits directional beacons within a sector. Each beacon contains the locator's position and the angles of the sector boundary lines. A normal node collects the beacons from all locators it hears. Second, it determines an approximate search area within which it is located based on the coordinates of the locators heard. Third, it computes the overlapping sector region using a majority vote scheme. Finally, SeRLoc determines a node location as the center of gravity of the overlapping region. These steps are shown in Algorithm 2.

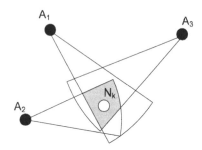

Fig. 6. SeRLoc Localization

Algorithm 2 SeRLoc

1: Receive beacons from locators; each beacon contains the position of locator and the angles of sector boundary.

2: Find four values: $(X_{min}, Y_{min}, X_{max}, Y_{max})$ among all the locator positions heard.

3: Set the search area as the rectangle $(X_{min} - R, Y_{min} - R, X_{max} + R, Y_{max} + R)$, where R is the radio range.

4: Partition the search area into grids.

5: **for** each beacon received **do**

6:    Increase the value of a grid point by one if this grid point is within the sector defined in this beacon.

7: **end for**

8: Estimated Position = CenterOfGravity(the grid points with the largest values)

We note that SeRLoc is unique in its secure design. It can deal with various kinds of attacks including wormhole and Sybil attacks. We do not describe its security features here except to note that the authors prove in [15] that their approach is more secure, robust and accurate in the presence of attacks, compared with other state-of-the-art solutions that largely ignore this issue.

2.4 Multidimensional Scaling

The MDS-MAP algorithm proposed by Shang et al. in [28] is based on a data analysis technique, called MultiDimensional Scaling (MDS), extensively used in psychometrics. MDS attempts to provide a visualization (2D or 3D) of the original data, and preserving the essential information present in the data set (i.e., similarities in a multidimensional space). The MDS-MAP algorithm uses the classical metric scaling, the simplest case of the MDS technique, which has a closed form solution, enabling a relatively efficient computation (requires no iterations).

An important concept for MDS is how to compute the distance between two points. If we denote by $\mathbf{X}$ the matrix of coordinates of points ($n \times m$ matrix of n points in m dimensions), and by $\mathbf{D} = [d_{ij}]$ the matrix of distances between points, it can be shown that the matrix of squared distances between points, i.e., $\mathbf{D}^{(2)}$, can be written as follows:

$$\mathbf{D}^{(2)} = \mathbf{c1}' + \mathbf{1c}' - 2\mathbf{XX}' = \mathbf{c1}' + \mathbf{1c}' - 2\mathbf{B}$$

where $\mathbf{c}$ is a vector with elements the diagonal elements of $\mathbf{XX}'$. The matrix $\mathbf{B} = \mathbf{XX}'$ is the scalar product matrix. So the questions becomes, given the matrix of squared distances $\mathbf{D}^{(2)}$ how can one obtain $\mathbf{B}$, and implicitly $\mathbf{X}$? It can be shown [1] that by double-centering $\mathbf{D}^{(2)}$, one can obtain $\mathbf{B}$:

$$-\frac{1}{2}\mathbf{JD}^{(2)}\mathbf{J} = \mathbf{B}$$

where $\mathbf{J} = \mathbf{I} - \mathbf{11}'/n$ (called the centering matrix), $\mathbf{I}$ the identity matrix and $\mathbf{1}$ the n-dimensional column vector with all elements one. Once $\mathbf{B}$ is obtained, the coordinates $\mathbf{X}$ can are computed by eigendecomposition.

The steps of classical scaling are summarized as follows:

1. Compute the squared distances matrix $\mathbf{D}^{(2)} = [d_{ij}^2]$
2. Double-center the $\mathbf{D^2}$ matrix:

$$\mathbf{B} = -\frac{1}{2}\mathbf{JD}^{(2)}\mathbf{J}$$

3. Compute the singular value decomposition of $\mathbf{B} = \mathbf{VAV^T}$
4. Compute the coordinate matrix:

$$\mathbf{X} = \mathbf{V}_+\mathbf{A}_+^{1/2}$$

where $\mathbf{A}_+$ is the matrix of the first m eigenvalues greater than zero and $\mathbf{V}_+$ the first m columns of $\mathbf{V}$.

The MDS-Map algorithm that uses the classical metric scaling technique is shown in Algorithm 3.

Algorithm 3 MDS-MAP

1: Compute the shortest paths d_{ij}, $1 \le i, j \le n$. This gives the distances matrix $\mathbf{D}$.
2: Compute the relative positions (map), by applying classical MDS to the distance matrix $\mathbf{D}$, and retain the largest 2 (for a 2D space) or 3 (for a 3D space) eigenvalues and eigenvectors.
3: Transform the relative map, into an absolute map, based on the absolute positions of anchor nodes.

The main drawbacks of the MDS-MAP algorithm, the need for global information and centralized computation are addressed in subsequent work by the authors [27].

2.5 Gradient

In the Gradient algorithm, proposed by Nagpal et al. in [20], the anchor nodes initiate a gradient that self-propagates and allows a sensor node to infer its distance from the anchor. After estimating distances to three anchors a sensor node infers its own location through multilateration.

The steps of the algorithm are as follows:

- Each anchor node A_i initiates a flood of the network by broadcasting a packet containing its position and a counter with the initial value set to one.
- Each sensor node N_j keeps track of the shortest path (in terms of radio hop counts, h_{j,A_i}) to an anchor A_i from which it has received a beacon. A distance estimate, between the sensor node and anchor is obtained by:

$$d_{ji} = h_{j,A_i} d_{hop}$$

where d_{hop} is the estimated Euclidian distance covered by one radio hop, and it is given by the Kleinrock-Silvester formula [11]:

$$d_{hop} = r \left(1 + e^{-n_{local}} - \int_{-1}^{1} e^{- \frac{n_{local}}{\pi} \left(\arccos t - t\sqrt{1-t^2} \right)} \right)$$

- Each node N_j computes its coordinates such that the total error is minimized:

$$E_j = \sum_{i=1}^{n} (d_{ji} - \hat{d}_{ji})$$

where $d_{ji} = \sqrt{(x_i - x_j)^2 + (y_i - y_j)^2}$ and $\hat{d}_{ji}$ is the estimated distance computed through gradient propagation, as shown above.

The coordinate are incrementally updated in the following way:

$$\Delta x = -\alpha \frac{\partial E_j}{\partial x_j} \text{ and } \Delta y = -\alpha \frac{\partial E_j}{\partial y_j}$$

where:

$$\frac{\partial E_j}{\partial x_j} = \sum_{i=1}^{n} (x_j - x_i) \left(1 - \frac{d_{ji}}{\hat{d}_{ji}} \right) \text{ and } \frac{\partial E_j}{\partial x_j} = \sum_{i=1}^{n} (x_j - x_i) \left(1 - \frac{d_{ji}}{\hat{d}_{ji}} \right)$$

Sources for errors in location estimation arise in the Gradient scheme from: incorrect estimation of the one-hop distance (d_{hop}), and the multilateration procedure.

2.6 Ad-Hoc Positioning System

In a similar manner with the Gradient method, the Ad-Hoc Positioning System (APS) proposed by Niculescu and Nath [22], uses the hop-by-hop propagation of distances to known anchors (a set of anchors is assumed to be present in the WSN). After obtaining distance estimates to three or more anchors, a sensor node employs a multi-lateration (similar with that of GPS) for iteratively improving its location estimation. The main difference resides in how a sensor node N_j estimates its distance to an anchor A_i (d_{ji} from the Gradient method, presented before).

The steps of the APS localization scheme algorithm are the following:

- Each anchor node A_i initiates a flood of the network by broadcasting a packet containing its position and a counter with the initial value set to one.
- Each sensor node N_j keeps track of the shortest path (in terms of radio hop counts, h_{j,A_i}) to an anchor A_i from which it has received a beacon. In [22] the authors propose four methods for propagating the distances from anchors to sensor nodes: DV-Hop, DV-Distance, Euclidian and Coordinate. The method that does not assume ranging, DV-Hop is described below. An example of the DV-Hop scheme is shown in Figure 7. At the end of this phase, node N_j knows that it is 3 hops, 2 hops and 1 hop from A_1, A_2 and A_3, respectively.

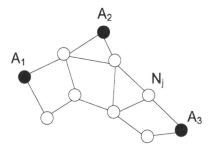

Fig. 7. The DV-Hop localization scheme

- Once an anchor node A_i obtains distances to other anchors, it computes a correction factor c_i (the estimated 1 radio hop Euclidean distance), which it propagates in the network. Corrections are propagated through controlled flooding, i.e., after a node receives and forwards the first correction, it will stop forwarding subsequent corrections. The correction factor is computed as follows:

$$c_i = \frac{\sum \sqrt{(x_i - x_j)^2 + (y_i - y_j)^2}}{\sum h_i}$$

for all anchors $A_j \neq A_i$ from which it has received a beacon (anchor A_j is positioned at (x_j, y_j) and h_i is the number of hops between the sensor node and anchor A_i).

For the example shown in Figure 7, if the distances $\overline{A_1A_2}$, $\overline{A_2A_3}$ and $\overline{A_3A_1}$ are 30m, 40m and 50m, respectively, the correction factor for anchor A_3 is: $c_3=(50+40)/(4+3)=12.9$m/hop.

- A least square method (the authors used the Householder method) is employed for solving the non-linear system of equations:

$$\begin{bmatrix} \Delta\rho_1 \\ \Delta\rho_2 \\ \Delta\rho_3 \\ ... \\ \Delta\rho_n \end{bmatrix} = \begin{bmatrix} \hat{1}_{1x} & \hat{1}_{1y} \\ \hat{1}_{2x} & \hat{1}_{2y} \\ \hat{1}_{3x} & \hat{1}_{3y} \\ ... & ... \\ \hat{1}_{nx} & \hat{1}_{ny} \end{bmatrix} \begin{bmatrix} \Delta x \\ \Delta y \end{bmatrix}$$

where $\Delta\rho_i = \hat{\rho}_i - \rho_i$, $\hat{\rho}_i$ and ρ_i are the estimated and the real distances between a sensor node and an anchor A_i, $\hat{1}_{ix}$ is the unit vector of $\hat{\rho}_i$ in the x direction and Δx and Δy are the corrections in the position estimate for the node N_j.
For the example shown in Figure 7, the estimated distances between node N_j and anchors A_1, A_2 and A_3 are $\hat{\rho}_1=4*12.9=51.6$m, $\hat{\rho}_2=38.7$m and $\hat{\rho}_3=12.9$m.

2.7 Probability Grid

In a similar manner with the DV-Hop, a localization scheme that can be used in scenarios where the topology of deployment is known a priori to be a grid, is proposed in [31].

The steps of the localization scheme are the following:

- Each anchor node A_m initiates a flood of the network by broadcasting a packet containing its position and a counter with the initial value set to one.
- Each sensor node N_k keeps track of the shortest path (in terms of radio hop counts) to each of the anchors A_l from which it has received beacons.
- Once an anchor node A_m obtains distances to other anchors, it computes a correction factor c_m (the estimated radio range), and it propagates it in the network.
- After receiving hop-count estimates to three or more anchor nodes, and a correction factor c_m a sensor node N_k evaluates the probability of being located at any position in the grid (labeled (i, j)). For this, it computes an expected hop count:

$$\lambda = d_{(i,j),l}/c_m$$

where $d_{(i,j),l}$ is the Euclidian distance between anchor A_l and the point (i, j) being evaluated. It then computes the probability of it (node N_k) to be positioned at (i, j):

$$p_{k,(i,j)} = \prod_{l=1}^{|A|} P_{(i,j)}^{h_{k,l}}$$

where $P_{(i,j)}^{h_{k,l}}$ is the probability of node N_k, positioned at (i, j), to be $h_{k,l}$ hops from anchor A_l.

- A node N_k chooses as its location, the position in the grid (i,j) with the maximum probability $p_{k,(i,j)}$.

The authors make the observation that $h_{k,l}$ is a discrete random variable that represents the number of radio hops between one anchor and the point of interest, i.e. (i,j). The main features that the distribution function for $h_{k,l}$ needs to exhibit are: to have one parameter λ (defined above) to be narrow and skewed positively for small values of λ and become broader and relatively symmetric for larger values of λ. This is illustrated in Figure 8:

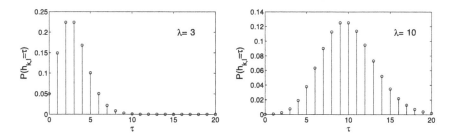

Fig. 8. The intuition behind the PMF of $h_{k,l}$

These requirements follow the intuition that for smaller values of the parameter λ (i.e., grid points closer to the anchor) the number of hops (call it τ) has a limited range of possible values with higher and higher values being less and less probable (positively skewed). As the distance between the anchor and the node increases (λ increases), the number of possibilities for the hop count (τ) increases and the distribution becomes bell-shaped, i.e., smaller and larger hop counts are equally probable. The authors found through simulations that a Poisson distribution is a good approximation for the $h_{k,l}$ discrete random variable. The distribution is given by:

$$P_{(h_{(k,l)}=\tau)} = \frac{\lambda^{\tau-1}e^{-\lambda}}{(\tau-1)!}$$

where $\tau = 1, 2, \ldots$.

3 Anchor-Free Solutions

Anchor-free localization schemes exploit the proximity to an event with a known location: a light event in [30] [24] or a nearby radio packet in [29]. One common characteristic for these schemes is the moving of the complexity (hardware and computational, associated with an accurate localization) from the sensor node to a central, more sophisticated device. By controlling well the spatio-temporal properties of the events (light and radio packets), a much higher accuracy in localization (when compared with the anchor-based schemes) can be obtained. While anchor nodes are not required for any of the following schemes, anchor nodes can be beneficial for extensions of the proposed schemes.

3.1 Spotlight

The main idea of the Spotlight localization system [30] is to generate controlled events in the field where the sensor nodes are deployed. An event could be, for example, the presence of light in an area. Using the time when an event is perceived by a sensor node and the spatio-temporal properties of the generated events, spatial information (i.e. location) regarding the sensor node can be inferred. The system architecture for the Spotlight localization system is shown in Figure 9.

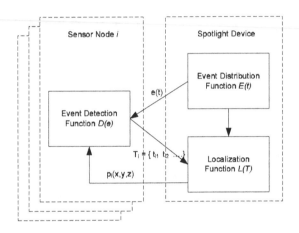

Fig. 9. Spotlight system architecture

With the support of these three functions, the localization process goes as follows:

- A Spotlight device distributes events $e(t)$ in the space A over a period of time.
- During the event distribution, sensor nodes record the time sequence $T_i = \{t_{i1}, t_{i2}, ..., t_{in}\}$ at which they detect the events.
- After the event distribution, each sensor node sends the detection time sequence back to the Spotlight device.
- The Spotlight device estimates the location of a sensor node i, using the time sequence T_i and the known $E(t)$ function.

The Event Distribution Function $E(t)$ is the core technique used in the Spotlight system and the authors propose three designs for it, with different tradeoffs/costs. These designs are presented below.

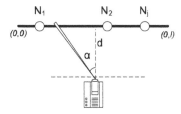

Fig. 10. The implementation of the Point Scan EDF

Point Scan

The Point Scan EDF is applicable to a simple sensor system where a set of nodes are placed along a straight line ($A = [0, l] \subset R$). The Spotlight device generates point events (e.g., light spots) along this line with constant speed s, as shown in Figure 10. The set of timestamps of events detected by a node i is $T_i = \{t_{i1}\}$. The Event Distribution Function $E(t)$ is:

$$E(t) = \{p \mid p \in A, p = t * s\}$$

where $t \in [0, l/s]$. The resulting localization function is:

$$L(T_i) = E(t_{i1}) = \{t_{i1} * s\}$$

Line Scan

Some devices, e.g. lasers, can generate an entire line of events simultaneously. With these devices, the Line Scan Event Distributed Function can be supported. Assuming that the sensor nodes are placed in a two dimensional plane ($A = [l \times l] \subset R^2$) and that the scanning speed is s. The set of timestamps of events detected by a node i is $T_i = \{t_{i1}, t_{i2}\}$.

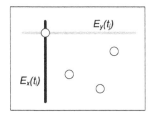

Fig. 11. The implementation of the Line Scan EDF

The Line Scan EDF, depicted in Figure 11, is defined as follows:

$$E_x(t) = \{p_k \mid k \in [0, l], p_k = (t * s, k)\} \text{ for } t \in 0, l/s$$

$$E_y(t) = \{p_k \mid k \in [0, l], p_k = (k, t * s - l)\} \text{ for } t \in l/s, 2l/s$$

and $E(t) = E_x(t) \cup E_y(t)$.

The location of a node can be calculated from the intersection of the two event lines, as shown in Figure 11. More formally:

$$L(T_i) = E(t_{i1}) \cup E(t_{i2})$$

Area Cover

Other devices, such as light projectors, can generate events that cover an area. This allows the implementation of the Area Cover EDF. The idea of Area Cover EDF is to partition the space A, where the sensor nodes are deployed, into multiple sections and assign a unique binary identifier, called code, to each section. Let's suppose that the localization is done within a plane ($A \in R^2$). Each section S_k within A has a unique code k.

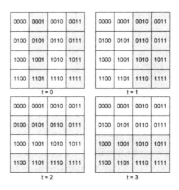

Fig. 12. The implementation of the Area Cover EDF

The Area Cover EDF, with its steps shown in Figure 12 is then defined as follows:

$$BIT(k, j) = \begin{cases} \text{true if } j^{th} \text{ bit of } k \text{ is } 1 \\ \text{false if } j^{th} \text{ bit of } k \text{ is } 0 \end{cases}$$

$$E(t) = \{p \mid p \in S_k, BIT(k, t) = true\}$$

and the corresponding localization algorithm is:

$$L(T_i) = \{p \mid p = COG(S_k), BIT(k, t) = true \text{ if } t \in T_i,$$
$$BIT(k, t) = false \text{ if } t \in T - T_i\}$$

where $COG(S_k)$ denotes the center of gravity of S_k.

Cost Comparison

Although all three proposed techniques are able to localize the sensor nodes, they differ in the localization time, communication overhead and energy consumed by the Event Distribution Function (call it Event Overhead). Assume that all sensor nodes are located in a square with edge size D, and that the Spotlight device can generate N events (e.g. Point, Line and Area Cover events) every second and that the maximum tolerable localization error is r. Table 1 presents the execution cost comparison of the three different Spotlight techniques.

Criterion	Point Scan	Line Scan	Area Cover
Localization Time	$D^2/r^2)/N$	$D^2/r^2)/N$	$\log_r(D/N)$
# Detections	1	2	$\log_r D$
# Time Stamps	1	2	$\log_r D$
Event Overhead	D^2	$2D^2$	$D^2 \log_r(D/2)$

Table 1. Execution Cost Comparison

Table 1 indicates that the Event Overhead for the Point Scan method is the smallest - it requires a one-time coverage of the area, hence the D^2. However the Point Scan takes a much longer time than the Area Cover technique, which finishes in $\log_r D$ seconds. The Line Scan method trades the Event Overhead well with the localization time. By doubling the Event Overhead, the Line Scan method takes only $r/2D$ percentage of time to complete, when compared with the Point Scan method. From Table 1, it can be observed that the execution costs do not depend on the number of sensor nodes to be localized. It is important to remark the ratio "Event Overhead"/"Localization Time", which is indicative of the power requirement for the Spotlight device. This ratio is constant for the Point Scan ($r^2 N$) while it grows linearly with area, for the Area Cover ($D^2 N/2$). If the deployment area is very large, the use of the Area Cover EDF is prohibitively expensive, if not impossible. For practical purposes, the Area Cover is a viable solution for small to medium size networks, while the Line Scan works well for large networks.

For the Spotlight system evaluation, the authors deployed 10 XSM [6] motes in a football field. The Spotlight device consisted of diode lasers, a computerized telescope mount, connected to a laptop. The Event Distribution Function investigated was the Point Scan. The range between the Spotlight device and the sensor nodes was approximately 170m.

Figure 13 shows the average localization errors versus the size of the event (diameter of the laser beam, on the ground), for different scanning speeds s. Localization errors of 10-20cm are reported.

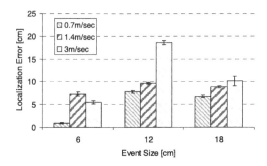

Fig. 13. Localization Error vs. Event Size for Spotlight system.

3.2 Lighthouse

In a similar way to the Spotlight localization system, the Lighthouse scheme, proposed by Römer [24], makes use of the free-space optical channel between a device (called Lighthouse in this case) and sensor nodes.

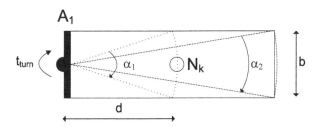

Fig. 14. Lighthouse Localization

The main idea of the Lighthouse system is exemplified in Figure 14. A parallel light beam of width b, emitted by anchor A_1 rotates with a certain period t_{turn}. A sensor node N_k detects this light beam for a period of time t_{beam}, which is dependent on the distance d between the Lighthouse device and the sensor node, in the following way:

$$d = \frac{b}{2\sin(\alpha_1/2)} = \frac{b}{2\sin(\pi t_{beam}/t_{turn})}$$

From measuring t_{beam} and knowing b and t_{turn}, one can compute the distance between the sensor node and the lighthouse device d. By constructing a device with three mutually perpendicular light emitting Lighthouses, a 3D location can be obtained.

The main difficulty encountered by the authors in the implementation of the Lighthouse prototype is ensuring that the light beam is perfectly parallel (zero divergence), having a width b. Instead, two laser beams of widths b_i and angle orientations β_i, γ_i and δ_i $i = 1, 2$, are used. To account for the misalignments, the authors develop a better approximation for the resulting beam width b:

$$b \approx C^b + \sqrt{d^2 + h^2}C^\beta + hC^\gamma + dC^\delta$$

where $C^b = b_1 + b_2$, $C^\beta = \sin\beta_1 + \sin\beta_2$, $C^\gamma = \tan\gamma_1 + \tan\gamma_2$ and $C^\delta = \sin\delta_1 + \sin\delta_2$. These parameters are constant for a particular Lighthouse system, and they are obtained through calibration, by localizing four points with known locations.

The experiments use 22 nodes placed in a $5x5m^2$ area, with the Lighthouse device positioned at the coordinate (0,0). The accuracy of the localization algorithm is presented relative to the distance between the Lighthouse device and the sensor node (i.e., $|\hat{x} - x|/x$). The mean relative error (difference between the computed location and ground truth) in localization is 1.1% in one direction and 2.8% in the second direction (the difference is attributed to the calibration). This translates in localization errors of a few centimeters.

3.3 Walking-GPS

In many applications it is envisioned that WSN will be deployed from Unmanned Aerial Vehicles. In the meantime, manual deployments have been prevalent and the employed localization solutions have used some variant of associating the sensor node ID with prior knowledge of that ID's position in the field.

In [29] the authors propose a solution, called Walking GPS, in which the deployer (either person or vehicle) carries a GPS device that periodically broadcasts its location. The sensor nodes being deployed, infer their position from the location broadcast by the GPS device. The proposed solution is simple, cost effective and has very little overhead.

In the Walking GPS architecture the system is decoupled into two software components: the GPS Mote and the Sensor Mote. The GPS Mote runs on a Mica2 mote. The mote is connected to a GPS device, and outputs its location information at periodic intervals. The Sensor Mote component runs on all sensor nodes in the network. This component receives the location information broadcast by the GPS Mote and infers its position from the packets received. The proposed architecture pushes all complexity derived from the interaction with the GPS device to a single node, the GPS Mote, and to significantly reduce the size of the code and data memory used on the sensor node. Through this decoupling, a single GPS Mote is sufficient for the localization of an entire sensor network, and the costs are thus reduced.

A relatively simple design for the GPS Mote would have been to periodically broadcast the actual GPS location received from the GPS device. In order to reduce the overhead incurred when exchanging data containing global GPS coordinates, the Walking GPS system uses a local, Cartesian, coordinate system. The conversion between coordinate systems is performed by the GPS mote. A local coordinate system of reference is better suited for WSN, than a global coordinate system.

The localization scheme that makes use of the Walking GPS solution has two distinct phases:

1. The first phase is during the deployment of the sensor nodes. This is when the Walking GPS solution takes place. The deployer has a GPS-enabled mote attached to it; the GPS-enabled mote periodically beacons its location; the sensor nodes that receive this beacon infer their location based on the information present in this beacon.
2. The second phase is during the system initialization. If at that time, a sensor node does not have a location, it asks its neighbors for their location information. The location information received from neighbors is used in a triangulation procedure by the requester, to infer its position. This second phase enhances the robustness of the scheme.

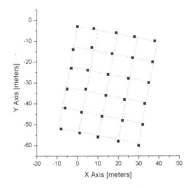

Fig. 15. Walking GPS system evaluation. Nodes deployed in a grid.

The experimental evaluation of the entire system, consisted of 30 MICA2 motes that were deployed in a 5x6 grid (for ease of measuring the localization error). The experimental results are shown in Figure 15. The average localization error obtained from fitting a grid to the experimental data is 0.8m with a standard deviation of 0.5m.

4 Open Problems

4.1 Security

Recently, several research groups have started to address robust and secure localization. For example, SeRLoc [15] demonstrates robustness against wormhole, Sybil and compromise of network nodes attacks. However, this work assumes a particular two-tier architecture and special hardware they call locators. In addition, the jamming of the wireless medium is not considered. It is an excellent start, but a lot more needs to be done especially for military domains and to meet various reality assumptions.

For securing localization, robust statistical methods (e.g. least median square) have been proposed [16]. The assumption is that an attacker selectively alters distance estimates to known anchor locations. The highest contamination ratio (i.e., affected readings) that the mathematical model supports is 50%, with significant degradation even at 35%. If an attacker possesses the capability of affecting distance estimates easily, then it is very likely that all distance estimates will be affected, making this set of solutions, less effective. The idea, however, of using robust statistical models, is a very good one.

In a similar approach, in [17], two solutions are proposed for secure localization: an attack resistant minimum mean square error (MMSE) which suffers from an unbounded localization error (the attack can result in an arbitrarily large localization error), and a voting-based scheme, which corrects the unbounded localization error, at a higher computational and storage cost.

Distance bounding has also recently been proposed as a technique for secure verification of localization. In [25] a combination of ultra-sound and radio communication is used for bounding the location claimed by a node, to a region. In this region, called region of interest, a set of trusted verifiers has to exist. This scheme is robust against attackers that can not be physically present in the region of interest. Similarly, [4] proposes a Verifiable Multilateration, that also relies on the distance bounding technique. The basic idea is to use the Time of Flight (ToF) of radio communication. Since the speed of light can not be exceeded, the location to be verified can not be closer than it actually is. It can only be further. However, claiming a longer distance would require a shorter distance to an even further positioned verifier. The main drawback is the hardware requirements (with nanosecond accuracy) imposed on the sensor nodes. In addition, [4] requires a relatively large number of anchor nodes.

In order to address some of the deficiencies of SeRLoc (e.g. jamming is not considered) and the Verifiable Multilateration (e.g., relatively high number of anchor nodes), a new scheme is proposed in [14]. This scheme can be used for both, location determination and location verification. The main idea is to fully utilize the strengths of both solutions: SeRLoc's use of sectored antennas and the distance bounding properties of Verifiable Multilateration. The deficiencies of both schemes, are still present.

The most recent effort on secure localization [5], attempts to depart from the aforementioned, "traditional", approaches, which require high speed hardware, sectored antennas or statistics, with a limited robustness. The idea is to use covert, hidden base stations (their position is known only to an authority), in addition to the "public", known base stations. The role of covert base stations is to perform TDoA (between radio and ultra-sound) ranging and verify the location computed and claimed by a node. For the effectiveness of this solution, the covert base stations communicate with a central location verification authority either in a wired manner or infra-red, to reduce the risk of being detected by the attacker. The authors also propose mobile base station assisting with the verification of location. While this

direction for secure localization is novel, in its current form, has demanding requirements for the infrastructure of covert base stations.

4.2 Impact of Localization on Protocols

In localization for WSN, achieving better results (usually with regard to location accuracy) requires increasing the relative cost of the localization scheme via additional
hardware, communication overhead, or the imposition of constraints and system requirements. Although more accurate location information is preferable, the desired
level of granularity should depend on a cost/benefit analysis of the protocols that utilize this information. In this section, we investigate the impact of localization error
on other communication protocols and proposed sensor network applications. Designers of sensor network systems with certain performance requirements can use
this analysis to aid in their architectural design and in setting system parameters.
Although requirements are expected to vary between deployments, we found that in
the general case for the protocols studied, performance degradation is moderate and
tolerable when the average localization error is less than 0.4R.

Routing Performance

A localization service is critical for location-based routing protocols such as geographic forwarding (GF) [21], [10], [12] and [34]. In these protocols, individual
nodes make routing decisions based on knowledge of their geographic location.
While most work in location-based routing assumes perfect location information, the
fact is that erroneous location estimates are virtually impossible to avoid. Problems
arise as error in the location service can influence location-based routing to choose
the best next hop (the neighbor closest to the destination), or can make a node inad-
vertently think that the packet could not be routed because no neighbors are closer to
the final destination.

To investigate the impact of localization error on routing, the authors of [8] studied the GF [21] routing protocol under the low traffic network conditions so that
network congestion does not influence the results. The baseline was the perfect localization, the protocol where every sensor node knows its correct physical location.

Figure 16 shows the delivery ratio (the percentages of packets that reach destination over all packets sent) with regard to node density for various levels of location
error. From this graph, we see that for average localization errors of 0.2R and 0.4R,
the delivery ratios of GF are very close to the baseline (no error). Beyond these numbers, the results diminish with increased error; a trend that could be problematic and
costly depending on the implemented architecture, reliability semantics, tolerance of
message loss, and application requirements. For example, when localization error is
the same as the node radio range, even with high node density (20 nodes per radio
range), the delivery ratio still falls below 60%.

Another metric affected by localization error is the route path length. Figure 17
measures the hop count increase (in percentage) due to location error to assess the
cost in communication overhead of this error. We see from this graph that for low

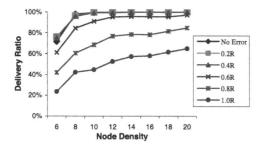

Fig. 16. Delivery ratio with different localization errors, changing node density

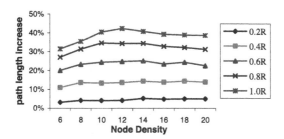

Fig. 17. Path length overhead with different localization errors under varying node density

localization error (less than 0.4R), this routing overhead remains moderate (less than 15%). However, as was the case for the delivery ratio metric, when localization error grows beyond 0.4R, the routing overhead increases to as high as 45%. We also note that this trend occurs regardless of the network node density, a fact that was not true for our previous metric. We acknowledge here that GF was chosen as a representative protocol, and an in depth study about localization's impact on various routing protocols and its implications on the design of location-dependent systems is future work.

Target Estimation Performance

Many of the most frequently proposed applications for WSN utilize target position estimations for tracking, search and rescue, or other means. In these proposed applications, when a target is identified, some combination of the nodes that sensed that target report their location to a centralized node (leader or base station). This node then performs aggregation on the received data to estimate the actual location of the target. Because target information could be used for locating survivors during a disaster, or identifying an enemy's position for strategic planning, the accuracy of this estimation is crucial to the application that uses it.

Intuitively an increase in localization error directly leads to target estimation error. To better understand the degree to which this error propagates to other protocols,

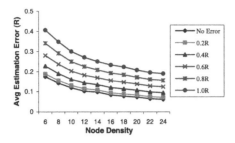

Fig. 18. Target estimation error with different localization errors under varying node density

the authors of [8] investigate the average estimation error under different node densities for varying degrees of location error. For these experiments, a simple and widely used target estimation algorithm is used: the average X and Y coordinates of all reporting nodes are taken as the target location estimation. The sensing range is set to be equal to the node radio range so that the node density is equivalent to the average number of sensors involved in target estimation. The results of various experiments are depicted in Figure 18. This graph shows that target estimation error due to location error is dampened during the aggregation process. As before, the baseline occurs when no localization error exists. Aside from showing varying degrees of estimation error with respect to node location error, Figure 18 also shows that the absolute target estimation error decreases with increased node density. For example, when localization error is equal to 1.0R, and node density reaches 12 nodes per radio range, the estimation error is only about 67% as large as when the node density is 6 nodes per radio range. From this chart we see that more nodes participating in estimation results in more random estimation error being ameliorated through aggregation.

Object Tracking Performance

In [8], the authors further evaluate the performance of target estimation by simulating a tracking application that uses estimation in context. In this experiment, a mobile evader randomly walks around the specified terrain while a pursuer attempts to catch it. In this simple experiment, the pursuer is informed of the current location of the evader periodically via sensing nodes in the terrain that detect the evader, coordinate to estimate the targets position with regard to their own positions, and periodically report this result to the mobile pursuer. When receiving a report, the pursuer readjusts its direction in an attempt to intercept the evader. When the pursuer comes within the node communication radius of the evader, the evader is considered caught and the simulation ends. For this experiment, the average tracking time (the time from pursuer take-off to when the evader is caught) under different localization errors is compared to the tracking time in the case of no localization error. Figure 19 shows normalized tracking time in relation to the pursuer speed for various degrees of localization error.

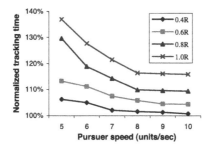

Fig. 19. Normalized tracking time with different localization errors varying pursuer speed. Terrain size 1000x1000m, Radio range = 40m, density = 8 nodes/radio range. Evader speed = 5m/s

From Figure 19 we see that the tracking time overhead decreases with increased pursuer speeds. More importantly, Figure 19 shows that the tracking time increases as localization error increases. This result implies that it is important for tracking applications with real-time requirements to take localization error into consideration. For example, when the average localization error is known to be 0.8R, and the pursuer speed is 5 units per second, the pursuer requires 30% more time in comparison to the ideal situation in which no localization error exists. To reduce this overhead to 10%, either the pursuers speed must be increased to 10 units per second, or the estimation error must be reduce to 0.4R. Again, Figure 19 shows that 0.4R is a tolerable bound for estimation error since tracking time only increases by 7% in the worst case.

4.3 Impact of Environment on Localization

The problem of range-free localization is further complicated by the diverse types of environments, where a WSN system can be deployed. Outdoor, real deployment environments very little resemble typical lab environments. Hence, issues like calibration, mobility (if nodes are "moved" by the environment, or the WSN is designed to be mobile), the lack of line-of-sight, the existence of obstructions and multipath effects often arise in realistic, outdoor environments.

Some preliminary work on the aforementioned issues are the following:

Calibration

Whitehouse formulates the calibration problem in WSN as a parameter estimation problem [33]. Each device in the WSN is parameterized and the values of the parameters are chosen such that the system performance is maximized (higher accuracy in location estimation). The author propose a macro-calibration procedure, called joint-calibration, that calibrates each device, by optimizing the overall system performance, instead of individual nodes. The steps of the joint calibration are the following:

1. Model the overall system, by using individual, device specific parameters.
2. Collection data.
3. Tune the parameters of individual devices, such that the overall system performance is improved.

The key insight into how to choose parameters to be tuned, such that the overall system performance improves is to look at trends in the transmitter/receiver pairs, and identify individual nodes for which the chosen parameters are problematic.

The proposed joint calibration is a good solution where manual calibration is possible. Obviously, in rugged, remote outdoor environments, auto-calibration (i.e., no manual intervention) is highly desirable.

Mobility

Hu and Evans [9] propose a sequential Monte-Carlo (SMC) localization algorithm for WSN in which sensor nodes and anchors are all mobile. The authors show that mobility can be used to enhance localization accuracy, a rather counterintuitive result - one would expect to be a significant impediment for an accurate positioning.

The proposed algorithm is an adaptation of the Sequential Monte Carlo localization scheme, frequently used in robot localization, target tracking and computer vision, to the domain of WSN. The main idea of the SMC localization algorithm is to represent the posterior distribution of possible locations using a set of weighted samples and to update them recursively in time.

From simulations of a 10Rx10R WSN, with an average number of nodes per transmission range of 10, the authors report localization errors of approximately 0.5R, when both sensor nodes and anchors move at a speed of R meters/sec. The localization error starts from high values (1.9R) and decreases rapidly, with the accumulation of new observations (nodes entering the ranges of new anchor nodes).

Line-of-Sight and Multipath

Real, outdoor environments pose significant challenges for range-free localization. Localization schemes designed and evaluated in "friendly" environments frequently fail to produce encouraging results in real deployments. When line-of-sight is a main assumption of the scheme [30], and does not always hold, or when obstructions and multipath for acoustic and radio waves are not considered [35], the performance of the localization scheme is degraded.

In order to address this, a potential direction to pursue is multimodal localization. In a multimodal localization system, more than one localization scheme is executed, in an attempt to reduce the impact the assumptions of a single localization scheme could have on the overall localization accuracy. By using Bayesian inference, and the knowledge (even if partial) obtained during the execution of one localization scheme, a finer, more accurate positioning can be obtained from the execution of subsequent localization schemes. For example, if a WSN is localized using the Line

Scan scheme of the Spotlight system, described before, and due to some environmental conditions one of the two events created in the network is not detected (the Spotlight localization scheme fails to produce a location in this case), the knowledge gained from the detection of the other event can be used to initialize a subsequently executed localization scheme.

5 Conclusions

In this chapter we presented a suite of range-free localization schemes for WSN. We define ranging, in the context of sensor networks, as the ability of a sensor node to infer distances to its neighbor sensor nodes, either through localization specific hardware (e.g., ultrasound transceivers) or the strength of the received radio signal. Hence, the localization schemes presented here (i.e., range-free schemes) do not posses sophisticated hardware and do not rely on the received signal strength for inter-node ranging. The sensor nodes we consider have simple radio communication and sensing capabilities.

The taxonomy that we adopt for categorizing the range-free localization schemes is based on the (non)existence of an infrastructure of anchor nodes (i.e., at least three nodes, for a 2D localization, with known locations) in the WSN. An anchor-free localization scheme exploits the proximity to an event with a known location: a light event in Spotlight [30] and Lighthouse [24] and a nearby radio packet in Walking GPS [29].

One main observation is the high accuracy in localization of the anchor-free, event based, localization schemes, at a reduced, per node, cost. It is remarkable to obtain location accuracies of tens of centimeters, at zero dollar cost (if the sensor node is equipped with a photo sensor for the mission it was deployed for) and relatively low communication overhead (reduced energy cost). Characteristic to the anchor-free localization schemes, is the moving of the complexities associated with the localization from the sensor node to a capable, sophisticated device. While the cost of such device is not negligible, the possibility of its reuse make the event-based, anchor-free solutions very attractive. The anchor-free, event based, class of localization schemes seems a very promising direction for high accuracy, low cost localization in WSN.

Despite the extensive attention the range-free localization has received, several open problems remain. Among these are how to secure the radio communication and sensing channels that sensor nodes posses, how to make range-free localization more robust against attacks, node or protocol failures (possibly due to its strict assumptions), understand the impact of localization schemes on other protocols and how to design more robust, cost efficient, calibration techniques. The breadth and depth of all these issues present interesting opportunities for future research in the domain of range-free node localization in WSN.

References

1. BORG, I., AND GROENEN, P. *Modern Multidimensional Scaling*. Series in Statistics. Springer, 1997.
2. BULUSU, N., HEIDEMANN, J., AND ESTRIN, D. GPS-less low cost outdoor localization for very small devices. *IEEE Personal Communications Magazine 7*, 5 (October 2000), 28–34.
3. BULUSU, N., HEIDEMANN, J., AND ESTRIN, D. Adaptive beacon placement. In *International Conference on Distributed Computing Systems (ICDCS)* (2001).
4. CAPKUN, S., AND HUBAUX, J. Secure positioning of wireless devices with application to sensor networks. In *IEEE Conference on Computer Communications (Infocom)* (2005).
5. CAPKUN, S., SRIVASTAVA, M., AND CAGALJ, M. Securing localization with hidden and mobile base stations. Tech. rep., NESL-UCLA, 2005.
6. DUTTA, P., GRIMMER, M., ARORA, A., BIBYK, S., AND CULLER, D. Design of a wireless sensor network platform for detecting rare, random, and ephemeral events. In *International Symposium on Information Processing in Sensor Networks (IPSN)* (2005).
7. FANG, L., DU, W., AND NING, P. A beacon-less location discovery scheme for wireless sensor networks. In *IEEE Conference on Computer Communications (Infocom)* (2005).
8. HE, T., HUANG, C., BLUM, B., STANKOVIC, J. A., AND ABDELZAHER, T. Range-Free localization schemes in large scale sensor networks. In *ACM International Conference on Mobile Computing and Networking (Mobicom)* (2003).
9. HU, L., AND EVANS, D. Localization for mobile sensor networks. In *ACM International Conference on Mobile Computing and Networking (Mobicom)* (2004).
10. KARP, B., AND KUNG, H. T. Gpsr: Greedy perimeter stateless routing for wireless networks. In *ACM International Conference on Mobile Computing and Networking (Mobicom)* (2000).
11. KLEINROCK, L., AND SILVESTER, J. Optimum tranmission radii for packet radio networks or why six is a magic number. In *IEEE National Telecommunication Conference* (1978).
12. KO, Y. B., AND VAIDYA, N. H. Location-aided routing (lar) in mobile ad hoc networks. In *ACM International Conference on Mobile Computing and Networking (Mobicom)* (1998).
13. KWON, Y., MECHITOV, K., SUNDRESH, S., KIM, W., AND AGHA, G. Resilient localization for sensor networks in outdoor environments. In *IEEE International Conference on Distributed Computing Systems (ICDCS)* (2005).
14. LAZOS, L., CAPKUN, S., AND POOVENDRAN, R. Rope: Robust position estimation in wireless sensor networks. In *International Workshop on Information Processing in Sensor Networks (IPSN)* (2005).
15. LAZOS, L., AND POOVENDRAN, R. SeRLoc: Secure range-independent localization for wireless sensor networks. In *ACM Workshop on Wireless Security (WiSe)* (2004).
16. LI, Z., TRAPPE, W., ZHANG, Y., AND NATH, B. Robust statistical methods for securing wireless localization in sensor networks. In *International Workshop on Information Processing in Sensor Networks (IPSN)* (2005).
17. LIU, D., NING, P., AND DU, W. Attack-resistant location estimation in sensor networks. In *International Workshop on Information Processing in Sensor Networks (IPSN)* (2005).
18. MARÓTI, M., KUSÝ, B., BALOGH, G., VÖLGYESI, P., NÁDAS, A., MOLNÁR, K., DÓRA, S., AND LÉDECZI, Á. Radio interferometric geolocation. In *ACM Conference on Embedded Networked Sensor Systems (SenSys)* (2005).

19. MOORE, D., LEONARD, J., RUS, D., AND TELLER, S. Robust distributed network localization with noisy range measurements. In *ACM Conference on Embedded Networked Sensor Systems (SenSys)* (2004).

20. NAGPAL, R., SHROBE, H., AND BACHRACH, J. Organizing a global coordinate system from local information on an ad hoc sensor network. In *International Workshop on Information Processing in Sensor Networks (IPSN)* (2003).

21. NAVAS, J. C., AND IMIELINSKI, T. Geographic addressing and routing. In *ACM International Conference on Mobile Computing and Networking (Mobicom)* (1997).

22. NICULESCU, D., AND NATH, B. Ad-hoc positioning system. In *IEEE Global Communications Conference (GLOBECOM)* (2001).

23. PRIYANTHA, N. B., BALAKRISHNAN, H., DEMAINE, E., AND TELLER, S. Mobile-assisted localization in wireless sensor networks. In *IEEE Conference on Computer Communications (Infocom)* (2005).

24. RÖMER, K. The lighthouse location system for smart dust. In *ACM/USENIX International Conference on Mobile Systems, Applications, and Services (MobiSys)* (2003).

25. SASTRY, N., SHANKAR, U., AND WAGNER, D. Secure verification of location claims. In *ACM Workshop on Wireless Security (WiSe)* (2003).

26. SAVARESE, C., RABAEY, J., AND LANGENDOEN, K. Positioning algorithms for distributed ad-hoc wireless sensor networks. In *USENIX Annual Technical Conference* (2002).

27. SHANG, Y., AND RUML, W. Improved mds-based localization. In *IEEE Conference on Computer Communications (Infocom)* (2004).

28. SHANG, Y., RUML, W., ZHANG, Y., AND FROMHERZ, M. P. J. Localization from mere connectivity. In *ACM International Symposium on Mobile Ad Hoc Networking and Computing (Mobihoc)* (2003).

29. STOLERU, R., HE, T., AND STANKOVIC, J. A. WalkingGPS: A practical localization system for manually deployed wireless sensor networks. In *IEEE Workshop on Embedded Networked Sensors (EmNetS)* (2004).

30. STOLERU, R., HE, T., STANKOVIC, J. A., AND LUEBKE, D. A high-accuracy low-cost localization system for wireless sensor networks. In *ACM Conference on Embedded Networked Sensor Systems (SenSys)* (2005).

31. STOLERU, R., AND STANKOVIC, J. A. Probability grid: A location estimation scheme for wireless sensor networks. In *IEEE International Conference on Sensor and Ad Hoc Communications and Networks (SECON)* (2004).

32. VICAIRE, P., AND STANKOVIC, J. A. Elastic localization. Tech. Rep. CS-2004-35, University of Virginia, 2004.

33. WHITEHOUSE, K., AND CULLER, D. Calibration as parameter estimation in sensor networks. In *ACM Intenational Workshop on Sensor Networks and Applications(WSNA)* (2002).

34. XU, Y., HEIDEMANN, J., AND ESTRIN, D. Geography-informed energy conservation for ad hoc routing. In *ACM International Conference on Mobile Computing and Networking (Mobicom)* (2001).

35. ZHANG, J., YAN, T., STANKOVIC, J. A., AND SON, S. H. Thunder: A practical acoustic localization scheme for outdoor wireless sensor networks. Tech. Rep. CS-2005-13, University of Virginia, 2005.

A Beacon-Less Location Discovery Scheme for Wireless Sensor Networks

Lei Fang[1], Wenliang Du[1], and Peng Ning[2]

[1] Department of Electrical Engineering
and Computer Science
Syracuse University
Email: {lefang,wedu}@ecs.syr.edu
[2] Department of Computer Science
North Carolina State University
Email:pning@ncsu.edu

Summary. In wireless sensor networks (WSNs), sensor location plays a critical role in many applications. Having a GPS receiver on every sensor node is costly. In the past, a number of location discovery schemes have been proposed. Most of these schemes share a common feature: they use some special nodes, called beacon nodes, which are assumed to know their own locations (e.g., through GPS receivers or manual configuration). Other sensors discover their locations based on the information provided by these beacon nodes.

In this paper, we show that efficient location discovery can be achieved in sensor networks without using beacons. We propose a beacon-less location discovery scheme. based on the following observations: in practice, it is quite common that sensors are deployed in groups, i.e., sensors are put into n groups, and sensors in the same group are deployed together at the same deployment point (the deployment point is different from the sensors' final resident location). Sensors from the same group can land in different locations, and those locations usually follow a probability distribution that can be known a priori. With this prior deployment knowledge, we show that sensors can discover their locations by observing the group memberships of its neighbors. We model the location discovery problem as a statistical estimation problem, and we use the Maximum Likelihood Estimation method to estimate the location. We have conducted experiments to evaluate our scheme.

Keyword: System Design.

1 Introduction

Sensor networks have been proposed for various applications. In many of these applications, nodes need to find their locations. For example, in rescue applications, rescue personnel can perform their tasks only if location of the hazardous event (reported by sensors) is known. Location is also important for geographic routing protocols, in which the location information (in the form of coordinates) is used to select the next forwarding host among the sender's neighbors [11–13, 20, 23]. Because of the constraints on sensors, finding location for sensors is a challenging problem. The location discovery problem is referred to as *localization* problem in the literature.

The Global Positioning System (GPS) [10] solves the problem of localization in outdoor environments for PC-class nodes. However, due to cost, it is highly undesirable to have a GPS receiver on every sensor node. This creates a demand for efficient and cost-effective location discovery algorithms in sensor networks. In the past several years, a number of location discovery protocols have been proposed to reduce or completely remove the dependence on GPS in wireless sensor networks [3,6,9,15–18,21,22]. Most of these schemes share a common feature: they use some special nodes, called beacon nodes, which are assumed to know their own locations (e.g., through GPS receivers or manual configuration). Other sensors discover their locations based on the information provided by these beacon nodes.

Although the overall cost of beacon-based location discovery schemes is significantly less than the GPS-like schemes, the cost for each beacon node is still expensive. To have a more robust and accurate positioning system, the number of beacon nodes tend to increase. Therefore, it is appealing to achieve location discovery without using beacon nodes.

In general, a positioning system consists of two components: one is the reference points, whose coordinates are known; the other is the spatial relationship between sensors and the reference points. For example, in Global Positioning System, the satellites are the reference points, and the time of arrival reveals the relationship between a GPS receiver and the satellites. In beacon-based positioning system, beacons are reference points, and relationships between a sensor and the reference points include time of arrival, time difference of arrival, angle of arrival, received signal strength, and hop-based distance, etc. For a positioning system that does not use beacon nodes, we still need to find some type of reference points with which sensors can find their locations.

We have observed that when sensors are deployed, the coordinates of the deployment points are usually known. Let us look at a deployment method that uses an airplane to deploy sensor nodes. The sensors are first pre-arranged in a sequence of smaller groups. These groups are dropped out of the airplane sequentially as the plane flies forward. This is analogous to parachuting troops or dropping cargo in a sequence. The positions where each sensor group are dropped out of the airplane are referred to as *deployment points*; their coordinates can be pre-determined and stored in sensors' memories prior to the deployment. Then during the deployment, using the GPS receivers on the airplane, we can ensure that the actual deployment points are the same as the pre-determined coordinates. We will use these deployment points as the reference points.

Next, we need to find a way to allow each sensor to establish a spatial relationship with the reference points, so that sensors can use this relationship (along with the coordinates of the reference points) to find their own locations. We have observed the following facts: after the deployment, sensors usually do not land in locations that are uniformly random across the whole deployment area, they tend to be distributed in areas around their deployment points. Therefore, sensors in different locations

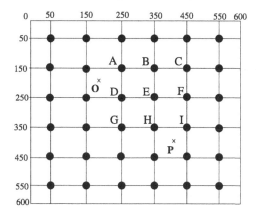

Fig. 1. An Example of Group-based Deployment (each dot represents a deployment point).

will observe different types of neighbors (i.e. neighbors from different groups). For example, assume that the deployment points are arranged in a grid style depicted in Figure 1, and a group of nodes are deployed at each deployment point. After the deployment, a node at location O will find out that more of its neighbors are from group A and D than from group H and I; on the contrary, a node at location P has more neighbors from group H and I than from A and D. This means, knowing how many of its neighbors are from each deployment group, a sensor can derive its spatial relationship with the deployment points.

To derive such a spatial relationship, we need to have a prior knowledge about how sensors from each group are distributed after the deployment, i.e., how likely can they land in a location z meters away from the their deployment points? In practice, given the methods and the conditions of the deployment, such knowledge can usually be modeled using a probability distribution function (pdf).

Based on the prior knowledge about the deployment points and the pdf of the deployment, we propose a beacon-less location discovery scheme, KPS (deployment Knowledge-based Positioning System). In our scheme, each sensor first finds out the number of its neighbors from each group. We call this the *observation* of a sensor. With this observation, a sensor estimates a location based on the principle that the estimated location should maximize the probability of the observation. This is exactly the principle of the *maximum likelihood estimation (MLE)*. Therefore, we use the MLE method to conduct the location estimation. Our results have shown that KPS can achieve a decent accuracy.

The rest of the paper is organized as follows: the next section overviews the existing work on location discovery. Section 3 presents the modeling of deployment knowledge. Section 4 describes our beacon-less scheme. Section 5 presents the evaluation results. Section 6 compares the beaconless scheme with the existing localization schemes. Finally we conclude and lay out some future work in Section 8.

2 Related Work

In the past several years, a number of location discovery protocols have been proposed to reduce or completely remove the dependence on GPS in wireless sensor networks [1, 3, 4, 6, 8, 9, 15–19, 21, 22].

Most solutions for location discovery in sensor networks require a few nodes called beacons (they are also called anchors or reference points), which already know their absolute locations via GPS or manual configuration. The density of the anchors depends on the characteristics and probably the budget of the network since GPS is a costly solution. Anchors are typically equipped with high-power transmitters to broadcast their location beacons. The remainder of the nodes then compute their own locations from the knowledge of the known locations and the communication links. Based on the type of knowledge used in location discovery, localization schemes are divided into two classes: range-based schemes and range-free schemes.

Range-based protocols use absolute point-to-point distance or angle information to calculate location between neighboring sensors. Common techniques for distance/angle estimation include Time of Arrival (TOA) [10], Time Difference of Arrival (TDOA) [1, 8, 19], Angle of Arrival (AOA) [17], and Received Signal Strength (RSS) [1]. While producing fine-grained locations, range-based protocols remain cost-ineffective due to the cost of hardware for radio, sound, or video signals, as well as the strict requirements on time synchronization and energy consumption.

Alternatively, coarse-grained range-free protocols are cost-effective because no distance/angle measurement among nodes is involved. In such schemes, errors can be masked by fault tolerance of the network, redundancy computation, and aggregation [9]. A simple algorithm proposed in [3] and [4] computes location as the centroid of its proximate anchor nodes. It induces low overhead, but high inaccuracy as compared to others. An alternate solution, DV-Hop [18], extends the single-hop broadcast to multiple-hop flooding, so that sensors can find their distance from the anchors in terms of hop counts. Using the information about the average distance per hop, sensors can estimate their distance from the anchors. Amorphous positioning scheme [15] adopts a similar strategy as DV-Hop; the major difference is that Amorphous improves location estimates using offline hop-distance estimations through neighbor information exchange.

Another existing range-free scheme is APIT algorithm [9]. APIT resolves the localization problem by isolating the environment into triangular regions between anchor nodes. A node uses the point-in-triangle test to determine its relative location with triangles formed by anchors and thus narrows down the area in which it probably resides. APIT defines the center of gravity of the intersection of all triangles that a node resides in as the estimated node location.

Our proposed scheme is significantly different from the existing schemes. The major advantage of our scheme is the removal of the dependency on the expensive beacon (or anchor) nodes. However, our scheme does not intend to replace the existing beacon-based schemes, because there are situations when the accurate deployment knowledge is difficult to obtain prior to deployment. If indeed the deployment knowledge can be obtained, our scheme can substantially reduce the cost associated with the expensive beacon nodes.

Rao et al. also proposed a localization scheme without beacons [20]. In this scheme, nodes flood the network to discover the distance (hops) between perimeter nodes. Compared to this flooding scheme, our scheme is more efficient in communications, because in our scheme, nodes only need to communicate with their neighbors once.

3 Modeling of the Deployment Knowledge

We assume that sensor nodes are static once they are deployed. We define *deployment point* as the point location where a sensor is to be deployed. This is not the location where this sensor finally resides. The sensor node can reside at points around this deployment point according to a certain probability distribution. As an example, let us consider the case where sensors are deployed from a helicopter. The deployment point is the location of the helicopter. We also define *resident point* as the point location where a sensor finally resides.

3.1 Group-based Deployment Model

In practice, it is quite common that nodes are deployed in groups, i.e., a group of sensors are deployed at a single deployment point, and the probability distribution functions of the final resident points of all the sensors from the same group are the same.

In this work, we assume such a group-based deployment, and we model the deployment knowledge in the following (we call this model the *group-based deployment model*):

1. N sensor nodes to be deployed are divided into n equal size groups so that each group, G_i, for $i = 1, \ldots, n$ is deployed from the deployment point with index i. To simplify the notion, we also use G_i to represent the corresponding deployment point, and let (x_i, y_i) represent its coordinates.
2. Locations of the deployment points are pre-determined prior to deployment. Their coordinates are stored in each sensor's memory. The deployment points can form any arbitrary pattern. For example, they can be arranged in a square grid pattern (see Figure 1), a hexagonal grid pattern, or other irregular patterns.
3. During deployment, the resident point of a node k in group G_i follows a probability distribution function $f_k^i(x, y \mid k \in G_i) = f(x - x_i, y - y_i)$. An example of the pdf $f(x, y)$ is a two-dimensional Gaussian distribution. Figure 2 shows an example of two-dimensional Gaussian distribution at the deployment point $(150, 150)$.

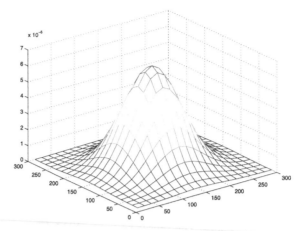

Fig. 2. Deployment distribution for one group.

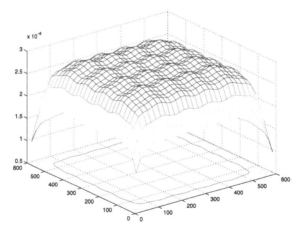

Fig. 3. The overall deployment distribution over the entire region.

3.2 Deployment Distribution

There are many different ways to deploy sensor networks, for example, sensors could be deployed using an airborne vehicle. The actual model for deployment distribution depends on the deployment method.

In this paper, we model the sensor deployment distribution as a Gaussian distribution (also called Normal distribution). Gaussian distribution is widely studied and used in practice. Although we only employ the Gaussian distribution in this paper, our methodology can also be applied to other distributions.

We assume that the deployment distribution for any node k in group G_i follows a two-dimensional Gaussian distribution, which is centered at the deployment point (x_i, y_i). Namely, the mean of the Gaussian distribution μ equals (x_i, y_i), and the pdf for node k in group G_i is the following [14]:

$$f_k^i(x,y \mid k \in G_i) = \frac{1}{2\pi\sigma^2} e^{-[(x-x_i)^2+(y-y_i)^2]/2\sigma^2}$$
$$= f(x - x_i, y - y_i),$$

where σ is the standard deviation, and $f(x,y) = \frac{1}{2\pi\sigma^2} e^{-(x^2+y^2)/2\sigma^2}$. Without loss of generality, we assume that the pdf for each group is identical, so we use $f_k(x,y \mid k \in G_i)$ instead of $f_k^i(x,y \mid k \in G_i)$ throughout this paper.

Although the distribution function for each single group is not uniform, we still want the sensor nodes to be evenly deployed throughout the entire region. By choosing a proper distance between the neighboring deployment points with respect to the value of σ in the pdf, the probability of finding a node in each small region can be made approximately equal. Assuming that a sensor node is selected to be in a given group with an equal probability, the average deployment distribution (pdf) of any sensor node over the entire region is:

$$f_{overall}(x,y) = \frac{1}{n} \sum_{i=1}^{n} f_k(x,y \mid k \in G_i). \tag{1}$$

To see the overall distribution of sensor nodes over the entire deployment region, we have plotted $f_{overall}$ in Eq. (1) for $6 \times 6 = 36$ groups over a $600m \times 600m$ square region with the deployment points $2\sigma = 100m$ apart (assuming $\sigma = 50$). We use the grid strategy to arrange the deployment points as depicted in Figure 1. Figure 3 shows the overall distribution. From Figure 3, we can see that the distribution is almost flat (i.e. nodes are fairly evenly distributed) in the whole region except near the boundaries.

4 A Beacon-Less Location Discovery Scheme

After sensors are deployed, each sensor broadcasts its group id to its neighbors, and each sensor can count the number of neighbors from G_i, for $i = 1, \ldots, n$. Assume that a sensor finds out that it has $a_1, \ldots, a_n$ neighbors from group $G_1, \ldots, G_n$, respectively. The question is whether this information, along with the deployment knowledge, can help the sensor estimate its own location.

Intuitively speaking, the observation of the neighbors' group ids is helpful. For example, if a sensor sees many of its neighbors from group G_j but zero neighbors from group G_k, we will know that the sensor is close to the deployment point of G_j, and it is far away from the deployment point of G_k. However, we need a systematic method to use this neighborhood information to calculate the sensor's location directly.

Assume that the location of the sensor of concern is $\theta = (x,y)$. Given the number (m) of nodes deployed in each group and the pdf function of the deployment, we can compute the probability that $a_1, \ldots, a_n$ nodes (from group $G_1, \ldots, G_n$, respectively) can be observed by a node at the location θ. Let X_i be the random variable that represents the number of nodes from group G_i that are neighbors to the node at location θ. Let $a = (a_1, \ldots, a_n)$ be a vector representing the observation. The probability that a is observed by a node at θ is the following:

$$f_n(a \mid \theta) = \Pr(X_1 = a_1, \ldots, X_n = a_n \mid \theta).$$

Note that, given θ, all X_i are mutually independent. Therefore,

$$
\begin{aligned}
f_n(a \mid \theta) \\
= \Pr(X_1 = a_1 \mid \theta) \cdots \Pr(X_n = a_n \mid \theta).
\end{aligned}
\tag{2}
$$

The above probability indicates how likely it is to observe $X_1 = a_1, \ldots, X_n = a_n$ at location θ. The function $f_n(a \mid \theta)$ describes the joint pdf for every observed vector $a = (a_1, \ldots, a_n)$ in the sample. When $f_n(a \mid \theta)$ is regarded as a function of θ for a given vector a, in statistics, it is called the *likelihood function*.

The goal of the location discovery now becomes an estimation problem, namely, we need to select the parameter θ from the parameter space Ω. We should certainly not consider any value of $\theta \in \Omega$ for which it would be impossible to obtain the vector a that was actually observed. Instead it would be natural to try to find a value of θ for which the probability density $f_n(a \mid \theta)$ is large, and to use this value as an estimate of θ. For each possible observed vector a, we are led by this reasoning to consider a value of θ for which the likelihood function $f_n(a \mid \theta)$ is a maximum and to use this value as an estimate of θ. This is the concept of *maximum likelihood estimation* (abbreviated as MLE).

The method of MLE was introduced by R. A. Fisher in 1912, it is by far the most widely used method of estimation in statistics. The principle of MLE is simple. That is to find the parameter values that make the observed data most likely. In other words, MLE is a method by which the probability distribution that makes the observed data most likely is sought. Details of MLE can be found in most of the statistics textbooks [5].

Let us see how to compute the likelihood function $f_n(a \mid \theta)$ when the vector a is observed. Let $g_i(\theta)$ be the probability that a sensor from group G_i can land within the neighborhood of point θ (we will show how to compute $g_i(\theta)$ later in this section). Therefore, the probability that exactly a_i sensors are within the neighborhood of point θ is the following (where m is the number of sensors deployed at each deployment point):

$$f(X_i = a_i \mid \theta) = \binom{m}{a_i} (g_i(\theta))^{a_i} (1 - g_i(\theta))^{(m-a_i)}$$

Therefore, according to Equation (2), the likelihood function $f_n(a \mid \theta)$ can be computed using the following equation:

$$
\begin{aligned}
f_n(a \mid \theta) &= \prod_{i=1}^{n} f(X_i = a_i \mid \theta) \\
&= \prod_{i=1}^{n} \binom{m}{a_i} (g_i(\theta))^{a_i} (1 - g_i(\theta))^{(m-a_i)}.
\end{aligned}
$$

The value of θ that maximizes the likelihood function $f_n(G \mid \theta)$ will be the same as the value of θ that maximizes $\log f_n(G \mid \theta)$, because logarithm is an increasing function. Therefore, it will be more convenient to determine the MLE by finding the value of θ that maximizes

$$L(\theta) = \log f_n(G \mid \theta)$$
$$= \sum_{i=1}^{n} \log \binom{m}{a_i} + \sum_{i=1}^{n} a_i \log g_i(\theta))$$
$$+ \sum_{i=1}^{n} (m - a_i) \log(1 - g_i(\theta)). \tag{3}$$

There are various ways to find the θ that maximizes $L(\theta)$. When $L(\theta)$ is differentiable and the maximal exists, it must satisfy the following partial differential equations known as the *likelihood equations*:

$$\frac{\partial L(\theta)}{\partial x} = 0 \quad \text{and} \quad \frac{\partial L(\theta)}{\partial y} = 0.$$

This is because the definition of maximum or minimum of a continuous differentiable function implies that its first derivatives vanish at such points.

If the first derivative has a simple analytic form, we can solve the above likelihood equations to find the value for $\theta = (x, y)$. However in practice, often we cannot derive an equation with a simple analytic form for its first derivative. This is especially likely if the model is complex and involves many parameters and/or complex probability functions. As we will show later when we describe how to compute $g_i(\theta)$, $L(\theta)$ is indeed very complicated. In such situations, the MLE estimate must be sought numerically. We will describe several numerical methods in the next subsection.

4.1 Finding Maximum

Gradient Descent

Gradient descent [7], also known as the method of steepest descent, is a common method in numerical analysis. The key idea of gradient descent is to find the maximum of a function based on the information of its gradient. Intuitively, we can imagine that a two-dimensional function is represented as a surface in a three-dimensional space, and the maximum point (also called peak) holds a zero gradient. The goal of the gradient descent method is to find a shortest path to reach the peak from a selected starting point. Usually the path consists of many iteration steps, and at each step, the choice of the direction is where the function increases most quickly. The whole process is like hill climbing, and the goal is to reach the top of the hill using the minimal amount of steps. To reduce the computation cost, numerous optimization schemes have been proposed to find a shortest path to the maximum point. One method is the conjugate gradient method [2], which usually converges faster to the maximum point than the gradient descent method. These optimizations are beyond the scope of this paper. In our work, we only focus on the most basic gradient descent method.

The gradient descent method, if used improperly, can be computationally intensive, and thus not suitable for resource-constrained sensor nodes. The cost of the gradient descent method in our scheme can be significantly affected by the selection of the starting point and the computation of the likelihood function $L(\theta)$ and its first derivatives. In section 4.3 and 4.4, we will show how to simplify $L(\theta)$ and its first derivatives using approximation and table-lookup approaches.

In the next subsections, we describe two algorithms that achieve a much better efficiency, however, at the cost of the accuracy. These two algorithms can be used as a stand-alone approach to estimate the location when the accuracy requirement is not high. Moreover, they can also be used to find the starting point for the gradient descent method.

A Geometric Approach

From a geometric perspective, if a sensor can get its distance from at least three deployment points, it can calculate its position. We will give a much simplified scheme to estimate a sensor's distance from a deployment point. Assume that the sensor has observed a_i neighbors from the deployment group G_i, it can use the MLE to find the distance z, such that the probability to observe a_i neighbors from group G_i is maximized.

We use $L_i(\theta)$ to represent the log likelihood function, where θ represents the location of the sensor. Based on Equation (3), we have

$$L_i(\theta) = \log f(X_i = a_i \mid \theta)$$
$$= \log \binom{m}{a_i} + a_i \log g_i(\theta)$$
$$+ (m - a_i) \log(1 - g_i(\theta)).$$

Let us use z to represent the distance from θ to the deployment point of G_i. Let $g(z)$ represent the probability that a sensor from group G_i can land within a circle (with radius R), the center of which is z distance from the deployment point of G_i. Because $g(z) = g_i(\theta)$, we can use z to replace θ in the above equation:

$$L_i(z) = \log \binom{m}{a_i} + a_i \log g(z)$$
$$+ (m - a_i) \log(1 - g(z)).$$

To find z, such that $L_i(z)$ is maximized, we let the first derivate of $L_i(z)$ be zero:

$$\frac{dL_i(z)}{dz} = 0$$

Therefore, we get the following result:

$$g(z) = \frac{a_i}{m}.$$

When $g(z)$ is complicated, we can use the table-lookup approach to find z given a_i and m, namely, we pre-calculate $g(z)$ for various values of z, and store the table of results in sensor's memory. Once $\frac{a_i}{m}$ is known after the deployment, a sensor can find z by looking up the value of z from the table. If the accuracy requirement on z is not so high, the amount of memory needed for such a table is not so large.

As we will see from the next subsection, using regression, we can approximate $g(z)$ using a Gaussian distribution. Therefore, finding z from a_i and m is quite simple.

Once we get the distance of the sensor from three deployment points, we can find the location of the sensor using the coordinates of these deployment points. The computation of this scheme is quite efficient.

Small area search approach

In the above geometric approach, because we only considered three deployment points, the accuracy might not be desirable, especially when the number of the neighbors from the other deployment points is not negligible. That is, the location we find might not be the maximum point of the likelihood function $L(\theta)$. To improve the accuracy, we use the point (X_0, Y_0) found by the geometric approach as an initial point, and then conduct a search in the nearby locations to find the maximum of $L(\theta)$, i.e., the value of $L(\theta)$ will be computed on the following points:

$$X = X_0 + i * LEN \quad -RG \leq i \leq RG$$
$$Y = Y_0 + j * LEN \quad -RG \leq j \leq RG,$$

where LEN is the length of each step (e.g. we can set it to 2 to 5 meters), and RG determines the search range. We use the number of steps along each direction to represent the range.

We will pick the point the has the maximum of $L(\theta)$ as the node's estimated location. The computational cost of the search depends on the number of steps and the step length. This approach will bring better result than the simple geometric scheme at the cost of computations. A performance comparison will be given later in Section 5.

4.2 Computing $g_i(\theta)$

We use z to represent the distance from point θ to the deployment point of group G_i. We define Ψ as the set of all deployment groups in the KPS scheme. We draw two circles. The first circle has a radius ℓ, and is centered at i, the deployment point of group G_i. We call this circle the i-circle. The second circle has a radius R, and is centered at $\theta = (x, y)$. We call this circle the θ-circle. When two circles intersect, we call the i-circle's arc within the θ-circle the L_{arc}, and we use $L_{arc}(\ell, z, R)$ to represent the length of the arc. We now consider an infinitesimal ring area $L_{arc}(\ell, z, R) \cdot d\ell$. The bold areas in Figure 4.a and 4.b show the infinitesimal ring areas.

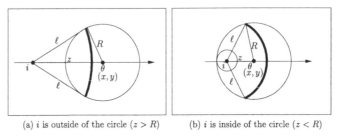

(a) i is outside of the circle ($z > R$)   (b) i is inside of the circle ($z < R$)

Fig. 4. Probability of nodes residing within a circle.

Based on the two-dimensional Gaussian distribution, the probability that a node n_i from group $i \in \Psi$ with deployment point (x_i, y_i) resides within this small ring area is

$$\frac{1}{2\pi\sigma^2} e^{-\frac{\ell^2}{2\sigma^2}} \cdot L_{arc}(\ell, z, R) \cdot d\ell$$
$$= f_R(\ell \mid n_i \in G_i) \cdot L_{arc}(\ell, z, R) \cdot d\ell,$$

where $f_R(\ell \mid n_i \in G_i)$ is defined as the following Gaussian distribution:

$$f_R(\ell \mid n_i \in G_i) = \frac{1}{2\pi\sigma^2} e^{-\frac{\ell^2}{2\sigma^2}}.$$

Using geometry knowledge, it is not difficult to derive the following equation for $L_{arc}(\ell, z, R)$:

$$L_{arc}(\ell, z, R) = 2\ell \cos^{-1}\left(\frac{\ell^2 + z^2 - R^2}{2\ell z}\right).$$

We define $g(z \mid n_i \in G_i)$ as the probability that the sensor node n_i from group i resides within the θ-circle, where z is the distance between θ and the deployment point of group G_i.

To calculate $g_i(z \mid n_i \in G_i)$, we integrate the probabilities over all the ring areas (for different ℓ) within the θ-circle. Therefore, when $z \rangle R$ (as shown in Figure 4.a),

$$g(z \mid n_i \in G_i)$$
$$\int_{z-R}^{z+R} f_R(\ell \mid n_i \in G_i) \cdot L_{arc}(\ell, z, R) \, d\ell.$$

When $z \langle R$ (as shown in Figure 4.b),

$$g(z \mid n_i \in G_i)$$
$$= \int_0^{R-z} \ell \cdot 2\pi f_R(\ell) \, d\ell$$
$$\int_{R-z}^{z+R} f_R(\ell \mid n_i \in G_i) \cdot L_{arc}(\ell, z, R) \, d\ell.$$

Putting both $z \rangle R$ and $z \langle R$ cases together, we have the following:

$$
g(z \mid n_i \in G_i)
$$
$$
= 1\{z \langle R\} \left[1 - e^{-\frac{(R-z)^2}{2\sigma^2}}\right]
$$
$$
\oint_{|z-R|}^{z+R} f_R(\ell \mid n_i \in G_i) \cdot L_{arc}(\ell, z, R) \, d\ell, \qquad (4)
$$

where $1\{\cdot\}$ is the set indicator function[3].

Therefore, $g_i(\theta)$, the probability that a node from the deployment group G_i can land within the neighborhood of point θ, can be computed in the following:

$$
g_i(\theta) = g(\sqrt{(x - x_i)^2 + (y - y_i)^2} \mid n_i \in G_i).
$$

For the sake of simplicity, we use $g(z)$ to represent $g(z \mid n_i \in G_i)$ in the rest of this paper, when it is obvious to see from the context that we are referring to the nodes in group G_i.

The formula for $g(z)$ is quite complicated, and we cannot afford to compute it using Equation (4) in sensor networks. Simplifying the analytical representation of $g(z)$, if possible, is difficult and beyond the scope of this paper. In this paper, we propose two approaches to improve the computations. The first is the table-lookup approach, and the second is the regression approach.

4.3 Simplifying $g(z)$: Table-lookup Approach

Since $g(z)$ only depends on R and σ, which are known prior to the deployment, we can pre-calculate $g(z)$ offline for each z value, and store the results as a table in sensor's memories. When a sensor needs the result for a specific value, e.g., z_0, it can use z_0 as the index to look up the value of $g(z_0)$ from the table. The computation takes only constant time.

Although the range of z is from 0 to $+\infty$, the values of $g(z)$ beyond certain range is negligible (our analysis shows that $g(z)$ is an exponentially decreasing function). Let α represent the size of the range, in which $g(z)$ has non-negligible values. We divide this range into ω equal-size sub-ranges, and store the $\omega + 1$ dividing points into a table. When a sensor needs to compute $g(z_0)$, it first finds the sub-range that contains z_0 by looking up the table; then it treats the two end-points of the sub-range as the two ends of a straight line, and finds the value corresponding to z_0 on that line. The sensor uses this value for $g(z_0)$.

As we can see that the precision of this approach depends on the size of the sub-range, the smaller the size is, the better. However, smaller sub-range also means more memory is needed for the whole table. Assume each value of $g(z)$ can be represented by two bytes, then we need 2000 bytes of memory to store the table if we divide the

---

[3] The value of $1\{\cdot\}$ is 1 when the evaluated condition is true, 0 otherwise.

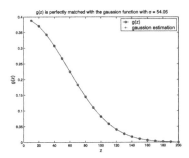

Fig. 5. Gaussian function with Ω=54.05 really matches the g(z)

range into 1000 pieces. In fact, in our experiments, when the range is divided into 200 pieces (i.e., using 400 bytes of memory), the accuracy is almost not affected.

Note that $g(z)$ does not depend on the deployment points; therefore, as long as the deployment follows the same p.d.f., the same $g(z)$ table can be used, regardless of how the deployment points are arranged.

4.4 Simplifying $g(z)$: Regression Approach

In the regression approach, we want to find a much simple representation for $g(z)$. Such representation does not need to produce the exact same values as the original $g(z)$, as long as it is a reasonable approximation.

After plotting $g(z)$, we have observed that the shape of $g(z)$ is very much like a Gaussian distribution with mean zero. Therefore we use the following Gaussian distribution to conduct the regression (the Gaussian distribution is adjusted by multiplying πR^2):

$$g(z) = (\frac{1}{2\pi\Omega^2}e^{-z^2/2\Omega^2}) \cdot \pi R^2.$$

The goal of the regression is to find out the standard deviation Ω of the regressed Gaussian distribution, such that the error between $g(z)$ and the regressed distribution function is minimized. We get the following relationships:

$$\Omega = 6.328\frac{R^2}{\sigma^2} + \sigma$$

For example, when $R = 40$, $\sigma = 50$, the value of $\Omega = 54.05$. We plot both $g(z)$ and our regression result in Figure 5. The results show that the regression is very accurate for $R = 40$ and $\sigma = 50$. We also plot the mean difference between the original $g(z)$ values and the regression results for various values of R (Figure 6). The figure shows that when R is not too large, the regression is quite accurate.

The above simplification can significantly reduce the costs for computing $L(\theta)$; however, being able to compute $L(\theta)$ efficiently is not sufficient. If the gradient descent method is to be used to find the maximum of $L(\theta)$, we should also be able to

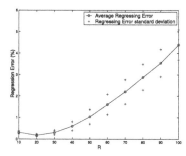

Fig. 6. Regression Errors

compute the first derivative of $L(\theta)$ efficiently.[4] Let $z^2 = (x - x_i)^2 + (y - y_i)^2$, where (x_i, y_i) is the deployment point of group G_i. The first derivative on x, $\frac{\partial L(\theta)}{\partial x}$, can be derived in the following (the first derivative on y can be similarly derived):

$$\frac{\partial L(\theta)}{\partial x} = \sum_{i=1}^{n} \frac{a_i \frac{\partial g_i(\theta)}{\partial x}}{g_i(\theta)} - \sum_{i=1}^{n} \frac{(m - a_i) \frac{\partial g_i(\theta)}{\partial x}}{1 - g_i(\theta)},$$

where $\frac{\partial g_i(\theta)}{\partial x}$ can be calculated in the following:

$$\frac{\partial g_i(\theta)}{\partial x} = \frac{R^2}{-2\Omega^4} e^{-((x-x_i)^2 + (y-y_i)^2)/2\Omega^2} (x - x_i)$$

$$= g_i(\theta) \frac{-1}{\Omega^2} (x - x_i).$$

Combining the above two equations together (and also applying the similar method to y), we get the following:

$$\frac{\partial L(\theta)}{\partial x} = \frac{-1}{\Omega^2} \sum_{i=1}^{n} \frac{a_i - mg_i(\theta)}{1 - g_i(\theta)} (x - x_i),$$

$$\frac{\partial L(\theta)}{\partial y} = \frac{-1}{\Omega^2} \sum_{i=1}^{n} \frac{a_i - mg_i(\theta)}{1 - g_i(\theta)} (y - y_i),$$

Therefore, once we know how to compute $g(z)$, we can also compute the first derivative of $L(\theta)$. To further improve the performance, we can use the table-lookup approach to store the table of the Gaussian distribution into sensor's memory. However, our experiments show only 10% of the performance improvement. This is because the computation on $g(z)$ is not the major cost.

---

[4] Although we can approximately calculate the first derivative of $g(z)$ at point z_0 by using $\frac{g(z_1) - g(z_0)}{z_1 - z_0}$, where z_1 is another point close to z_0, the computation is less accurate than the direct calculation.

5 Evaluation

This section provides a detailed quantitative analysis evaluating the performance of our beacon-less location discovery scheme. The obvious metric for the evaluation is the location estimation error. We have conducted a variety of experiments to cover different system configurations including varying the node density and varying the transmission range. We have also investigated how the boundary effects affect the accuracy of the location estimation. Moreover, we have compared the performance of the three approaches described in Section 5.4.

In our experiments, the deployment area is a square plane of 1000 meters by 1000 meters. In this paper, we only use the square grid pattern for our deployment: namely, the plane is divided into 10×10 grids of size $100m \times 100m$; centers of these grids are chosen as deployment points. Figure 1 shows our deployment strategy. Similar experiments can be conducted for other deployment patterns.

We still use m to represent the number of nodes in each group, R to represent the transmission range. We set the σ of the Gaussian distribution to 50 in all of the experiments. We then randomly generate the sensor networks based on the deployment model.

In the experiments we calculate ΔZ, the average distance between a node's actual position and estimated position. We use ΔZ as the average estimation error of KPS. In our simulation, we estimate the locations for all the nodes in the plane, and then we calculate the average errors. The numerical approach used in all the experiments is gradient descent unless it says otherwise.

5.1 Estimation Error when Varying Node Density

Because KPS is based on statistical methods, the sample size is critical to the accuracy of the estimation. In KPS, the sample size is decided by the density of the network, which is equivalent to m, the number of nodes deployed in each group (because we have fixed the deployment area and the number of deployment groups). Therefore, in this experiment, we investigate how the estimation error changes when m changes.

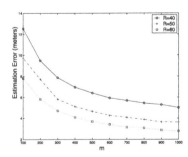

Fig. 7. Estimation errors vs. m ($R = 40$, 50, and 60).

For each experiment, we fix $\sigma = 50$, and then change m from 100 to 1000. We repeat the same experiment for $R = 40, 50$, and 60. The simulation results are depicted in Figure 7. The figure shows that our estimation is quite accurate. For example, when $m = 400$ and $R = 40$, the estimation error is only 8 meters, which equals $0.2R$. The figure also shows that the accuracy of the location estimation becomes better when m increases, i.e., each sensor can observe more nodes in its neighborhood.

In practice, if we do not have enough sensors to deploy to reach the desired node density, we can still achieve the desired density by deploying dummy nodes along with the sensor nodes. A dummy node is a low-cost node, whose only functionality is to broadcast its group identity to its neighbors. A dummy node does not need to find its own location, nor does it need to carry out sensing or computing tasks. Its only goal is to increase the sample size, such that the sensors in its neighborhood can estimate their location more accurately. Therefore, the cost of a dummy node can be much lower than a sensor node.

5.2 Estimation Error when Varying Transmission Range

Another way to increase the sample size is to increase the transmission range R. When R increases, the number of neighbors for each sensor will increase. In this experiment, we investigate how R affects the estimation accuracy. We fix $m = 400$, and vary R from 40 meters to 120 meters. The simulation results are depicted in Figure 8.

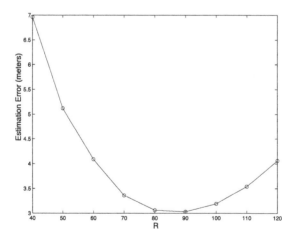

Fig. 8. Estimation errors vs. R ($m = 400$).

The figure shows an interesting trend: when R increases from 40 to 90, the estimation error decreases without a surprise. However, starting from $R = 90$, the estimation error increases. This can be intuitively explained using an extreme-case example: assume that $R = \infty$, which means that all the sensor nodes can observe exactly the same set of neighbors (i.e., all the other nodes in the network). Therefore, the estimated location for all the sensors will be the same (actually it will be the center of the deployment area).

The extreme-case example indicates that when R increases to infinite, the estimation error will increase, and eventually will converge to a constant, when the center of the deployment area is selected as every sensor's estimated location. This is largely due to the boundary effects that we are going to discuss in the next experiment. Namely, when R increases, more and more nodes will be affected by the boundary effects because their neighboring areas cover the areas outside of the deployment area, where the node density is close to zero. Therefore, for those nodes, the difference of their observations becomes smaller and smaller while the difference of their locations is still constant. We will further investigate the boundary effects in our experiments.

The fact that Figure 8 has a minimum point tells us we should choose the proper R in practice; just increasing R won't always give us better results.

5.3 Estimation Error vs. Boundary Effects

Boundary is also a factor we must consider. Because there are less nodes on the boundaries, the variance of a node's neighbors is large compared to the nodes near the center. So it's less accurate for nodes to determine their positions with their observations.

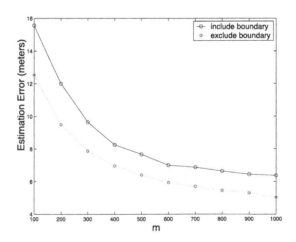

Fig. 9. Boundary ($R = 40$ and $\sigma = 50$).

In the experiment, we calculate the errors in two different ways: one includes the nodes on the boundary, and the other does not. The boundary nodes are defined as those that are within 50 meters of any of the four borders (50 is chosen because $\sigma = 50$). The results are shown in Figure 9. It is clear that nodes deployed near the boundary will make the estimation error larger.

5.4 Comparison of Three Find-Maximum Methods

As we have discussed in Section 4.1, we propose to use three different approaches to find the maximum point of the likelihood function. Among these three approaches, gradient descent can provide the best accuracy, but its computation cost is most expensive. The geometric approach is the least expensive one, but it produces the worst estimation error. The small-area-search approach is in the middle. In this experiment, we quantitatively compare the computation cost and the accuracy of these three approaches.

Computation Cost

The algorithms for the three approaches are tested on a PC with Intel P4 2.8G Hz CPU and 1G memory. We set $R = 40$ and $m = 100$. We measure the average time for a sensor to find its location. Their performance comparisons are shown in Table 1.

Table 1. Comparison of the Three Numerical Approaches.

Algorithm	Computation Expense
Geometric Method	0.02ms
Small Area Search (1 step)	0.05ms
Small Area Search (2 steps)	0.14ms
Small Area Search (3 steps)	0.26ms
Small Area Search (4 steps)	0.32ms
Gradient Decent	0.68ms

The relative comparison among these algorithms is more important than their absolute values. We can find that the computational cost of the gradient descent is 34 times more expensive than the geometric method. Given the fact that the geometric method is very simple (its cost is almost negligible), and the location discovery is only conducted once, the gradient descent method is also affordable for sensor networks. Also as we mentioned before, implementing the optimization technologies such as table-lookup in the sensor system will make it more realistic to use the gradient descent algorithm.

Estimation Accuracy

From Figure 10, we see that the gradient descent approach supplies the best results and the geometric approach produces the worst results. The small-area-search scheme becomes more and more accurate when the number of steps increases. The figure shows that the accuracy of the 2-step method is already close to the gradient decent method.

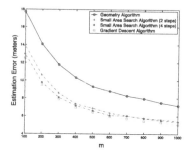

Fig. 10. Comparison of different numerical approach ($R = 40$).

6 Comparisons with existing schemes

In this section, we compare our KPS scheme with the existing beacon-based location schemes. Because KPS and its assumptions are significantly different from the existing beacon-based schemes, comparing the localization accuracy is not much meaningful. Therefore our comparisons mainly focus on cost, robustness, security, and mobility.

6.1 Cost analysis

Communication Cost

Communication cost is a major concern in sensor networks. For the beacon-based localization schemes, communication cost is very low, because sensor nodes only need to receive signals from the beacon nodes; there is no interaction between sensor nodes. The KPS scheme depends on the knowledge of neighbors, it requires each node to broadcast (only one-hop broadcasting) a message to its neighbors. However, this broadcasting is necessary for neighbor discovering that is required by other functionalities of sensor networks, such as routing. Therefore, the KPS localization scheme does not introduce extra communication cost.

Computation and Storage Cost

Compared with the beacon-based location scheme, the calculation of KPS is more complex, so the computation cost of KPS is much higher than most of the beacon-based localization schemes. Most computation burden comes from the Find-Maximum methods. However, our simulation results have shown that the computation cost is still realistic.

As we mentioned in 4.3, to reduce the computation cost, we can store some precalculated table in sensor's memory. The size of the table can be limited to several kilobytes.

Device Cost

The cost of device on beacon-based schemes is much higher than the KPS scheme. KPS is sensitive to node density. If the node density of the sensor networks is high enough, no extra device is needed; if the node density is too low, cheap dummy nodes can be deployed to help achieve acceptable localization accuracy. However, beacon-based schemes must depend on special beacon nodes, which are much more expensive than normal sensor nodes.

6.2 Robustness and Security

In the beacon-based schemes, the localization accuracy largely depends on a small number of beacon nodes. When some of these nodes fail to function or when they are tampered with by adversaries (for example, some compromised beacon nodes might report false positions), a significant number of sensors can be affected, i.e., their derived locations can be much far away from their actual locations.

In contrast, the beacon-less KPS scheme is much more robust and secure. In KPS, each sensor depends on its neighbors to find its own location. When one or a few neighbors fail, the localization results will not be affected much. When some compromised neighbors intentionally send out false group memberships, their lies cannot be arbitrary, because a lie that deviates too much from the deployment knowledge can reveal anomalies. Therefore, the KPS scheme can even tolerate node compromise to certain degree. Further analysis regarding this property is undergoing.

6.3 Limitation of the KPS Scheme

Although KPS achieves localization without using expensive beacon nodes, it does have its limitations. First, Beacon-based schemes support mobile sensor networks. Namely nodes can obtain their locations even if they are mobile. However, KPS depends on the distribution of the node deployment; once a node moves, the distribution cannot be maintained. Therefore, KPS can only be used in a static sensor networks. Second, locations of deployment points are critical. They must be estimated with high accuracy. Although this can be easily achieved for an airborne deployment because GPS can be used on an airplane, the goal is hard to achieve for other types of deployment. In addition, KPS also requires an accurate modeling of deployment knowledge. In our future work, we will study the accuracy of localization if the actual deployment deviates from the model. Due to these limitations, we do not claim that KPS can replace the existing beacon-based localization schemes in all applications. Our KPS scheme provides an less-expensive alternative in those applications that satisfy our assumptions. We believe such assumptions are reasonable in many sensor network applications.

6.4 Summaries

The comparisons of the KPS scheme and beacon-based localization schemes are summarized in Table 2

Table 2. Comparison of KPS and beacon-based schemes.

	KPS	beacon-based
Communication overhead	Low	Low
Computation cost	High	Low
Device cost	Low	High
Robustness/Security	High	Low
Mobility	None	Good

7 Conclusion and Future Work

In sensor networks, traditional localization schemes use beacons as the reference points to help sensors find their locations. We present KPS, a beacon-less localization scheme, in which sensors use the deployment distribution and the position of deployment points to find the locations. The major advantage of the KPS scheme is that we do not need the expensive beacon nodes, while achieving comparable location discovery results. We have conducted extensive evaluation. Our results show that when the node density is high, the location estimation error achieved by KPS can be less than a few meters. These results show that the accuracy provided by KPS is sufficient to support various applications in sensor networks. In our future work, we plan to study how the inaccuracy of the deployment knowledge can affect the accuracy of the location discovery. Our motivation is that in practice, the deployment knowledge that we know prior to the deployment might not be quite accurate. It will be interesting to know how KPS is affected by that. We also plan to provide more analytical evaluation results on KPS.

8 Acknowledgment

Du's work was supported by Grants ISS-0219560 and CNS-0430252 from the US National Science Foundation, and also by Grant W911NF-05-1-0247 from the US Army Research Office (ARO). Ning's work was supported by Grants CNS-0430223 and CAREER-0447761 from the US National Science Foundation.

References

1. P. Bahl and V. N. Padmanabhan. RADAR: An in-building RF-based user location and tracking system. In *Proceedings of the IEEE INFOCOM*, pages 775–784, March 2000.
2. R. Barrett, M. Berry, T. F. Chan, J. Demmel, J. Donato, J. Dongarra, V. Eijkhout, R. Pozo, C. Romine, and H. Van der Vorst. *Templates for the Solution of Linear Systems: Building Blocks for Iterative Methods, 2nd Edition.* SIAM, Philadelphia, PA, 1994.
3. N. Bulusu, J. Heidemann, and D. Estrin. GPS-less low cost outdoor localization for very small devices. In *IEEE Personal Communications Magazine*, pages 28–34, October 2000.
4. N. Bulusu, J. Heidemann, and D. Estrin. Density adaptive algorithms for beacon placement, April 2001.

5. M. H. DeGroot and M. J. Schervish. *Probability and Statistics*. Addison Wesley, 3rd edition, 2002. Chapter 6.

6. L. Doherty, K. S. Pister, and L. E. Ghaoui. Convex optimization methods for sensor node position estimation. In *Proceedings of INFOCOM'01*, 2001.

7. R. Hamming. *Numerical Methods for Scientists and Engineers*. Dover Pubns, 2nd edition, 1987.

8. A. Harter, A. Hopper, P. Steggles, A. Ward, and P. Webster. The anatomy of a context-aware application. In *Proceedings of MOBICOM'99*, Seattle, Washington, 1999.

9. T. He, C. Huang, B. M. Blum, J. A. Stankovic, and T. F. Abdelzaher. Range-free localization schemes in large scale sensor networks. In *Proceedings of the Ninth Annual International Conference on Mobile Computing and Networking (MobiCom '03)*, 2003.

10. B. Hofmann-Wellenhof, H. Lichtenegger, and J. Collins. *Global Positioning System: Theory and Practice*. Springer Verlag, 4th ed., 1997.

11. X. Hong, K. Xu, and M. Gerla. Scalable routing protocols for mobile ad hoc networks. *IEEE Network magazine*, (4), 2002.

12. B. Karp and H. T. Kung. GPSR: Greedy perimeter stateless routing for wireless networks. In *Proceedings of ACM MobiCom 2000*, 2000.

13. Y. B. Ko and N.H. Vaidya. Location-aided routing (lar) in mobile ad hoc networks. In *Proceedings ACM/IEEE MOBICOM 98)*, pages 66–75, October 1998.

14. A. Leon-Garcia. *Probability and Random Processes for Electrical Engineering*. Reading, MA: Addison-Wesley Publishing Company, Inc., second edition, 1994.

15. R. Nagpal, H. Shrobe, and J. Bachrach. Organizing a global coordinate system from local information on an ad hoc sensor network. In *IPSN'03*, 2003.

16. A. Nasipuri and K. Li. A directionality based location discovery scheme for wireless sensor networks. In *Proceedings of ACM WSNA'02*, September 2002.

17. D. Niculescu and B. Nath. Ad hoc positioning system (APS) using AoA. In *Proceedings of IEEE INFOCOM 2003*, pages 1734–1743, April 2003.

18. D. Niculescu and B. Nath. Dv based positioning in ad hoc networks. In *Journal of Telecommunication Systems*, 2003.

19. N. B. Priyantha, A. Chakraborty, and H. Balakrishnan. The cricket location-support system. In *Proceedings of MOBICOM*, Seattle, Washington'00, August 2000.

20. A. Rao, S. Ratnasamy, C. Papadimitriou, S. Shenker, and I. Stoica. Geographic routing without location information. In *Proceedings of ACM MOBICOM 2003*, pages 96–108, September 2003.

21. A. Savvides, C. Han, and M. Srivastava. Dynamic fine-grained localization in ad-hoc networks of sensors. In *Proceedings of ACM MobiCom '01*, pages 166–179, July 2001.

22. A. Savvides, H. Park, and M. Srivastava. The bits and flops of the n-hop multilateration primitive for node localization problems. In *Proceedings of ACM WSNA '02*, September 2002.

23. Y. Xu, J. Heidemann, and D. Estrin. Geography-informed energy conservation for ad hoc routing. In *Proceedings of ACM MobiCom 2000*, Rome, Italy, July 2001.

Learning Sensor Location from Signal Strength and Connectivity*

Neal Patwari[1], Alfred O. Hero III[1], and Jose A. Costa[2]

[1] Dept. of Electrical Engineering and Computer Science
University of Michigan, Ann Arbor, USA
npatwari@umich.edu and hero@umich.edu

[2] Center for the Mathematics of Information
California Institute of Technology, Pasadena, USA
jcosta@caltech.edu

Received signal strength (RSS) or connectivity, *i.e.*, whether or not two devices can communicate, are two relatively inexpensive (in terms of device and energy costs) measurements at the receiver that indicate the distance from the transmitter. Such measurements can either be quickly dismissed as too unreliable for localization, or idealized by ignoring the non-circular nature of a transmitter's coverage area. This chapter finds a middle ground between these two extremes by using measurement-based statistical models to represent the inaccuracies of RSS and connectivity.

While a particular RSS or connectivity measurement may be hard to predict, a statistical model for RSS and connectivity can in fact be well-characterized. Many numerical examples are used to provide the reader with intuition about the variability of real-world RSS and connectivity measurements.

This chapter then gives a description of three sensor localization algorithms which are based on 'manifold learning', a class of non-linear dimension reduction methods. These algorithms include Isomap [1], the distributed weighted multidimensional scaling (dwMDS) algorithm [2], and the Laplacian Eigenmap adaptive neighbor (LEAN) algorithm [3]. The performance of these estimators is compared via simulation using the RSS and connectivity measurement models.

The results show that while RSS and connectivity measurements are highly variable, these manifold learning-based algorithms demonstrate their robustness by achieving location estimates with low bias and often variance close to the lower bound. Due to their desirability as low cost, low complexity measurements, system designers should consider RSS and connectivity for sensor network applications.

* This work was partially funded by the DARPA Defense Sciences Office under Office of Naval Research contract #N00014-04-C-0437. Distribution Statement A. Approved for public release; distribution is unlimited.

1 Introduction

As indicated in the scope of this edited volume, localization is a key technology in wireless sensor networks. However, there is a tradeoff between accurate sensor location and simple, low cost and energy efficient devices. In many applications, device cost will be a more severe constraint than accuracy.

Application Example: Logistics. As an example, consider deploying a sensor network in a warehouse. Pallets, boxes and parts to be warehoused are tagged with wireless sensors when first brought into the facility. The sensors will allow both monitoring of storage conditions (such as humidity and temperature) and determination of the object location at all times. Compared to radio-frequency identification (RFID) tags, which are only located when they pass within a few feet of a reader, networked wireless sensors can be queried and located as long as they are within range of the closest other wireless sensor in the network.

The objective in warehouse logistics is allowing a human to more quickly locate a particular object in the warehouse; thus being able to locate the device to within a few meters will likely be acceptable. The high density of objects to be tracked will require that each sensor must have extremely low costs, on the order of cents, in order to make it cost effective. Connectivity and RSS become very desirable methods.

Application Example: Traffic Monitoring. RSS and connectivity would be valuable in vehicle traffic monitoring. If you're stuck on the highway in a traffic jam, and can't get a traffic report on the radio, it would be very valuable to know how far the backup extends, in order to plan an alternate route if necessary. If a small percentage of cars had a short-range wireless sensor (with accelerometer and a compass), a message could hop forward, find the point at which the traffic eases, and then measure backwards, via RSS or connectivity, the distance to your car. Its possible that messages could propagate backwards to alert the drivers of cars before they hit the jam. Low device cost and long battery life would be critical to achieve a high enough density of sensors to make this application useful.

1.1 Relevant Research

There is now a large literature of algorithms which use RSS or connectivity (a.k.a. proximity) for sensor localization. These algorithms are either distributed or centralized. A number of distributed localization algorithms are compared in [4]. RSS-based localization has been implemented on motes and Besides the papers within this volume, the references within [5, 6] also provide links to the extensive literature.

This chapter focuses on the subset of algorithms which propose manifold learning-based algorithms for sensor localization, which are introduced in Section 4. In general, manifold learning approaches can lead to distributable algorithms that extract the information perceived to be most accurate. The communication can often be limited to just the nearest neighbors, and there are manifold learning algorithms guaranteed to find the global optimal solution.

1.2 Outline of Chapter

Prior to discussing the manifold learning algorithms themselves, this chapter explores in detail statistical measurement models and their derivation from empirical radio channel studies. Section 2.1 presents RSS measurements and Section 2.2 presents connectivity measurements. Then, to provide more intuitive understanding of the probabilities involved, Section 3 presents two numerical examples and graphically presents results which can be derived directly from the statistical models of Section 2. Finally, Section 4 presents a general overview and comparison of three manifold learning-based algorithms. Section 5 makes those comparisons quantitative by using the statistical models from Section 2 in simulations. Section 6 discusses the results and concludes.

2 Measurement Models

The key to developing reliable sensor localization systems which use pair-wise measurements is to accurately represent the statistics of the measurements. The system designer must design sensor networks that will be deployed in many places and many environments, none of which are known to the designer. Over the ensemble of these deployments, the environment-dependent errors in the measurements are unpredictable and must be modeled as random. In this section, we discuss what past measurements have indicated about these statistical models.

2.1 Received Signal Strength

It is assumed that a system designer will do two things to attempt to reduce the variability in RSS measurements. In particular, a pair of sensors will:

1. Make multiple RSS measurements over time, and
2. Use a wideband measurement of RSS.

The first assumption is that in applications in which sensors are mostly stationary, it would be acceptable to trade off some time delay for increased accuracy. When there are moving objects such as people or vehicles in the environment, the time averaging possible with the multiple measurements allows us to reduce their effects.

The latter assumption helps reduce the RSS variability to frequency selective fading. In multipath channels, multipath signals add constructively or destructively at the receiver as a function of the frequency. A wideband measurement, such as using frequency hopping spread-spectrum (FH-SS) or orthogonal-frequency division multiplexing (OFDM), can reduce the variability by averaging RSS over a large frequency band.

The RSS measurement that has been averaged is largely subject to shadowing effects. Shadowing is the signal attenuation caused by stationary objects in the radio channel, such as walls, furniture, and buildings, and can't be averaged out by time or frequency averaging.

Statistical Model

Typically, the ensemble mean received power in a real-world, obstructed channel decays proportional to d^{-n_p}, where n_p is the 'path-loss exponent', typically between 2 and 4 [7, 8]. The ensemble mean power at distance d is typically modeled as

$$\bar{P}(d) = \Pi_0 - 10 n_p \log_{10} \frac{d}{\Delta_0} \tag{1}$$

where Π_0 is the received power (dBm) at a short reference distance Δ_0.

The difference between a measured received power and its ensemble average, due to the randomness of shadowing, is modeled as log-normal (*i.e.*, Gaussian if expressed in dB). The log-normal model is based on many years of radio channel measurement results [7–9] and analytical evidence [10]. This model has further been tested via experimental measurements in wireless sensor networks in both indoor and outdoor environments, at both 900 MHz and at 2.4 GHz [11, 12]. These measurements have verified the ensemble mean power model from (1) and that the variation around the ensemble mean is log-normal. Results from one of these measurement campaigns, reported in [12], are shown in Fig. 1. This campaign measured the RSS between each pair of sensors in a 44-node wireless sensor network.

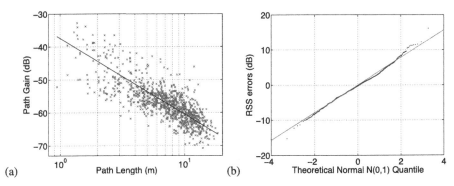

Fig. 1. RSS measurements in a wireless sensor network in [12] show that (a) mean RSS $\bar{P}(\|\mathbf{z}_i - \mathbf{z}_j\|)$ decays linearly with log distance as in (1) with $n_p = 2.3$, and $\sigma_{dB} = 3.92$; and (b) the quantile-quantile (QQ) plot of RSS errors, $P_{i,j} - \bar{P}(\|\mathbf{z}_i - \mathbf{z}_j\|)$ (dB), vs. the Gaussian distribution.

As a result, the received power (dBm) at sensor i transmitted by j, $P_{i,j}$, is distributed as

$$f\left(P_{i,j} = p | \{\mathbf{z}_i\}_{i=1}^N\right) = \mathcal{N}\left(p; \bar{P}(\|\mathbf{z}_i - \mathbf{z}_j\|), \sigma_{dB}^2\right), \tag{2}$$

where $\mathcal{N}(x; \mu, \sigma)$ is our notation for the value at x of a Gaussian p.d.f. with mean μ and variance σ, and the actual transmitter-receiver separation distance $\|\mathbf{z}_i - \mathbf{z}_j\|$ is given by

$$\|\mathbf{z}_i - \mathbf{z}_j\| = \sqrt{(x_i - x_j)^2 + (y_i - y_j)^2}, \tag{3}$$

for a two-dimensional location coordinate $\mathbf{z}_i = [x_i, y_i]^T$. Note that the standard deviation of received power (when received power is expressed in dBm), σ_{dB}, has units of (dB) and is relatively constant with distance. Typically, σ_{dB} is as low as 4 and as high as 12 [7].

Estimating Range from RSS

The 'range', *i.e.*, the estimated distance between devices i and j, can be estimated from $P_{i,j}$. First, the maximum likelihood estimate of range is presented. The log-likelihood of $P_{i,j}$ given $d_{i,j} = \|\mathbf{z}_i - \mathbf{z}_j\|$ is,

$$\log f \left(P_{i,j} | \{\mathbf{z}_i\}_{i=1}^N \right) = c_1 - \frac{\left[P_{i,j} - \bar{P}(\|\mathbf{z}_i - \mathbf{z}_j\|) \right]^2}{2\sigma_{dB}^2} \tag{4}$$

where log indicates the natural logarithm, and c_1 is a constant independent of $\{\mathbf{z}_i\}_{i=1}^N$. Because of the quadratic form, it is clear that the maximum of the log-likelihood occurs when $P_{i,j} = \bar{P}(\|\mathbf{z}_i - \mathbf{z}_j\|)$, where the $\bar{P}$ is given in (1). As a direct result, the distance $\delta_{i,j}^{MLE}$ which best estimates $\|\mathbf{z}_i - \mathbf{z}_j\|$ in the maximum-likelihood sense is,

$$\delta_{i,j}^{MLE} = \Delta_0 10^{\frac{\Pi_0 - P_{i,j}}{10 n_p}} \tag{5}$$

Consider what happens if we write $P_{i,j} = \bar{P}(\|\mathbf{z}_i - \mathbf{z}_j\|) + \eta_{i,j}$, where $\eta_{i,j}$ is 'noise' in the measurement which is zero-mean, Gaussian, with variance σ_{dB}^2. In this case,

$$\delta_{i,j}^{MLE} = \Delta_0 10^{\frac{\Pi_0 - \bar{P}(\|\mathbf{z}_i - \mathbf{z}_j\|) - \eta_{i,j}}{10 n_p}}$$

$$\delta_{i,j}^{MLE} = \|\mathbf{z}_i - \mathbf{z}_j\| 10^{-\frac{\eta_{i,j}}{10 n_p}}. \tag{6}$$

The expected value of the MLE distance estimate is,

$$E \left[\delta_{i,j}^{MLE} \right] = C \|\mathbf{z}_i - \mathbf{z}_j\|, \tag{7}$$

where

$$C = \exp \frac{1}{2\gamma}, \quad \text{where } \gamma = \left(\frac{10 n_p}{\sigma_{dB} \log 10} \right)^2. \tag{8}$$

The parameter C is a multiplicative bias factor, a function of the ratio, σ_{dB}/n_p. For $\sigma_{dB}/n_p = 1.70$, as measured in [12], $C = 1.08$, and for $\sigma_{dB}/n_p = 2.45$, as measured in [11], $C = 1.18$. So, depending on the channel parameters, this bias can be a significant factor.

Motivated by (7), a bias-corrected estimator (a pseudo-MLE) of distance can be defined just by dividing the MLE by C,

$$\delta_{i,j}^{BC} = \frac{\Delta_0}{C} 10^{\frac{\Pi_0 - P_{i,j}}{10 n_p}}. \tag{9}$$

The most important result of the log-normal model is that RSS-based range estimates (from either estimator above) have standard deviation proportional to their actual range. Consider the variance of the MLE distance estimator, which can be calculated to be,

$$\text{var}\{\delta_{i,j}^{MLE}\} = (C^4 - C)\|z_i - z_j\|^2$$

This is why RSS errors are referred to as multiplicative. In comparison, errors in distance estimates based on time-of-arrival (TOA) are additive. Clearly, RSS is most valuable between nearby sensors.

2.2 Connectivity Measurements

It is common for localization research to consider connectivity (a.k.a. proximity) measurements as a simple, inexpensive, low-bandwidth, and backward-compatible location measurement. Whether or not devices have accurate RSS measurement circuitry on their receivers, two devices can determine whether or not they can communicate. Two sensors are *not* considered to be connected solely based on the distance between them – two sensors are connected if the receiving sensor can successfully demodulate packets transmitted by the other sensor. The receiver fails to successfully demodulate packets when the received signal strength (RSS) is too low. Since RSS is a random variable due to the unpredictability of the fading channel, and connectivity is a function of RSS, connectivity is also a random variable.

Binary Quantization Model

Specifically, the connectivity measurement of sensors i and j, $Q_{i,j}$, is modeled as a binary quantization of RSS,

$$Q_{i,j} = \begin{cases} 1, & P_{i,j} \geq P_1 \\ 0, & P_{i,j} \langle P_1 \end{cases} \tag{10}$$

where $P_{i,j}$ is the received power (dBm) at sensor i transmitted by sensor j, and P_1 is the receiver threshold (dBm) under which packets cannot be demodulated. Note that we can both talk about a receiver power threshold P_1 and also the threshold distance d_1 at which the mean received power is P_1. From (1), this threshold distance is

$$d_1 = \Delta_0 10^{\frac{\Pi_0 - P_1}{10 n_p}}. \tag{11}$$

Noise-free Connectivity Model

To generate simulation results for connectivity-based localization, the 'noise-free' connectivity model is sometimes used. In this model, radio coverage is assumed to be a perfect circle around the transmitter. Thus, pairs of devices will have exact knowledge of whether or not they are separated by more or less than the coverage radius. Since this is not complete knowledge of distance itself, such a model would still result in localization uncertainty or 'error'. However, these errors are likely to be just a small contribution to the localization errors in a real system. While such a model might be appropriate for the formulation or visualization of localization algorithms, it is not a means for accurate simulation of estimator variance.

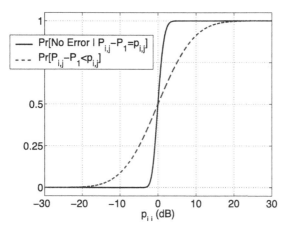

Fig. 2. Two plots relating to the variation in proximity measurements when sensors i and j are separated by the threshold distance d_1 from (11): (- - - - -) the CDF of $P_{i,j}$ in dB above the receiver threshold P_1 from (10), and (——) the probability of no packet error given $P_{i,j}$ in dB above P_1 (for a packet of 200 bits and a coherent BPSK receiver without FEC).

More Detailed Connectivity Models

In reality, being in-range of another device (transmitting a packet which the other device correctly demodulates) is not a step function of received power. Two additional sources of variation in connectivity measurements are:

1. The randomness of frame errors given the received power level, and
2. The possibility of multiple-access interference during a particular transmission.

To discuss issue (1), consider that if we are given received power $P_{i,j}$, connectivity $Q_{i,j} \in \{0, 1\}$ is a binary random variable, such that

$$\mathcal{P}[Q_{i,j} = 1 | P_{i,j}] = \mathcal{P} \, [\text{No Packet Error} | P_{i,j}] \qquad (12)$$

where the probability of a packet error is a function of the type of signalling and forward error correction (FEC) used, packet length, and whether the receiver is coherent or non-coherent. If all of these details of the transceiver implementation were known, a more accurate model could be defined using (12). However, note that the uncertainty in the RSS given distance is much more significant than the uncertainty in error-free packet reception given RSS. In typical digital receivers, there is a large range of received powers for which the probability of packet error is very close to zero, and a large range of power for which the probability is very close to one. The range of power for which the probability of packet error is neither close to one or zero is small in comparison. Fig. 2 plots $\mathcal{P}[\text{No Packet Error} | P_{i,j}]$ from (12) under the following conditions: a packet of 200 bits, a coherent BPSK receiver, with no FEC. For comparison, Fig. 2 also plots the CDF of received power for sensors separated by d_1 under a log-normal model with $\sigma_{dB} = 7.38$ dB (as measured in [11]).

Regarding issue (2), outside interference or multiple-user interference (MUI) from other sensors will cause packets to be lost at random times. The interference power raises the noise floor and increases the required power threshold for the desired signal. However, if sensors send multiple packets over time, especially for networks of mostly stationary sensors, it will be likely that a packet received with RSS greater than P_1 will be received without MUI during some transmission, and thus the sensors will measure that they are connected.

Statistical Model

Given the binary quantization model for connectivity in (10) and the log-normal model for $P_{i,j}$ in (2), it can be shown that the probability mass function of $Q_{i,j}$ given the coordinates of devices i and j is

$$
\mathcal{P}[Q_{i,j} = s | \mathbf{z}_i, \mathbf{z}_j] = \begin{cases} \Phi\left[\sqrt{\gamma} \log \frac{\|\mathbf{z}_i - \mathbf{z}_j\|}{d_1}\right], & s = 0 \\ 1 - \Phi\left[\sqrt{\gamma} \log \frac{\|\mathbf{z}_i - \mathbf{z}_j\|}{d_1}\right], & s = 1 \end{cases} \tag{13}
$$

where $\Phi[\cdot]$ is the cumulative distribution function (CDF) of the univariate zero-mean unit-variance Gaussian distribution, and d_1 is the threshold distance given in (11), and γ is given in (8).

3 Numerical Examples

For the purposes of obtaining an intuitive understanding of the variability of RSS and connectivity measurements, it is valuable to show some numerical values for particular cases. The following examples show that RSS and connectivity are in fact highly variable measurements. While these numbers might scare away system designers, this is certainly not the desired effect. In fact, these models are used to generate RSS and connectivity in Section 5 and demonstrate the accuracy of location estimation algorithms. The ability to achieve localization, even given the variability of the measurements, indicates the robustness of the 'cooperative' localization concept.

3.1 RSS-based Distance Estimates

Consider the log-normal RSS measurement $P_{i,j}$ between devices i and j. Consider two different channels, with channel parameters $\sigma_{dB}/n_p = 1.70$ and $\sigma_{dB}/n_p = 2.48$, which correspond to results from measurement campaigns reported in [12] and [11], respectively. (The lower channel parameter was a result of using a very wideband measurement of RSS, while the higher measured the RSS of a continuous wave (CW) signal.)

As a numerical example, let's calculate the probability that the bias-corrected estimate of distance, $\delta_{i,j}^{BC}$, is within an interval around the correct distance. Specifically, consider the interval,

$$0.5\|\mathbf{z}_i - \mathbf{z}_j\| \langle \delta_{i,j}^{BC} \langle 2.0\|\mathbf{z}_i - \mathbf{z}_j\| \tag{14}$$

From (6), we can rewrite the interval from (14) as,

$$0.5\|\mathbf{z}_i - \mathbf{z}_j\| \langle \frac{\|\mathbf{z}_i - \mathbf{z}_j\|}{C} 10^{-\frac{\eta_{i,j}}{10 n_p}} \langle 2.0\|\mathbf{z}_i - \mathbf{z}_j\|$$

$$\frac{\log(0.5C)}{\log 10} \langle -\frac{\eta_{i,j}}{10 n_p} \langle \frac{\log(2.0C)}{\log 10}$$

$$-10 n_p \frac{\log(2.0C)}{\log 10} \langle \eta_{i,j} \langle -10 n_p \frac{\log(0.5C)}{\log 10}$$

Since $\eta_{i,j}$ is zero-mean Gaussian with standard deviation σ_{dB}, the probability that $\eta_{i,j}$ falls in this interval is just the area under a normal curve, excluding the tails. Specifically,

$$\mathcal{P}\left[\frac{\|\mathbf{z}_i - \mathbf{z}_j\|}{2} \langle \delta_{i,j}^{BC} \langle 2\|\mathbf{z}_i - \mathbf{z}_j\|\right] = \Phi\left[\sqrt{\gamma}\log\frac{2}{C}\right] - \Phi\left[\sqrt{\gamma}\log\frac{1}{2C}\right] \tag{15}$$

where γ is given in (8).

Numerical Solutions

For the two measured channel parameters, $\sigma_{dB}/n_p = 1.70$ and $\sigma_{dB}/n_p = 2.48$, the numerical solutions to (15) are 0.92 and 0.76, respectively.

If instead of using half and twice the actual distance as the interval min and max in (14) we had used 2/3 and 3/2 of the actual distance, specifically,

$$0.667\|\mathbf{z}_i - \mathbf{z}_j\| \langle \delta_{i,j}^{BC} \langle 1.5\|\mathbf{z}_i - \mathbf{z}_j\|,$$

then these probabilities would be reduced to 0.69 and 0.51 for the $\sigma_{dB}/n_p = 1.70$ and $\sigma_{dB}/n_p = 2.48$ channels, respectively.

What is the difference between the results using the bias-corrected estimator and the MLE estimator of distance? Deriving from (5), we would arrive at the same formula as (15), but with $C = 1$. The probabilities that the MLE distance estimate would be in either the wider or narrower interval are slightly higher than for the bias-corrected estimate (by about 0.01), using either of the two channel parameter ratios. Thus bias correction may help with aligning the estimator mean to the true mean, but it doesn't help reduce the confidence intervals.

Clearly, if RSS is to be used as an estimate of distance, system robustness must be considered. Even in the better channel, the probability of getting a decent estimate of distance (within 2/3 and 3/2 of the actual distance) is 69%. A sensor localization system in a wireless network must be designed to make many RSS measurements between many pairs of devices, such that the worst of the errors can be discarded or down-weighted.

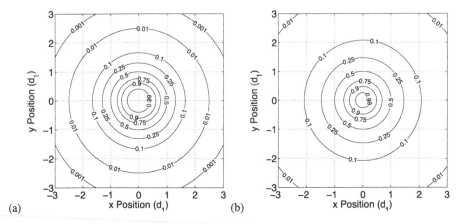

Fig. 3. For a transmitter at the center $(0,0)$, the probability that the received power will be above the receiver threshold, as a function of the location of the receiver, given a channel parameter σ_{dB}/n_p of (a) 1.70 [12] and (b) 2.48 [11]. Units are in terms of the threshold distance d_1.

3.2 Connectivity and Coverage Area

Equation 13 gives the probability of connectivity for a pair of devices separated by distance $\|z_i - z_j\|$, given the channel parameter σ_{dB}/n_p and the distance threshold d_1. To make this more concrete, two examples are provided in Fig. 3.

As indicated by the figure, regardless of the channel parameter, the probability of having received power higher than the threshold when separated by the threshold distance is 50%. But when the ratio σ_{dB}/n_p is higher, it is more likely that a distant receiver *will* be connected. As a comparison, in Fig. 3(a), the 10% probability contour line is about $1.65d_1$ away from the transmitter, while in 3(b), the 10% probability contour line is about $2.08d_1$ away. Similarly, when the ratio σ_{dB}/n_p is higher, it is more likely that a nearby receiver will *not* be connected. To show this, in Fig. 3(a), the 90% probability contour line is about $0.60d_1$ away from the transmitter, while in 3(b), this line is about $0.48d_1$ away. In sum, as the ratio σ_{dB}/n_p increases, we are less certain of *either* nearby sensors measuring that they are connected or distant sensors measuring that they are not connected.

4 Manifold Learning Localization Algorithms

In cooperative sensor localization, we have a network of N sensors, and we want to estimate the coordinates $\{z_i\}_{i=1}^n$ of sensors $1\ldots n$, which have unknown location. A few sensors, $n+1\ldots N$ are assumed to have perfect *a priori* knowledge of their coordinates, $\{z_i\}_{i=n+1}^N$. We are given a mesh of many pair-wise distance or connectivity estimates, from RSS or connectivity, as discussed above. While not all pairs will make measurements, we assume that many neighboring sensors will.

Algorithms have been developed by applying manifold learning techniques to the sensor location estimation problem [1–3, 13–17]. Manifold learning is a class of non-linear dimension reduction methods. These are an extension of linear dimension reduction methods such as multi-dimensional scaling (MDS) or principle components analysis (PCA). In PCA, low dimensional coordinates are found by projecting the high dimensional data to the linear subspace which best fits the data. When the high dimensional data don't lie in a linear subspace, the results are inaccurate. For example, for the 3-D data in Fig. 5, PCA would attempt to find a 2-D plane that, when the data were projected to the plane, would best fit the data. Since the 3-D data actually lie on a curved surface, no single plane would serve to fit all of the data. In comparison, in manifold learning, the subspace is only assumed to be locally linear. When reducing the dimension of the data, only the relationships between neighboring high dimensional data points are preserved.

In sensor localization, the manifold learning framework is applicable for two reasons:

1. Measurements between nearby sensors are often more precise and less biased than those between further apart sensors, and
2. Using only measurements between nearby sensors reduces bandwidth, energy, and computational requirements.

Determination of the Neighbor Graph

The first step in manifold learning algorithms is to determine the neighbor graph. In the manifold learning literature, a pair are considered to be neighbors if the Euclidean distance between their high-dimensional data points is less than a threshold. For pair-wise RSS measurements, we might threshold the measured $\delta_{i,j}$ with a pre-determined radius R, or an adaptive threshold set to ensure at least K neighbors. These two neighbor selection methods are called the 'R-radius' and the 'K-nearest-neighbors' (KNN) methods, respectively. Note that the KNN method essentially sets a dynamic radius R for each device depending on its local sensor density, so in either case, we can refer to the threshold distance. For connectivity, the receiver's power threshold decides on neighbors – if $Q_{i,j} = 1$, then we believe that $\delta_{i,j} \langle d_1$ for the receiver threshold distance d_1 given in (11), and thus we consider i and j to be neighbors. For connectivity, our pre-determined radius R is equal to d_1, and for RSS, we may select $R \le d_1$.

In wireless sensor networks, distance or connectivity between sensors is measured in a noisy channel, so there is an additional complication – it isn't known which sensors are *actually* within R of each other. Neighbor selection in noise is discussed in Section 4.5 and in more detail in [2, 3].

In sum, the selection of neighbors determines a graph in which neighboring sensors' nodes are connected, and non-neighboring sensors' nodes are not. This neighborhood graph is the key input into the next step.

Low-dimensional Coordinate Estimation

The second step in manifold learning algorithms is to find the low-dimension coordinates which best represent the neighbor relationships. This search can be represented as the minimization a cost function or as a constrained optimization problem. Generally, these optimization approaches are of two types: *distance-based* and *similarity-based* approaches. These are contrasted by analogy in Fig. 4. As the name would indicate, the distance-based methods encode information regarding the distances between points in the graph. The similarity-based methods encode inverse distance, or some decreasing function of distance. These cost functions are described in detail in the next sections.

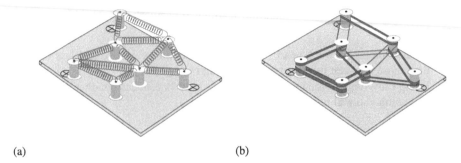

(a)   (b)

Fig. 4. Physical analogy of manifold learning algorithms based on (a) distance or (b) similarity between sensors. Sensors (spools) are connected by (a) springs which have natural length equal to the measured distance and can can push *and* pull their neighbors, or by (b) rubber bands with different weights (thickness of the band) which can *only* pull sensors together. Scaling and rotation are constrained to match the *a priori* known coordinates $\otimes$.

	Isomap / MDS-MAP	Laplacian Eigenmap / LEAN	dwMDS
Preserves	Distance	Similarity	Distance
Algorithm Basis	Eigen-decomposition	Eigen-decomposition	Iterative, distributed majorization
Minimize	$\sum_{i,j}(\|\mathbf{z}_i - \mathbf{z}_j\|^2 - \bar{\delta}_{i,j}^2)^2$	$\sum_{i,j} w_{i,j}\|\mathbf{z}_i - \mathbf{z}_j\|^2$	$\sum_{i,j} w_{i,j}(\|\mathbf{z}_i - \mathbf{z}_j\| - \delta_{i,j})^2$ $+r_i\|\mathbf{z}_i - \bar{\mathbf{z}}_i\|^2$
Computational Complexity	$\mathcal{O}(N^3)$	$\mathcal{O}(KN^2)$	$\mathcal{O}(KLN)$
Post-processing	Rotation	Rotation and Scaling	None
Notes	Sensitive to large range errors	Natural for connectivity	Directly incorporates prior info

4.1 Isomap

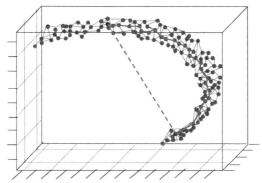

Fig. 5. In Isomap, the distance between two devices is not the direct path in the higher dimensional space (- - - -); rather, it is the shortest path distance between the two points along the edges of the nearest neighbor graph (——).

In Isomap [18], the distances $\delta_{i,j}$ measured between non-neighbors are ignored. Instead, Isomap replaces distances $\delta_{i,j}$ in (16) with $\tilde{\delta}_{k,l}$, which is set equal to the sum length along the shortest path on the neighbor graph between sensors k and l, for all pairs $(k,l) \in \{1, \ldots, N\}^2$. The general idea is that the shortest path on the neighborhood distance graph is a good approximation to the shortest distance on the manifold, as demonstrated in Fig. 5. Isomap then minimizes the cost,

$$S_{Isomap} = \sum_{i=1}^{N} \sum_{j=1}^{N} \left(\tilde{\delta}_{i,j}^2 - \|\mathbf{z}_i - \mathbf{z}_j\|^2 \right)^2 . \tag{16}$$

An algorithm called MDS-MAP, introduced by Shang et. al. [1, 13, 14], applies the Isomap algorithm, when measurements are connectivity, to sensor localization.

Computation: Because the distances are squared before taking the difference, the cost is a quadratic function of the individual coordinates. The minimum of S_{Isomap} can be found directly from the singular value decomposition of the appropriate transform of the $N \times N$ distance matrix $D^2 = [[\delta_{i,j}^2]]_{i,j}$, as derived in detail in [19]. This eigen-decomposition operation has computational complexity $\mathcal{O}(N^3)$. The Isomap algorithm also requires finding of shortest path between each pair of sensors in the network. Using Dijkstra's algorithm, this is an $\mathcal{O}(N^2)$ operation, and this calculation can be performed in a distributed manner in a wireless network.

Post-Processing: Isomap produces the relative, but not absolute map of all N devices in the network. As a post-processing step, all coordinate estimates are transformed (rotate, scale, and translate) by the transformation that makes the coordinate estimates of the known-location sensors, $\{\hat{\mathbf{z}}_i\}_{i=n+1}^{N}$, best match their *a priori* known coordinates in a least-squares sense.

4.2 Laplacian Eigenmap Adaptive Neighbor (LEAN)

The LEAN algorithm combines Laplacian Eigenmap (LE), a similarity-based manifold learning method, with an an adaptive neighbor weighting algorithm. This section first discusses the LE method given a set of weights $\{w_{i,j}\}$. Then, Section 4.4 discusses the initial selection of the weights, and Section 4.5 discusses the iterative adjustment of the weights in a 2-step adaptive algorithm.

The LE method considers the minimization of the cost S_{LE}:

$$S_{LE} = \sum_{i,j} w_{i,j} \|\mathbf{z}_i - \mathbf{z}_j\|^2 \tag{17}$$

subject to the translation and scaling constraints,

$$\sum_i \mathbf{z}_i = \mathbf{0} \quad \text{and} \quad \sum_i \|\mathbf{z}_i\|^2 = 1. \tag{18}$$

where $w_{i,j}$ are weights representing the 'similarity' of devices i and j. In the LEAN algorithm, these weights depend on the RSS or connectivity measurements, and are given explicitly in Sections 4.4 and 4.5. The key fact is that $w_{i,j} = 0$ when i and j are non-neighbors (*i.e.*, when $Q_{i,j} = 0$ or $\delta_{i,j} \rangle R$).

The minimum of cost S_{LE} without the constraints in (18) would occur when all the coordinates $\mathbf{z}_i$ were equal. The constraints remove the translation ambiguity by setting the origin as the center, and counteract the tendency to put all points at the origin by mandating a unit norm average coordinate.

Computation: The benefit of the formulation in (17) and (18) is that the globally optimum solution is found via eigen-decomposition. Defining the $N \times N$ weight matrix $W = [[w_{i,j}]]_{i,j}$ and its column sums (or row sums, since W is symmetric) $u_i = \sum_{j=1}^N w_{i,j}$, the graph Laplacian L_W is given by,

$$L_W = \text{diag}[u_1, \dots, u_N] - W, \tag{19}$$

where $\text{diag}[u_1, \dots, u_N]$ is the diagonal matrix with $\{u_i\}$ on its diagonal. Matrix L_W is sparse, since $w_{i,j} = 0$ for non-neighbors, and each row or column has at most $K_{max} + 1$ non-zero elements, where $K_{max} = \max_i K_i$, and K_i is the number of neighbors of sensor i. The eigen-decomposition of L_W is the set of $(\lambda_k, \mathbf{v}_k)$, for eigenvalues λ_k and eigenvectors $\mathbf{v}_k$, $k = 1 \dots N$. Here, it is assumed w.l.o.g. that the eigenvectors are sorted in increasing order by magnitude of eigenvalue. As presented in detail by Belkin and Niyogi in [20], the $\mathbf{v}_k$ for $i = 2 \dots r + 1$ provide the optimal lowest-cost, r-dimensional solution to (17). Specifically,

$$\hat{\mathbf{z}}_i = [\mathbf{v}_2(i), \dots, \mathbf{v}_{r+1}(i)], \tag{20}$$

where $\mathbf{v}_k(i)$ is the ith element of the kth eigenvector.

Finding the smallest eigenvalues and eigenvectors of a sparse and symmetric matrix is a computational problem which has been studied for decades for problems in physics and chemistry [21, 22], and can be solved using distributed algorithms for parallel processing. In particular, if sensors select local cluster-heads, the distributed algorithm can use data-distribution techniques and block-Jacobi preconditioning methods to reduce communication. Due to the sparsity of the graph Laplacian matrix, the computational complexity of the eigen-decomposition is $\mathcal{O}(KN^2)$, where K is the average number of neighbors of each sensor.

Post-Processing: As in the Isomap algorithm, the LE method produces a relative map of the coordinates. The same method as in the previous section is used to determine the best scaling, rotation, and translation of the coordinates based on the *a priori* coordinate information.

Related Research: The locally linear embedding (LLE) [23] and the Hessian-based LLE (HLLE) methods are also similarity-based manifold learning algorithms. The HLLE method [24] expands the optimization to attempt to preserve the local Hessian, *i.e.*, 2nd-order differences within local neighborhoods, within the final low-dimensional coordinate embedding.

4.3 Distributed Weighted Multi-dimensional Scaling (dwMDS)

The dwMDS method minimizes a cost function S_{dwMDS} in a distributed manner, by having each sensor i, for $i = 1 \dots n$, iteratively minimize its own local cost function, S_i [2, 16]. The global cost is additive, *i.e.*,

$$S_{dwMDS} = \sum_{i=1}^{n} S_i. \tag{21}$$

Using the dwMDS algorithm to optimize S_i at each unknown-location sensor $i = 1 \dots n$ acts to optimize S_{dwMDS}. In the dwMDS method, S_i are given by

$$S_i = r_i \|\mathbf{z}_i - \bar{\mathbf{z}}_i\|^2 + \sum_{j=1}^{N} \tilde{w}_{i,j} \left(\delta_{i,j} - \|\mathbf{z}_i - \mathbf{z}_j\| \right)^2, \tag{22}$$

where $\bar{\mathbf{z}}_i$ represents the mean coordinate of the *a priori* coordinate distribution for sensor i, r_i is the confidence in that mean coordinate, and $\tilde{w}_{i,j}$ is a weight corresponding to the expected accuracy in the $\delta_{i,j}$ measurement. When $r_i = 0$, it indicates no prior information exists for i, and any $0\langle r_i\langle\infty$ indicates imperfect but partial prior knowledge of i's location. Also, let $\tilde{w}_{i,j} = 2w_{i,j}$ if either $i\rangle n$ or $j\rangle n$, and $\tilde{w}_{i,j} = w_{i,j}$ otherwise. This is done so that each measurement $\delta_{i,j}$ is treated equally, whether or not it was measured between two unknown-location sensors or between and unknown-location and known-location node. Weights $w_{i,j}$ are similar to those used in Section 4.2 and their selection is discussed in Section 4.4.

Computation: In the dwMDS algorithm, sensors serially optimize their own coordinate, given their neighbors most recent coordinate estimate. Sensor i, during its turn, improves its estimate of $\mathbf{z}_i$. This improvement is done by optimizing a quadratic majorization function for S_i, which guarantees that each iteration of the optimization reduces the global cost S_{dwMDS}. The update function for i is simply a linear function of its neighbors' most recent coordinate estimates. We leave the detailed derivation and presentation of the calculation to the references [2, 16]. Each sensor requires calculation on the order of its number of neighbors K, in each of the L iterations required for convergence, so the total network-wide computational complexity is $\mathcal{O}(LKN)$. The dwMDS also has a slower increase in communication requirements than centralized localization algorithms as N increases [16]. Furthermore, since prior information is included directly in (22), there is no need to do post-processing.

4.4 Weight Selection

In this section, we describe the selection of weights $w_{i,j}$ in the LEAN and dwMDS methods. We consider separately measurements of RSS and connectivity.

RSS Measurements When distance estimates are available from RSS, weights $\{w_{i,j}\}$ in the LE and dwMDS methods are set as follows. The LOESS method, a non-parametric scheme, is used to set $w_{i,j}$:

$$w_{i,j} = \begin{cases} \exp\left\{-\delta_{i,j}^2/h_{i,j}^2\right\}, & \text{if } i \text{ and } j \text{ are neighbors} \\ 0, & \text{otherwise} \end{cases}, \qquad (23)$$

where $h_{i,j}$ is the maximum distance $\delta_{k,l}$ measured by either sensor i or j.

Connectivity Measurements Since no pure distance estimate is available when only connectivity is measured, it would seem that any $w_{i,j}$ between connected sensors i and j should have identical weight. However, this scheme is not the best approach because it tends to give too much 'pull' to sensors with many neighbors. This 'pull' serves to bias other sensors' coordinates towards itself, as described in [6]. Here, we use a simple symmetric weight scheme that sets weights so that the total weights for sensor i, $u_i = \sum_j w_{i,j} \approx 1$:

$$w_{i,j} = (1/K_i + 1/K_j)/2, \qquad (24)$$

where K_l is the total number of neighbors of sensor l. While u_i is not identical for all i, it is slightly higher for sensors with more neighbors than their neighbors. This behavior is desirable and helps to improve bias performance.

4.5 Adaptive Neighbor Selection:

Typically, we select neighbors closer than a threshold distance R. For RSS measurements, this distance R can be set less than d_1, the receiver threshold distance from (11), but in connectivity, note that $R = d_1$ is the only option. The key issue that this section addresses is that when distances estimated from noisy RSS or connectivity measurements are used to select neighbors, the neighbor selection process can be the source of significant bias. The act of selecting the neighbors with $\delta_{i,j}$ less than a threshold has a tendency to select the $\delta_{i,j}$ which are, on average, less than the actual distances $\|z_i - z_j\|$. And for connectivity, we know from Section 3.2 that sensors may include as neighbors many sensors which are further than R and may ignore sensors closer than R.

Both the dwMDS algorithm and the LE-based algorithm specifically counter this bias effect by using a two-stage adaptive algorithm:

Stage 1: First, distances $\{\delta_{i,j}\}$ are measured, and those (i, j) with $\delta_{i,j} \langle R$ (or $Q_{i,j} = 1$ for connectivity) are selected as neighbors. The dwMDS or LEAN algorithm then computes $\{\tilde{z}_i\}_{i=1}^n$, which are referred to as 'interim' coordinate estimates.

Stage 2: Next, the neighborhood is adjusted based on the interim coordinate estimates:

- In the dwMDS algorithm, the neighborhood graph is completely re-calculated based solely on the interim coordinates. Sensors with interim coordinates closer than R, i.e., $\|\tilde{z}_i - \tilde{z}_j\| \langle R$, are selected as neighbors. The weight matrix W is re-calculated as given in Section 4.4 using the new neighbor graph.
- In the LE-based localization algorithm, define K_i to be the number of neighbors of sensor i. Also define $\tilde{K}_i$ to be the number of sensors with interim coordinates closer than R to $\tilde{z}_i$, i.e., $|\{j : \|\tilde{z}_i - \tilde{z}_j\| \langle R\}|$. Next, set the new weights $w_{i,j}$ so that the new sensor weight sums $\tilde{u}_i$ are given by $\tilde{u}_i = u_i \sqrt{K_i/\tilde{K}_i}$, as detailed in [3]. Essentially, increase a sensor's pull if its estimate is in a less dense area than it should be.

Using the new neighborhood structure, the dwMDS algorithm or the LE algorithm is re-run to estimate final coordinate estimates $\{\hat{z}_i\}$. Sections 5.1 and 5.2) show the dramatic effects of the adaptive neighbor selection in sensor localization.

5 Simulation Results

Geometries: The performance of these three algorithms is first demonstrated on a grid network, of 7×7 sensors arranged on a uniform grid of unit area, as shown in Fig. 6, in which the four corner devices are reference nodes and the remaining 45 are unknown location devices, and $L = 1$ m. The grid geometry is chosen first because it shows the geometric bias effects very well. Subsequently, in Section 5.4, random geometries are explored via simulation. Note that the choice of $L = 1$m is arbitrary, since any scaling of L would proportionally scale the simulation errors and lower bound on standard deviation. Essentially, all distances, estimator biases and standard devations can be taken in units of L.

Simulation Input and Output: For all experiments, $R = 0.4$ or 0.3 is chosen as the threshold distance, and the channel parameter ratio σ_{dB}/n_p is set to either 1.70 or 2.48. Independent Monte Carlo trials (200) are run to determine uncertainty ellipses and bias performance (per sensor) of the location estimates. The results are displayed in many figures in this section. Each figure plots the estimator mean (▼) and 1-standard deviation uncertainty ellipse of the estimator (——), compared to the

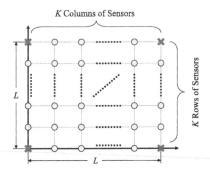

Fig. 6. Grid geometry layout of sensors.

actual device location (•). One-standard deviation uncertainty ellipses, as a coarse generalization, can be seen as a 2-D confidence interval, within which most of the coordinate estimates will lie.

Also plotted in the simulation figures are the Cramér-Rao lower bound (CRB) on the uncertainty ellipses (- - - - -), which were derived and presented for RSS and connectivity measurements in [12] and [25], respectively. Any unbiased estimator must have 1-σ uncertainty ellipse larger than that given by the CRB. Note that the CRB may only provide a loose lower bound on the best performance achievable by any unbiased estimator.

For comparing different estimators, let the mean bias $\bar{b}$ and the RMS standard deviation $\bar{\sigma}$ of the estimator be defined as:

$$\bar{b} = \frac{1}{n} \sum_{i=1}^{n} \|\bar{\mathbf{z}}_i - \mathbf{z}_i\|, \qquad \bar{\sigma} = \sqrt{\frac{1}{n} \operatorname{tr} \mathbf{C}} \qquad (25)$$

where n is the number of unknown-location sensors, $\bar{\mathbf{z}}_i$ is the mean of all of the estimates of sensor i over all trials of the simulation, $\mathbf{z}_i$ is the actual location of sensor i, and $\mathbf{C}$ is the covariance of the coordinate estimates over all trials.

5.1 RSS Results from dwMDS in Grid

To show the benefit of adaptive neighborhood selection from Section 4.5, we show in Fig. 7 the performance of the dwMDS algorithm with and without its 2nd adaptive stage, and with different values of σ_{dB}/n_p and threshold R. Fig. 7(a) stops the algorithm after Stage 1, using the interim coordinates as the final estimates. The biasing effect of neighborhood selection in noise results in $\bar{b} = 0.12$, by effectively shortening the average distance estimates and thus forcing a smaller sensor location estimate map. In contrast, Fig. 7(b) allows the completion of Stage 2, and is nearly unbiased with $\bar{b} = 0.02$, and has $\bar{\sigma} = 0.09$. Except at the edge nodes, the estimator variance is visibly close to the lower bound. The dwMDS algorithm (and also the lower bound) degrade with increasing ratio σ_{dB}/n_p as shown in Fig. 7(c), which has $\bar{\sigma} = 0.16$. The algorithm is shown to be robust the change to $R = 0.3$ in Fig. 7(d) which results in $\bar{\sigma} = 0.11$, and we note without showing additional plots that the error performance is robust to a wide range of R.

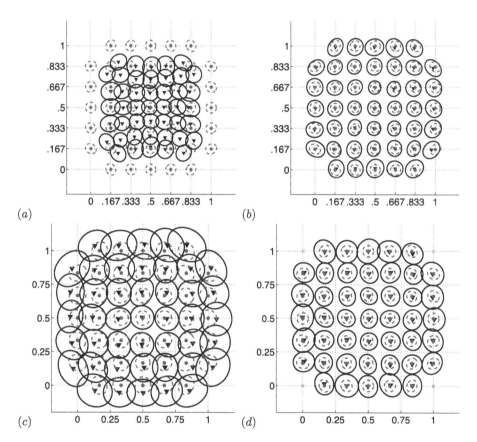

Fig. 7. Comparison of RSS-based dwMDS estimators using (a) single-stage or (b) adaptive, 2-stage neighbor selection, both with $\sigma_{dB}/n_p = 1.70$ and $R = 0.4$. Also shown is the adaptive dwMDS performance (c) when $\sigma_{dB}/n_p = 2.48$ and $R = 0.4$ or (d) when $\sigma_{dB}/n_p = 1.70$ and $R = 0.3$.

5.2 Connectivity Results from LEAN in Grid

The same tests as performed in Section 5.1 are now performed on the LEAN algorithm using measurements of connectivity. We should expect that the variance will increase, and we do see this in Fig. 8. Figs. 8(a) and 8(b) compare the LEAN performance with and without the second stage of the adaptive neighbor selection algorithm. In this comparison, the improvement is only marginal - both the bias and standard deviation decrease only slightly, so that $\bar{b} = 0.03$ and $\bar{\sigma} = 0.14$ in (b). The real benefit of the 2-stage adaptive LEAN algorithm is its ability to keep the bias very low over a wide range of σ_{dB}/n_p and R, a much wider range than possible without the adaptive weighting. Figs. 8(c) and 8(d) show the simulation results when the 2-stage adaptive LEAN algorithm is used with (c) $\sigma_{dB}/n_p = 2.48$ and $R = 0.4$; and (d) $\sigma_{dB}/n_p = 1.70$ and $R = 0.3$. The bias is nearly constant in all figures (a-d), but in (c), $\bar{\sigma} = 0.20$ and in (d), $\bar{\sigma}$ is the same as in (b) at 0.14.

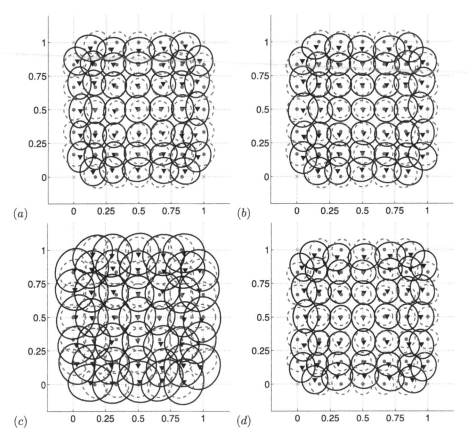

Fig. 8. Comparison of LEAN estimators using (a) single-stage or (b) adaptive, 2-stage neighbor selection, both with $\sigma_{dB}/n_p = 1.70$ and $R = 0.4$. Also shown is the adaptive dwMDS performance (c) when $\sigma_{dB}/n_p = 2.48$ and $R = 0.4$ or (d) when $\sigma_{dB}/n_p = 1.70$ and $R = 0.3$.

5.3 Connectivity and RSS Results from Isomap/MDS-MAP Method in Grid

Isomap can be run either from RSS or connectivity measurements. When RSS measurements are available, $\delta_{i,j}$ is calculated using the MLE in (5). When only connectivity measurements are available, $\delta_{i,j} = 1$ when i and j are connected (as in [1]). Since Isomap computes a relative map and then scales it to match the prior information, this choice of $\delta_{i,j} = 1$ for two connected devices is arbitrary and irrelevant. Simulation results are shown in Fig. 9(a-d), for the cases of measurements of connectivity or RSS, and for the channels with σ_{dB}/n_p equal to 1.70 and 2.48. These changing parameters and the simulation results are shown in Table 1.

	Simulation Inputs		Outputs	
	Measurements	σ_{dB}/n_p	$\bar{b}$	$\bar{\sigma}$
Fig. 9(a)	RSS	1.70	0.03	0.15
Fig. 9(b)	Connectivity	1.70	0.02	0.21
Fig. 9(c)	RSS	2.48	0.04	0.23
Fig. 9(d)	Connectivity	2.48	0.05	0.28

Table 1. Isomap simulation inputs and outputs.

In all simulations, $R = 0.4$. The estimator is largely unbiased, but the variance is significantly larger than the lower bound. In general, this can be attributed to the form of the cost in (16), which is a function of the squared difference between squared distances, rather than just the squared difference between distance itself. The squaring of the distance before taking the difference enables solution via eigen-decomposition as discussed in Section 4.1, but the algorithm is more sensitive to the tails of the density of $\delta_{i,j}$.

5.4 Simulation Results in a Random Deployment

Fig. 10 shows the performance of the dwMDS and LEAN algorithms in two different random deployments of sensors. The 49 coordinates for the random deployment were chosen independently from a uniform distribution on the square area. Then, the four sensors closest to the corners are selected as reference devices. All simulations use $\sigma_{dB}/n_p = 1.70$ and $R = 0.4$. In the first randomly-generated geometry, Fig. 10(a) shows dwMDS performance to be $\bar{b} = 0.03$ and $\bar{\sigma} = 0.10$; and Fig. 10(b) shows LEAN performance to be $\bar{b} = 0.04$ and $\bar{\sigma} = 0.12$. In the second randomly-generated geometry, Fig. 10(c) shows dwMDS performance to be $\bar{b} = 0.02$ and $\bar{\sigma} = 0.09$, while Fig. 10(d) shows the LEAN performance to be $\bar{b} = 0.05$ and $\bar{\sigma} = 0.12$.

Compared to the grid geometry, there are a few sensors with very high bias in the LEAN algorithm results. However, the dwMDS algorithm results show very low bias, even for the most isolated sensors. Variance is slightly higher for the dwMDS algorithm, while the LEAN algorithm has lower variance than in the grid geometry. The LEAN algorithm may actually benefit, in terms of variance, from the non-uniform density of sensors across the randomly-deployed network.

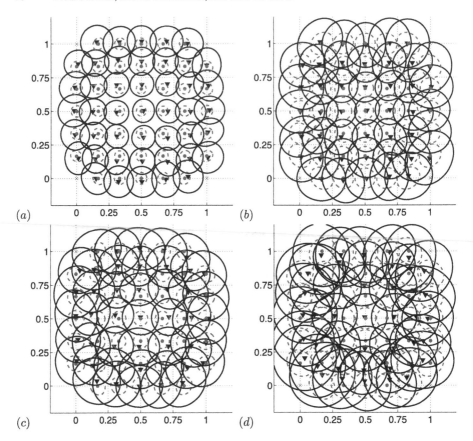

Fig. 9. Comparison of Isomap estimators using (a)&(c) RSS measurements and (b)&(d) connectivity measurements. All use $R = 0.4$, but for (a)&(b) $\sigma_{dB}/n_p = 1.70$ and for (c)&(d) $\sigma_{dB}/n_p = 2.48$.

6 Discussion and Conclusion

The simulation results show us particular advantages of the dwMDS and LEAN methods compared to the Isomap-based methods, when comparing estimator covariance. Furthermore, the standard deviation of localization using connectivity measurements (via LEAN) is almost 60% higher than with RSS measurements (via dwMDS). These increase in variance is approximately the same as the increase in the lower bounds from RSS to connectivity, as related in [25]. Both estimators, except for sensors at the edge of the networks, perform reasonably close to the lower bound.

One perspective on this analysis is that a high path loss exponent n_p is actually desirable in terms of localization accuracy. If a system operates in an environment in which n_p is high, but σ_{dB} is not proportionally as high, then more location information is possible. When antennas are very close to the ground, or when center frequencies are in oxygen absorption bands, path loss exponents are typically higher.

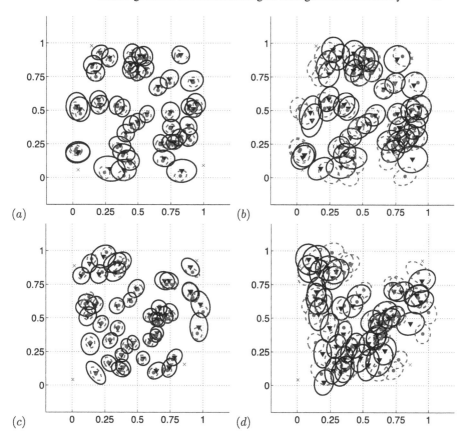

Fig. 10. Simulation of estimators (a)&(c) dwMDS with RSS and (b)&(d) LEAN with connectivity measurements, in two different non-uniform, random deployment of sensors. All use $R = 0.4$ and $\sigma_{dB}/n_p = 1.70$.

Of course, higher path loss exponents mean that transmit powers must be higher in order to achieve the same coverage.

Adaptive methods help dramatically improve both estimator performance (in the dwMDS algorithm) and help increase the variety of environments and network densities in which the estimator performs with low bias and variance close to the lower bound (as seen in the LEAN algorithm). These benefits of adaptive neighborhoods are demonstrated by the simulation results.

Opportunities for future research are many. While LEAN is distributable because of its reliance on finding a few of the smallest eigen-vectors of a sparse symmetric matrix, algorithms from the parallel processing literature must be optimized to perform well in a energy-limited sensor network. Other manifold learning-based methods are possible, and can be explored for their ability to reduce the computational or communication requirements of sensor localization.

Regardless, though, the statistics of RSS and connectivity measurements are critical to the determination of many measures of 'how well' a particular localization system will perform. This chapter has provided a thorough background in these models, and shown how they can be used in simulation of localization algorithms.

References

1. Yi Shang, Wheeler Ruml, Ying Zhang, and Markus P. J. Fromherz, "Localization from mere connectivity," in *Mobihoc '03*, June 2003, pp. 201–212.
2. Jose A. Costa, Neal Patwari, and Alfred O. Hero III, "Achieving high-accuracy distributed localization in sensor networks," in *IEEE Intl. Conf. Acoustic, Speech, & Signal Processing (ICASSP'05)*, March 2005, pp. 642–644.
3. Neal Patwari and Alfred O. Hero III, "Adaptive neighborhoods for manifold learning-based sensor localization," in *IEEE Workshop on Signal Processing Adv. Wireless Commun. (SPAWC'05)*, June 2005, pp. 1098–1102.
4. Koen Langendoen and Niels Reijers, "Distributed localization in wireless sensor networks: A quantitative comparison," *Elsevier Science*, May 2003.
5. Bhaskar Krishnamachari, *Networking Wireless Sensors*, Cambridge University Press, Cambridge UK, 2005.
6. Neal Patwari, Joshua Ash, Spyros Kyperountas, Robert M. Moses, Alfred O. Hero III, and Neiyer S. Correal, "Locating the nodes: Cooperative localization in wireless sensor networks," *IEEE Signal Processing Mag.*, vol. 22, no. 4, pp. 54–69, July 2005.
7. Theodore S. Rappaport, *Wireless Communications: Principles and Practice*, Prentice-Hall Inc., New Jersey, 1996.
8. Homayoun Hashemi, "The indoor radio propagation channel," *Proceedings of the IEEE*, vol. 81, no. 7, pp. 943–968, July 1993.
9. Greg Durgin, Theodore S. Rappaport, and Hao Xu, "Measurements and models for radio path loss and penetration loss in and around homes and trees at 5.85 GHz," *IEEE Journal on Sel. Areas in Comm.*, vol. 46, no. 11, pp. 1484–1496, Nov. 1998.
10. Alan J. Coulson, Allan G. Williamson, and Rodney G. Vaughan, "A statistical basis for lognormal shadowing effects in multipath fading channels," *IEEE Trans. on Veh. Tech.*, vol. 46, no. 4, pp. 494–502, April 1998.
11. Neal Patwari, Robert J. O'Dea, and Yanwei Wang, "Relative location in wireless networks," in *IEEE Vehicular Technology Conf. (VTC)*, May 2001, vol. 2, pp. 1149–1153.
12. Neal Patwari, Alfred O. Hero III, Matt Perkins, Neiyer Correal, and Robert J. O'Dea, "Relative location estimation in wireless sensor networks," *IEEE Trans. Signal Processing*, vol. 51, no. 8, pp. 2137–2148, Aug. 2003.
13. Yi Shang, Wheeler Ruml, Ying Zhang, and Markus P. J. Fromherz, "Localization from connectivity in sensor networks," *IEEE Trans. Parallel & Distr. Syst.*, vol. 15, no. 11, pp. 961–974, Nov. 2004.
14. A.A. Ahmed, Yi Shang, and Hongchi Shi, "Variants of multidimensional scaling for node localization," in *11th Int. Conf. Parallel & Distr. Syst.*, July 2005, vol. 1, pp. 140–146.
15. Xiang Ji and Hongyuan Zha, "Sensor positioning in wireless ad-hoc sensor networks with multidimensional scaling," in *IEEE INFOCOM 2004*, March 2004, vol. 4, pp. 2652–2661.
16. Jose A. Costa, Neal Patwari, and Alfred O. Hero III, "Distributed multidimensional scaling with adaptive weighting for node localization in sensor networks," *IEEE/ACM Transactions on Sensor Networks*, May 2006, (to appear).

17. Hyuk Lim and Jennifer C. Hou, "Localization for anisotropic sensor networks," in *IEEE INFOCOM 2005*, March 2005, vol. 1, pp. 138–149.
18. Joshua B. Tenenbaum, Vin de Silva, and John C. Langford, "A global geometric framework for nonlinear dimensionality reduction," *Science*, vol. 290, pp. 2319–2323, Dec 2000.
19. T. Cox and M. Cox, *Multidimensional Scaling*, Chapman & Hall, London, 1994.
20. Mikhail Belkin and Partha Niyogi, "Laplacian Eigenmaps for dimensionality reduction and data representation," *Neural Computation*, vol. 15, no. 6, pp. 1373–1396, June 2003.
21. Earnest R. Davidson, "The iterative calculation of a few of the lowest eigenvalues and corresponding eigenvectors of large real-symmetric matrices," *J. Comput. Phys.*, vol. 17, no. 1, pp. 87–94, Jan. 1975.
22. Luca Bergamaschi, Giorgio Pini, and Flavio Sartoretto, "Computational experience with sequential and parallel, preconditioned Jacobi Davidson for large, sparse symmetric matrices," *J. Computational Physics*, vol. 188, no. 1, pp. 318–331, June 2003.
23. Sam T. Roweis and Lawrence K. Saul, "Nonlinear dimensionality reduction by local linear embedding," *Science*, vol. 290, pp. 2323–2326, Dec 2000.
24. David L. Donoho and Carrie Grimes, "Hessian Eigenmaps: new locally linear embedding techniques for high-dimensional data," Tech. Rep. TR2003-08, Dept. of Statistics, Stanford University, March 2003.
25. Neal Patwari and Alfred O. Hero III, "Using proximity and quantized RSS for sensor localization in wireless networks," in *2nd ACM Workshop on Wireless Sensor Networks & Applications (WSNA'03)*, Sept. 2003, pp. 20–29.

Node Localization Using Mobile Robots in Delay-Tolerant Sensor Networks

Pubudu N Pathirana[1], Nirupama Bulusu[2], Andrey V Savkin[3], Sanjay Jha[4], and Thanh X Dang[5]

[1] Deakin University, Victoria 3217, Australia pubudu@deakin.edu.au
[2] Portland State University, Portland, OR 97207-0751, USA nbulusu@cs.pdx.edu
[3] University of New South Wales, Sydney 2052,Australia a.savkin@unsw.edu.au
[4] University of New South Wales, Sydney 2052,Australia sjha@cse.unsw.edu.au
[5] Portland State University, Portland, OR 97207-0751, USA dangtx@cs.pdx.edu

Summary. We present a novel scheme for node localization in a Delay-Tolerant Sensor Network (DTN). In a DTN, sensor devices are often organized in network clusters that may be mutually disconnected. Some mobile-robots may be used for data collection from the network clusters. The key idea in our scheme is to use this robot to perform location estimation for the sensor nodes it passes by based on the signal strength of radio messages received from them. Thus we eliminate the processing constraints of static sensor nodes and the need for static reference beacons. Our mathematical contribution is the use of a Robust Extended Kalman Filter (REKF) based state estimator to solve the localization. Compared to the standard extended Kalman filter, REKF is computationally efficient and also more robust, since it does not make any assumptions about the measurement noise. Finally, we have implemented our localization scheme on a hybrid sensor network testbed, and show that it can achieve node localization accuracy within 1m in a large indoor setting.

1 Introduction

Recent years have witnessed a boom in sensor networks research [3, 10] and commercial activities [8]. This has been motivated by the wide range of potential applications from environmental monitoring to condition-based maintenance of aircraft. Sensor networks are frequently envisioned to exist at large scale, and characterized by extremely limited end-node power, memory and processing capability.

The concept of a *delay-tolerant sensor network (DTN)* was first proposed by Fall [13]. A DTN would typically be deployed to monitor an environment over a long period of time, and characterized by non-interactive sensor data traffic. Sensors are randomly scattered and organize into one or more clusters that may be disconnected from each other. Each cluster has a cluster-head. Sensor information is typically aggregated at the cluster heads, which tend to have more resources and are responsible for communicating data to outside world. Wireless mobile robots (e.g. robomote [25]), unmanned aerial vehicles can roam around the network to collect data from cluster heads, or to dynamically reprogram or reconfigure the sensors. Examples of DTNs in existence are Sammi [5], Zebranet [12], and DataMules [24].

This chapter revisits the problem of *node localization, i.e.*, estimating sensor node positions for a delay-tolerant sensor network. A DTN has several distinguishing characteristics which motivate alternate approaches to node localization than those previously proposed. In a DTN, sensor nodes need neither be localized in real time, nor all at once. In this chapter, we propose a novel localization scheme for DTNs using received signal strength (RSSI) measurements from each sensor device at a data gathering mobile-robot. Our contributions are threefold:

- We motivate and propose a novel approach that allows one/more mobile robots to perform node localization in a DTN, eliminating the processing constraints of small devices. Mobility can also be exploited to reduce localization errors and the number of static reference location beacons required to uniquely localize a sensor network.
- We develop a novel Robust Extended Kalman Filter (REKF) [18], based state estimation algorithm for node localization in DTNs. Localization based on signal strength measurements is solved by treating it as on-line estimation in a nonlinear dynamic system (Section 3). Our model incorporates significant uncertainty and measurement errors and is computationally more efficient and robust in comparison to the extended Kalman filter implementation used to solve similar problem in cellular networks [15] [6] (simulations, section 4).
- We implement and validate our scheme on a novel hybrid sensor network testbed of motes, Stargates and Lego Mindstorm robots (section 5).

2 Related Work

In this section, we review research most relevant to our work — (i) delay-tolerant sensor networks, and (ii) sensor network localization.

2.1 Delay-tolerant Sensor Networks

Fall first proposed a Delay-tolerant Network architecture [13] for sensors deployed in mobile and extreme environments lacking an *always-on* infrastructure. These sensors are envisioned to monitor the environment over a long period of time. Herein, communication is based on an abstraction of message switching rather than packet switching. The abstraction of moderate-length message (known as *bundles*) delivery for non-interactive traffic can provide benefits for network management because it allows the network path selection and scheduling functions to have *a-priori* knowledge about the size and performance requirements of requested data transfers.

DTNs are already being used in practice. DataMules [24] uses a Mule that periodically visits sensor devices and collects information from these devices, in effect providing a message store-and-forward service, enabling low-power sensor nodes to conserve power. The Sammi Network [5] is a community of Sammi people, who are reindeer herders in Sweden who keep relocating their base. The Sammi communities do not have a wired or wireless communication infrastructure. Their relocation is controlled by an yearly cycle which depends on the natural behavior of reindeer. In the Zebranet wildlife tracking system [12], wireless sensor nodes attached to animals collect location data and opportunistically report their histories when they come within radio range of base stations. While previous research has focused on communication abstractions, we are investigating the challenges and opportunities that arise from mobile data collecting elements in DTNs.

2.2 Sensor Network Localization

Localization is one of the most widely researched topics within the area of sensor networks, and in robotics [2, 7, 19, 23]. Previous localization systems for sensor networks [19, 23] have been designed to *simultaneously* scale and *continuously* localize a large number of devices. To meet these requirements, devices usually compute their own location from distance (or other measurements) made to nearby reference beacons. However, localization requirements for these sensor networks are different from delay-tolerant sensor networks.

In a DTN, nodes neither need to be localized concurrently nor continuously. We tradeoff computational time for node localization for several other benefits. We can reduce the computational requirements for small sensor devices, by instead using one or more mobile robots to compute the location of sensors. We can employ more sophisticated algorithms since processing is performed by the robot rather than sensor devices. We can also reduce the number of static location reference beacons required, by exploiting the mobility of the robot. [6]

Previous research has also investigated RSSI-based localization schemes [1]. One of the main drawbacks in RSSI-based localization schemes is the RSSI measurement noise caused by short-scale and medium-scale fading. In our scheme, we reduce the impact of fading, to make RSSI-based localization more viable. Because the robot-receiver is mobile, over a period of time, we can statistically eliminate the fading noise in RSSI measurements (not possible with static transmitter-receiver pair).

Previously, Kalman filters and Bayesian filters have been applied to the localization problem (mainly in the context of robotics and cellular networks). In this chapter we propose using a Robust Extended Kalman Filter (REKF) as a state estimator in predicting sensor locations. These robust state estimation ideas emerged from the work of Savkin and Petersen [22]. It not only provides satisfactory results [17], but also eliminates the requirement of the knowledge of measurement noise in the more

---

[6] For instance, Eren *et al* [4] estimate that to uniquely localize a sensor network of n nodes in $O(\sqrt{(\log n)})$ steps, $O(\frac{n}{\log n})$ reference location beacons are required using the iterative trilateration scheme proposed in [23]. To localize with just 1 beacon, $O(n)$ steps will be required.

In our scheme, assume that the mobile robot can localize $O(log n)$ sensors in each step (a single Filter computation). This is not an unreasonable assumption to make since $log n \leq 10$, even for very large n. Using just 1 mobile robot, node localization can be achieved in $O(\frac{n}{log n})$ steps. To localize in $O(\sqrt{\log n})$ steps, our scheme requires $O(\frac{n}{\log n \sqrt{\log n}})$ steps.

commonly used standard Kalman filter implementation presented in [15]. It is significantly more computation and memory efficient than the more adaptive, but computationally complex and memory-intensive Bayesian filters; making it better suited to the sensor networks regime. In the next section, we describe in detail, our node localization scheme using mobile robots in a delay-tolerant sensor network.

3 Localization Methodology

To solve node localization based on RSSI measurements at a mobile-robot, we model it as an on-line estimation in a nonlinear dynamic system. In this section, we describe this system dynamic model and the nonlinear measurement model. We present the theoretical background for the Robust Extended Kalman filter used with this model in Appendix A.

3.1 System dynamic model

We use the terminology mobile $-$ robot for a mobile-node fitted with a wireless base-station. The sensors to be located are randomly distributed in an environment. The dynamic model for n sensors and the mobile-robot can be given in two dimensional cartesian coordinates as [21]

$$\dot{x}(t) = Ax(t) + B_1 u(t) + B_2 w(t) \tag{1}$$

where

$$A = \begin{bmatrix} \Theta & & 0 \\ & \ddots & \\ 0 & & \Theta \end{bmatrix}, -B_1 = \begin{bmatrix} \Phi \\ \vdots \\ \Phi \end{bmatrix},$$

$$B_2 = \begin{bmatrix} \Phi & & 0 \\ & \ddots & \\ 0 & & \Phi \end{bmatrix}$$

$$\Theta = \begin{bmatrix} 0 & 0 & 1 & 0 \\ 0 & 0 & 0 & 1 \\ 0 & 0 & 0 & 0 \\ 0 & 0 & 0 & 0 \end{bmatrix}, \Phi \begin{bmatrix} 0 & 0 \\ 0 & 0 \\ -1 & 0 \\ 0 & -1 \end{bmatrix}.$$

The dynamic state vector $x(t) = [x_1(t) \ldots x_i(t) \ldots x_n(t)]'$ with $x_i(t) = [X_i(t) \ Y_i(t) \ \dot{X}_i(t) \ \dot{Y}_i(t)]'$, where $i \in [1 \ldots n]$ and $X_i(t)$ and $Y_i(t)$ represent the position of the ith sensor (Sensor$_i$) with respect to the mobile $-$ robot at time t, and their first order derivatives $\dot{X}(t)$ and $\dot{Y}(t)$ represent the relative speed along the X and Y directions. In other words, if $x_c(t) = [X_c(t) \ Y_c(t) \ \dot{X}_c(t) \ \dot{Y}_c(t)]'$ represent the absolute state (position and velocity in order in the X and Y direction respectively) of the (mobile $-$ robot) and $x_s^i(t) = \left[X_s^i(t) \ Y_s^i(t) \ \dot{X}_s^i(t) \ \dot{Y}_s^i(t) \right]'$ denote the absolute state of the Sensor$_i$ in the same order, then $x_i(t) \triangleq x_c(t) - x_s^i(t)$. Furthermore, let $u(t)$ denote the two dimensional driving/acceleration command of the mobile $-$ robot from the respective accelerometer readings and $w(t)$ denote the unknown two-dimensional driving/acceleration command of the sensor if moving. Although it can be generalized for moving sensor case, as most applications rely on stationary sensors, here we consider the sensors as stationary and set $w(t) = 0$. This system can be represented in graphical form in the form of an input($u(t)$) and measurement (y) system as in Figure 1. We omitted B_2 as we only consider the case of stationary sensors. The basic idea in such a system is to estimate state x from measurement y. In the localization problem, as the sensor locations are unknown, we assume an arbitrary location (0,0). We show that this assumed state converges to the actual state and hence the unknown sensor location can be estimated (as the position/state of the mobile $-$ robot is known) within the prescribed time frame.

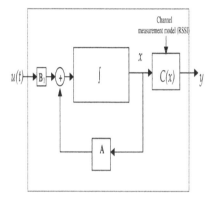

Fig. 1. Location estimation system

3.2 RSSI Measurement model

In wireless networks, the distance between two communicating entities is observable using the forward link RSSI (received signal strength indication) of the receiver. Measured in decibels at the mobile-robot for our case, RSSI can be modelled as a two folds effect: due to *path loss* and due to *shadow fading* [15]. Fast fading is neglected assuming that a low-pass filter is used to attenuate Rayleigh or Rician fade. Denoting the ith sensor as Sensor$_i$ (figure 2), the RSSI from the Sensor$_i$, $p_i(t)$ can be formulated as [27]

$$p_i(t) = p_{oi} - 10\varepsilon \log d_i(t) + v_i(t), \tag{2}$$

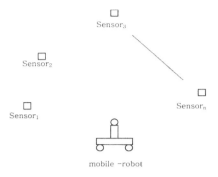

Fig. 2. Network Geometry

where p_{oi} is a constant determined by transmitted power, wavelength, and antenna gain of the mobile − robot. ε is a slope index (typically 2 for highways and 4 for microcells in the city), and $v_i(t)$ is the logarithm of the shadowing component, which is considered as an uncertainty in the measurement. $d_i(t)$ represents the distance between the mobile − robot and Sensor$_i$, which can be further expressed in terms of the position of the ith sensor with respect to the location mobile-robot i.e., $(x_i(t), y_i(t))$

$$d_i(t) = \left(x_i(t)^2 + y_i(t)^2\right)^{1/2} \tag{3}$$

For sensors within a network cluster, we uses measurements at a single mobile-robot as opposed to multiple ones [15]. The observation vector

$$y(t) = \begin{bmatrix} p_1(t) \\ \vdots \\ p_n(t) \end{bmatrix}, \tag{4}$$

is sampled progressively as the mobile − robot moves in the coverage area. The measurement equation for the measurements made by the mobile − robot for the n number of sensors are in the form of,

$$y(t) = C(x(t)) + v(t) \tag{5}$$

where $v(t) = [v_1(t) \cdots v_n(t)]'$ with

$$C(x(t)) = \begin{bmatrix} p_{oi} - 10\varepsilon \log\left(x_1(t)^2 + y_1(t)^2\right) \\ \vdots \\ p_{oi} - 10\varepsilon \log\left(x_n(t)^2 + y_n(t)^2\right) \end{bmatrix}. \tag{6}$$

We provide a brief intuitive explanation of REKF here (see Appendix for detailed theoretical background). We use the state space model (a representation of the dynamic system consisting of the mobile − robot and the n sensors using a set of differential equations derived from simple kinematic equations). Our dynamic system considers two noise inputs: (i) measurement noise (this is standard with any measurement), v in $y = C(x) + v$ and (ii) w - acceleration is also considered noise as it is unknown. In this application, the initial condition errors are quite significant as no knowledge is available regarding the sensor locations. This issue is directly addressed as the proposing algorithm is inherently robust against estimation errors of the initial condition (see equation 17 in the appendix). The two noise inputs and the initial estimation errors have to satisfy IQC equation (equation 17 in the appendix). If there exists a solution to the Ricati equation (see section A.2 in the appendix), then the IQC presented in a suitable form by the equation 17 is satisfied and the states can be estimated from the measurements using equation 19 which is a Robust version of the Extended Kalman Filter (REKF) .

In the application of REKF in a delay-tolerant network, the ith system (mobile $-robot$ and the Sensor$_i$) during a corresponding time interval is represented by the nonlinear uncertain system in (12) together with the following Integral Quadratic Constraint(IQC) (from equation 17) :

$$(x(0) - x_0)' N_i (x(0) - x_0)$$
$$+ \frac{1}{2} \int_0^s \left(w(t)' Q_i(t) w(t) \right) + v(t)' R_i(t) v(t) dt$$
$$\leq d + \frac{1}{2} \int_0^s z(t)' z(t) dt. \tag{7}$$

Here $Q_i\rangle 0, R_i\rangle 0$ and $N_i\rangle 0$ with $i \in \{1, 2, 3\}$ are the weighting matrices for each system i. The initial state(x_0) is the estimated state of respective systems at startup. It is essentially derived from the terminal state of the previous system together with other data available in the network (i.e robot position and speed) to be used as the initial state for the next system taking over the tracking. With an uncertainty relationship of the form of (7), the inherent measurement noise (see equation 5), unknown mobile robot acceleration and the uncertainty in the initial condition are considered as bounded deterministic uncertain inputs. In particular, the measurement equation with the standard norm bounded uncertainty can be written as (equation 5)

$$y = C(x) + \delta C(x) + v_0$$

where $|\delta| \leq \xi$ with ξ, a constant indicating the upper bound of the norm bounded portion of the noise. By choosing $z = \xi C(x)$ and $\nu = \delta C(x)$,

$$\int_0^T |\nu| dt \leq \int_0^T z' z dt.$$

Considering v_0 and the corresponding uncertainty in w as w_0 satisfying the bound

$$\Phi(x(0)) + \int_0^T [w_0(t)'Q w_0(t) + v_0(t)'R v_0(t)] dt \leq d,$$

it is clear that this uncertain system leads to the satisfaction of condition in inequality 13 and hence 17 (see [18]). This more realistic approach removes any noise model assumptions in algorithm development and guarantees the robustness.

3.3 Robust versus Optimal State Estimation

REKF tends to increase the robustness of the state estimation process and reduce the chance that a small deviation from the Gaussian process in the system noise causes a significant negative impact on the solution. However, we lose optimality and our solution will be just sub-optimal. To explain the connection between REKF and the standard extended Kalman Filter, consider the system 12 with

$$K(x, u) = \nu K_0(x, u) \tag{8}$$

where $K_0(x, u)$ is some bounded function, and $\nu \rangle 0$ is a parameter. Then, the REKF estimate $\tilde{x}(t)$ for the system 12, 8, 17 defined by 19, 20 converges to $\tilde{x}^0(t)$ as ν tends to 0. Here $\tilde{x}^0(t)$ is the extended Kalman state estimate for the system 12 with the Gaussian noise $\left[w(t)' \ v(t)' \right]$ satisfying

$$E\left\{ \begin{bmatrix} w(t) \\ v(t) \end{bmatrix} \left[w(t)' \ v(t)' \right] \right\} = \begin{bmatrix} Q(t) & 0 \\ 0 & R(t) \end{bmatrix};$$

The parameter ν in 8 describes the uncertainty in the system and measurement noise. For small ν, our robust state estimate becomes close to the Kalman state estimate with Gaussian noise. For larger ν, we achieve more robustness but less optimality. We show via simulation that for larger uncertainty (which is quite realistic) our robust filter still performs well whereas the standard extended Kalman estimate diverges.

4 Simulations

To examine the performance of the Robust Extended Kalman Filter for a sensor network, we simulate a mobile-robot equipped with a radio transceiver moving in the sensor coverage area. We assume the network knows the acceleration of the mobile-robot via GPS and accelerometer readings but has no information about the sensors. We simulate two scenarios — large sensors and small sensors.

4.1 Large Sensors

In Scenario I, we simulate large sensors scattered over a wide area. We expect sensors to be low cost and equipped with modest transmitters. To model this, we use a slow sampling rate of 2 sec per sample. We simulate a mobile robot and four sensors. The algorithm can be scaled to as many sensors as required by appropriately increasing the number of mobile robots. Simulation parameters are listed in Table 1. The mobile-robot measures the forward link signal from four sensors and estimates the state of the system from an arbitrary initial estimate (zero). Figure 3 shows how the estimated sensor location from an initial position converges to the actual sensor positions within the simulation time. Figure 4 shows the distance variation in X and Y directions separately for each sensor as well as the predicted distances approaching the actual distances. Figure 6 shows that the extended Kalman filter cannot be used with large uncertain instances as it diverges.

Parameter	Value	Comments
p_{oi}	20w	Base station transmission power
N	diag{10,10,10,1, 70,25,2,5, 100,1,20,1, 40,40,20,1}	Weighting on the Initial viscosity solution
Q	diag{2, 2, 11, 1, 1, 3, 1, 2}	Weighting on the uncertainty in the vehicle driving command
R	diag{$2 \times 10^3, 1.7 \times 10^3$,} $10 \times 10^3, 51 \times 10^3$	Weighting on the measurement noise
T	$5 mins$	Simulation time
A_{max}	$50 m/s^{-2}$	weighting on u(t)
Ts	2s	Sampling interval
$x_s^1(0)$	[500m 2500m 0 0]'	1st sensor initial state
$x_s^2(0)$	[2500m 500m 0 0]'	2nd sensor initial state
$x_s^3(0)$	[1000m 9000m 0 0]'	3rd sensor initial state
$x_s^3(0)$	[2500m 12500m 0 0]'	4th sensor initial state
$x_c(0)$	[2500m -2500m $20ms^{-1}$ $10ms^{-1}$]'	$mobile - robot$ initial state

Table 1. Simulation parameters for Scenario I.

4.2 Small Sensors

In Scenario II, we use sensors with much lesser signal strength (600mW) as in [11] with higher sampling rate as in commercially available systems. To demonstrate scalability, we increase both the time scale and the number of sensors in this scenario. The simulation parameters are given in Table 4. The arbitrary acceleration of the mobile robot is taken as $u(t) = A_{max} [-3 \sin(0.2t) + \phi_1 \ 0.9 \cos(0.05t) + \phi_2]$.

In the dynamical system simulation, we choose the functions given in Table 4 for the arbitrary mobile-robot acceleration (u), with ϕ_1, and ϕ_2 being uniform random distributions in the interval $[0 \ 0.1A_{max}]$. We consider uniformly distributed measurement noise in the interval $[0 \ 0.01\|y(t)\|]$ with $y(t)$ being the noise free measurement with $\xi = 0.05$. The equation for the state estimation and the corresponding Riccati Differential equation obtained from equation 19 and 20 are:

$$\dot{\tilde{x}}(t) = A\tilde{x}(t) + B_1 u_i(t)$$
$$+X^{-1}(t)[\beta^1 (\tilde{x}(t))' R_i (y(t) - \beta (\tilde{x}(t)))$$
$$+\xi^2 \beta^1 (\tilde{x}(t))' \beta^1 (\tilde{x}(t))]$$
$$\tilde{x}(t) = x_0. \tag{9}$$
$$\dot{X} + A'X + XA + XB_2 Q_i^{-1} B_2'X -$$
$$\beta^1 \tilde{x}(t)' R_i \beta^1 \tilde{x}(t) + \xi^2 \beta^1 (\tilde{x}(t))' \beta^1 (\tilde{x}(t)) = 0$$
$$X(0) = N, \tag{10}$$

where

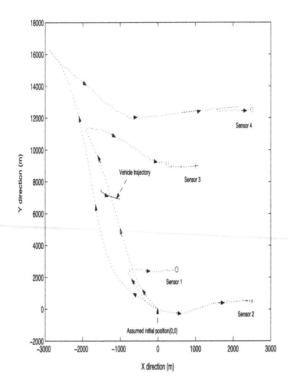

Fig. 3. Location estimation trajectories converging to the actual sensor locations.

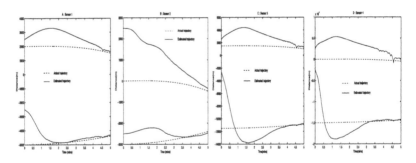

Fig. 4. Estimation convergence for sensors 1, 2, 3 and 4.

$$\beta(x) = C(x(t)$$

as shown in equation 6. Also here

$$\beta^1(x) = \nabla_x \beta(x), \qquad (11)$$

x_0 is $x(0)$, the relative initial dynamic state of the system. In the second scenario, ten sensors with lesser signal strength are used with the mobile robot. Figure 5(a) plots

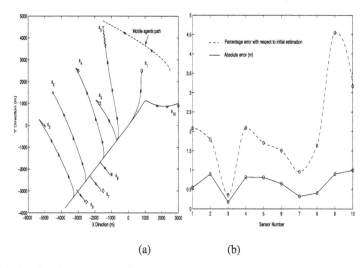

(a)   (b)

Fig. 5. (a) Estimation path for each sensor (Scenario 2). (b) Percentage error and the absolute error from the initial estimation.

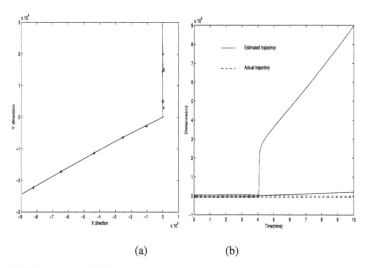

(a)   (b)

Fig. 6. (a) Using standard Kalman filter as the state estimator. (b) Divergence of state estimation when using standard Kalman filter.

the estimated trajectory approaching each respective sensor location from an initial estimate of each sensor location of (0,0). Figure 5(b) plots the percentage error in localizing each sensor with respect to the initial estimation error.

5 Implementation and Experimental Results

We have implemented our REKF-based localization scheme to verify its computational efficiency and estimation accuracy in a real environment. We now describe this implementation and report on preliminary experimental results.

5.1 Hybrid Sensor Platform

Our hybrid platform consists of three devices — (i) motes (ii) Stargates and (iii) Lego Mindstorm robots. They differ in processing, memory, battery, and mobility capabilities; and also in their operating systems software.

Motes

Motes [16], shown in Figure 7, are our resource impoverished devices and run the TinyOS event-driven operating system. The Mica2 mote sensors deployed in our experiment use the CC1000 radio from ChipCon, which provides an analog RSSI measurement that can be connected to an analog to digital converter (ADC) to produce digital signals. These RSSI measurements can be used for localization.

Stargate

Stargate [26], shown in Figure 7, is a resource-rich node that provides more capabilities than the MICA motes. It is a powerful Linux based single board computer with Intel 400MHz X-Scale processor (PXA255), Compact Flash, PCMCIA, Ethernet, USB Host, 64 MB SDRAM and an additional interface to communicate with a mote. We use a Stargate as the computational substrate for the mobile-robot. The Stargate runs the Robust Extended Kalman Filter which is implemented in Java. RSSI readings are measured for the mote interfaced with the Stargate, communicating with other sensors.

Platform	Core	Processor	Data Path	MHz	Memory
MICA2	Atmega 128L4	RISC	8 bits	4 MHz	128 KB program flash, 4 KB data
Stargate	Intel PXA255 Xscale	RISC	32 bits	400 MHz	32 MB flash, 64 MB SDRAM

Table 2. Stargate and MICA2 Comparison

Lego MindStorm: Mobility

We use the popular Lego MindStorm [14] platform, shown in Figure 7 to emulate a mobile robot. It is a programmable, non-maneuvring robot that constructed by connecting small Legos together. A Stargate is mounted onto the Lego. An Infra Red tower is used to program the RCX box, which controls the Lego MindStorm. At the core of the RCX is a Hitachi H8 microcontroller with 32K external RAM. The microcontroller is used to control three motors, three sensors, and an infrared serial communications port. Both the driver and firmware accept and execute commands from the PC through the IR communications port. To calculate the velocity, the Lego MindStorm is programmed to move at a constant speed (selected from eight power options) in a straight line.

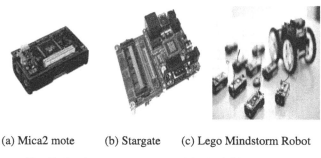

(a) Mica2 mote   (b) Stargate   (c) Lego Mindstorm Robot

Fig. 7. Hardware Components of the Hybrid Testbed.

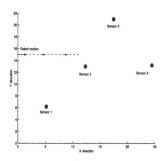

Fig. 8. Indoor sensor distribution and robot navigation topology.

5.2 Experimental Results

We placed the motes in the topology shown in Figure 8. and programmed the Lego Mindstorm robot to move in a straight line. We measured the distance travelled and time elapsed to accurately deduce the velocity of a mobile robot (sampling rate = 0.3s). The values of the weighting matrices are tuned with simulations using the modelled system parameters.

Computational Efficiency

To characterize computational efficiency, we measured the REKF computation time (for 10 sensors) as a function of the number of RSSI samples for the following cases:

- Java implementation (Sun's JVM) on Pentium IV 3GHz machine
- Matlab implementation on Pentium IV 3GHz machine
- Java implementation, with Open-Wonka on Stargate 400MHz machine

The computation time in milliseconds is shown in Table 3. The performance of Mat-

Number of Samples	Pentium-IV (Java) (seconds)	Pentium-IV (MATLAB) (seconds)	STARGATE (Java) (seconds)
90	-	-	115
2000	4	70	-
8000	15	27	-
16000	27	56	-

Table 3. Computation Times for REKF Implementations

lab and Java running on Pentium IV 3 GHz does not differ much (in the same order). However for Stargate, the performance degrades significantly with the number of samples. The Stargate has very limited memory (64 MB SDRAM) which makes Open-Wonka's garbage collector inefficient. To improve the performance on the Stargate, we can (i) break down the computation on Stargate to a smaller subsets of samples, or (ii) implement REKF in C instead of Java.

Estimation Accuracy

To evaluate estimation accuracy, we report on an experiment with four sensors, since the visualization of estimation convergence is clearer with a smaller number of sensors. The four sensors are positioned at $(6.1, 6)m$, $(12.2, 13.6)m$, $(18.3, 21.2)m$ and $(24.4, 13.6)m$. The mobile vehicle(robot) is initially located at $(0, 15.2)m$ and moves with a velocity of $2.52m/min$. For our indoor implementation, with the collected data and modelling the RSSI, we use $p_{oi} = 160mw$ and $\varepsilon = 3$(equation 2).

Figure 9 shows the localization of the four sensors with approximate error of $1m$. The results are very close to the real positions. The Lego Mindstorm is a non maneuvering robot (constant velocity). By incorporating vehicle maneuver further improvements to the localization error can obviously be made.

6 Algorithmic Improvements and Future Work

Experimental evaluation revealed the importance of finding the right parameters for the weighting matrices. Our next step is to implement automatic tuning of these parameters through machine learning techniques and enhance the localization scheme

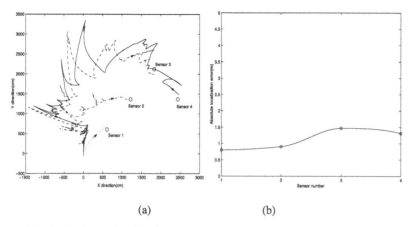

(a)   (b)

Fig. 9. (a) Sensor localization, and (b) localization error for the real system.

significantly. The algorithm stores all sensor states during the experiment period in order to validate the convergence of estimations. To improve computational efficiency, in a production system, where only the final estimation is needed, only two states for each sensor need to be stored (instead of nearly 500).

In general, the use of mobile robots in Delay-Tolerant Sensor Networks opens up a number of other interesting research possibilities. Once sensor localization has been performed, it is possible to create a topological map (or a path profile) that can be optimized so that the data collecting mobile-robot can follow this path for data collection in the shortest possible time, or to meet storage and power requirements. Mobile robots could act as relays between disconnected portions of the network, thereby forming a relay network. The trajectory of a mobile robot can be dynamically recalculated so that a mobile robot can slow down whenever it needs to download a lot of information. We intend to explore these aspects in future work.

7 Conclusions

In this chapter, we have provided a scheme for node localization using mobile robots in a delay-tolerant sensor network (DTN). To the best of our knowledge no other study has been done for such a network. DTNs are commonly deployed in long-term environmental monitoring applications. In a DTN, node localization need not happen in real time. Using one or more mobile robots to compute the location of sensors allows us to tradeoff computational time for node localization for several other benefits. First, we can eliminate processing constraints for small sensor devices. We can employ more sophisticated algorithms since processing is performed by the robot rather than sensor devices. Second, we can reduce the number of static location reference beacons required, by exploiting mobility of the robot. Third, it makes RSSI-based localization more viable. Because the robot-receiver is mobile, over a period of time, we can statistically eliminate the fading noise in RSSI measurements.

Parameter	Value	Comments
p_{oi}	600mw	Base station transmission power
N	diag$\{10^{-2}, 10^{-2}, 10^{-2}, 10^{-2},$ $10^{-8}, 10^{-8}, 10^{-8}, 10^{-8},$ $5,5,5,5,$ $5,5,5,5,$ $5,5,5,5,$ $10^{-4}, 10^{-4}, 10^{-4}, 10^{-4},$ $0.07, 0.07, 0.07, 0.07,$ $0.07, 0.07, 0.07, 0.07,$ $0.03, 0.03, 0.03, 0.03,$ $10^{-4}, 10^{-4}, 10^{-4}, 10^{-4}\}$	Weighting on the Initial viscosity solution
Q	diag$\{10^4, 10^4, 10^8, 10^8,$ $5 \times 10^3, 5 \times 10^3,$ $5 \times 10^3, 5 \times 10^3,$ $5 \times 10^3, 5 \times 10^3,$ $5 \times 10^{11}, 5 \times 10^{11},$ $10^{11}, 10^{11}, 10^{11}, 10^{11},$ $10^{14}, 10^{14}, 10^9, 10^9,\}$	Weighting on the uncertainty in the user driving command
R	diag$\{100, 3, 2 \times 10^9$ $2 \times 10^9, 2 \times 10^9$ $2 \times 10^9, 6 \times 10^5$ $64 \times 10^7, 64 \times 10^7$ $5 \times 10^5, 65\}$	Weighting on the measurement noise
T	$10mins$	Simulation time
A_{max}	$50m/s^{-2}$	weighting on u(t)
Ts	0.05s	Sampling interval
$x_s^1(0)$	[800m 2500m 0 0]'	1st sensor initial state
$x_s^2(0)$	[-5000m 0 0 0]'	2nd sensor initial state
$x_s^3(0)$	[-4500m 1500m 0 0]'	3rd sensor initial state
$x_s^4(0)$	[-3000m 2500m 0 0]'	4th sensor initial state
$x_s^5(0)$	[-1500m 4500m 0 0]'	5th sensor initial state
$x_s^6(0)$	[-2500m -3500m 0 0]'	6th sensor initial state
$x_s^7(0)$	[-1500m -3000m 0 0]'	7th sensor initial state
$x_s^8(0)$	[-1000m -2250m 0 0]'	8th sensor initial state
$x_s^9(0)$	[-1700m 1000m 0 0]'	9th sensor initial state
$x_s^{10}(0)$	[3000m 900m 0 0]'	10th sensor initial state
$x_c(0)$	$[2500m\ 2500m\ 2ms^{-1}\ 10ms^{-1}]'$	$mobile - robot$ initial state

Table 4. Simulation parameters for Scenario II.

We proposed applying a Robust Extended Kalman Filter based state estimator for node localization. It is computationally more efficient and robust to measurement noise compared to the more commonly used extended Kalman filter implementation. Real experiments in a large indoor area show that the localization accuracy is approximately 1m. This compares favorably with previously proposed RSSI localization schemes in an indoor setting [1] (accuracies within 3m), as well as with finer-grained acoustic time-of-flight localization schemes [19, 23] (accuracies vary 10cm - 25cm). Now that we have validated our ideas through simulation, implementation and experiment, we are working on further localization schemes refinements and on other mobile-robot uses in delay-tolerant sensor networks.

References

1. P. Bahl and V. N. Padmanabhan. Radar: An in-building user location and tracking system. In *Proceedings of IEEE Infocom 2000*, volume 2, pages 775–784, Tel-Aviv, Israel, March 2000. IEEE.
2. Nirupama Bulusu, John Heidemann, and Deborah Estrin. Gps-less low cost outdoor localization for very small devices. *IEEE Personal Communications Magazine*, 7(5):28–34, October 2000.
3. D.Estrin, R.Govindan, J.Heidemann, and S.Kumar. Next century challenges: Scalable coordination in sensor networks. In *Proceedings of the ACM MobiCom 99*, pages 263–270, Seattle, WA, USA, August " 15–20" 1999. ACM Press.
4. T. Eren, D. Goldenberg, W. Whitley, YR Yang, AS. Morse, BDO Anderson, and PN Belheumer. Rigidity, computation, and randomization of network localization. In *Proceedings of IEEE Infocom 2004*, March 2004.
5. Dealy Tolerant Networking Research Group. The sammi network. *http://www.cdt.luth.se/babylon/snc/*.
6. M. Hellebrandt and R. Mathar. Location tracking of mobiles in cellular radio networks. *IEEE Transcation on Vehicular Technology*, 48(5):1558–1562, September 1999.
7. Jeffrey Hightower and Gaetano Borriello. Location systems for ubiquitous computing. *IEEE Computer*, 34(8):57–66, August 2001.
8. Crossbow Technologies Incorporated. *http://www.xbow.com*.
9. M.R. James and I.R. Petersen. Nonlinear state estimation for uncertain systems with an integral constraint. *IEEE Transactions on Signal Processing*, 46(11):2926–2937, November 1998.
10. J.M.Kahn, R.H.Katz, and K.S.J.Pister. Next century challenges: Mobile networking for smart dust. In *Proceedings of ACM/IEEE MobiCom 99*, pages 271–278, Seattle, WA, USA, August " 15–20" 1999. ACM Press.
11. K. Joanna, R.W. Heinzelman, and H. Balakrishnan. Negotiation-based protocols for disseminating information in wireless sensor networks. *Wireless Networks*, 1999.
12. P. Juang, H. Oki, Y. Wang, M. Martonosi, L. Peh, and D. Rubenstein. Energy-efficient computing for wildlife tracking: Design tradeoffs and early experiences with zebranet. In *Proceedings of ASPLOS-X*, pages 96–107, San Jose, CA, October 2002. ACM.
13. K.Fall. A delay-tolerant network architecture for challenged internets. Technical Report IRB-TR-03-003, Intel Research Berkeley, February 2003.
14. Inc. Lego Mindstorms. *http://www.xbow.com*.
15. T. Liu, P. Bahl, and I. Chlamtac. Mobility modelling, location tracking, and trajectory prediction in wireless ATM networks. *IEEE Journal on Selected Areas in Communications*, 16(6):922–936, August 1998.
16. The Mica mote. *http://www.xbow.com/Products/productsdetails. aspx?sid=61*.
17. P.N. Pathirana and A.V. Savkin. Precision missile guidance with angle only measurements. In *Proceedings of the Information Decision and Control Conference*, Adelaide, Australia, February 2002.

18. I.R. Petersen and A.V. Savkin. *Robust Kalman Filtering for Signals and Systems with Large Uncertainities*. Birkhauser, Boston, 1999.

19. N. Priyantha, A. Chakraborty, and H. Balakrishnan. The cricket location support system. In *Proceedings of ACM MobiCom 2000*, pages 32–43, Boston, MA, USA, August 2000. ACM.

20. A.V. Savkin and R.J. Evans. *Hybrid dynamical systems :Controller and Sensor Switching Problems*. Birkhäuser, Boston, 2002.

21. A.V. Savkin, P.N. Pathirana, and F.A. Faruqi. The problem of precision missile guidance: LQR and H$^\infty$ control frameworks. *IEEE Transactions on Aerospace and Electronic Systems*, 39(3):901–910, July 2003.

22. A.V. Savkin and I.R. Petersen. Recursive state estimation for uncertain systems with an integral quadratic constraint. *IEEE Transactions on Automatic Control*, 40(6):1080–1083, 1995.

23. A. Savvides, C. Han, and MB. Srivastava. Dynamic fine-grained localization in ad-hoc networks of sensors. In *Proceedings of ACM MobiCom 2001*, pages 166–179, Rome, Italy, July 2001. ACM.

24. R. Shah, S. Roy, S. Jain, and W. Burnette. Datamules: Modeling a three-tier architecture for sparse sensor networks. In *Elsevier ad hoc networks journal*, volume 1, pages 215–233. Elsevier, 2003.

25. G.T. Sibley, M.H.Rahimi, and G.S.Sukhatme. Robomote: A tiny mobile robot platform for large-scale ad-hoc sensor networks. In *Proceedings of the IEEE International Conference on Robotics and Automation (ICRA-2002)*, volume 2, pages 1143–1148, Washington D.C, USA, May 11-15 2002. IEEE.

26. Stargate. *http://www.xbow.com/Products/productsdetails.as px?sid=85*.

27. H.H. Xia. An analytical model for predicting path loss in urban and suburban environments. In *PIRMC*, 1996.

Appendix

A Set-Value state estimation with a non-linear signal model

We consider a nonlinear uncertain system of the form

$$\dot{x} = A(x, u) + B_2 w$$
$$z = K(x, u) \qquad (12)$$
$$y = C(x) + v,$$

as a general form of the system given by equation 1 with measurement equation in the form of equation 5, and defined on the finite time interval $[0, s]$. Here, $x(t) \in \mathbb{R}^n$ denotes the *state* of the system $y(t) \in \mathbb{R}^l$ is the *measured output* and $z(t) \in \mathbb{R}^q$ is the *uncertainty output*. The uncertainty inputs are $w(t) \in \mathbb{R}^p$ and $v(t) \in \mathbb{R}^l$. Also, $u(t) \in \mathbb{R}^m$ is the known *control input*. We assume that all of the functions appearing in (12) are with continuous and bounded partial derivatives. Additionally, we assume that $K(x, u)$ is bounded. This was assumed to simplify the mathematical derivations and can be removed in practice [18] [9]. The matrix B_2 is assumed to be independent of x, and is of full rank.

The uncertainty in the system is defined by the following *nonlinear integral constraint* [18, 20]

$$\Phi(x(0)) + \int_0^s L_1(w(t), v(t)) \, dt \le d + \int_0^s L_2(z(t)) \, dt, \qquad (13)$$

where $d \ge 0$ is a positive real number. Here, Φ, L_1 and L_2 are bounded nonnegative functions with continuous partial derivatives satisfying growth conditions of the type

$$\|\phi(x) - \phi(x')\| \le \beta \left(1 + \|x\| + \|x'\| \right) \|x - x'\| \qquad (14)$$

where $\|\cdot\|$ is the euclidian norm with $\beta \rangle 0$, and $\phi = \Phi, L_1, L_2$. Uncertainty inputs $w(\cdot), v(\cdot)$ satisfying this condition are called *admissible uncertainties*. We consider the problem of characterizing the set of all possible states $\mathcal{X}_s$ of the system (12) at time $s \ge 0$ which are consistent with a given control input $u^0(\cdot)$ and a given output path $y^0(\cdot)$; i.e., $x \in \mathcal{X}_s$ if and only if there exists admissible uncertainties such that if $u^0(t)$ is the control input and $x(\cdot)$ and $y(\cdot)$ are resulting trajectories, then $x(s) = x$ and $y(t) = y^0(t)$, for all $0 \le t \le s$.

A.1 The State Estimator

The state estimation set $\mathcal{X}_s$ is characterized in terms of level sets of the solution $V(x, s)$ of the PDE

$$\frac{\partial}{\partial t} V + \max_{w \in \mathbb{R}^m} \left\{ \nabla_x V \cdot \left(A(x, u^0) + B_2 w \right) \right.$$
$$\left. - L_1 \left(w, y^0 - C(x) \right) + L_2 \left(K(x, u^0) \right) \right\} = 0$$
$$V(\cdot, 0) = \Phi. \qquad (15)$$

The PDE (15) can be viewed as a filter, taking observations $u^0(t), y^0(t), 0 \le t \le s$ and producing the set $\mathcal{X}_s$ as a output. The state of this filter is the function $V(\cdot, s)$; thus V is an information state for the state estimation problem.

Theorem 1. *Assume the uncertain system (12), (13) satisfies the assumptions given above. Then the corresponding set of possible states is given by*

$$\mathcal{X}_s = \{ x \in \mathbb{R}^n : V(x, s) \le d \}, \qquad (16)$$

where $V(x, t)$ is the unique viscosity solution of 15) in $C(\mathbb{R}^n \times [0, s])$.

Proof. see [18]. $\qquad \square$

A.2 A Robust Extended Kalman Filter

Here we consider an approximation to the PDE (15) which leads to a Kalman filter like characterization of the set $\mathcal{X}_s$. Petersen and Savkin in [18] presented this as a Extended Kalman filter version of the solution to the Set Value State Estimation problem for a linear plant with the uncertainty described by an Integral Quadratic Constraint (IQC). This IQC is also presented as a special case of equation 13. We consider uncertain system described by (12) and an integral quadratic constraint of the form

$$(x(0) - x_0)' X_0 (x(0) - x_0)$$
$$+\frac{1}{2} \int_0^s \left(w(t)' Q(t) w(t) \right) + v(t)' R(t) v(t) dt$$
$$\leq d + \frac{1}{2} \int_0^s z(t)' z(t) dt, \qquad (17)$$

where $N\rangle 0, Q\rangle 0$ and $R\rangle 0$. For the system (12), (17), the PDE (15) can be written as

$$\frac{\partial}{\partial t} V + \nabla_x V.A(x, u^0) + \frac{1}{2} \nabla_x V B_2 Q^{-1} B_2' \nabla_x V'$$
$$- \frac{1}{2} \left(y^0 - C(x) \right)' R \left(y^0 - C(x) \right)$$
$$+ \frac{1}{2} K(x, u^0)' K(x, u^0) = 0,$$
$$V(x, 0) = (x - x_0)' N (x - x_0). \qquad (18)$$

Considering a function $\hat{x}(t)$ defined as $\hat{x}(t) \triangleq \arg\min_x V(x, t)$, and the following equations (19),(20) and (21), define our approximate solution to the PDE (18) :

$$\dot{\tilde{x}}(t) = \quad A \left(\tilde{x}(t), u^0 \right)$$
$$+ X^{-1} [\nabla_x C \left(\tilde{x}(t) \right)' R \left(y^0 - C \left(\tilde{x}(t) \right) \right)$$
$$+ \nabla_x K \left(\tilde{x}(t), u^0 \right)' K \left(\tilde{x}(t), u^0 \right)],$$
$$\tilde{x}(t) = x_0. \qquad (19)$$

$X(t)$ is defined as the solution to the Riccati Differential Equation (RDE)

$$\dot{X} + \nabla_x A \left(\tilde{x}, u^0 \right)' X + X \nabla_x A \left(\tilde{x}, u^0 \right)$$
$$+ X B_2 Q^{-1} B_2' X - \nabla_x C(\tilde{x})' R \nabla_x C(\tilde{x})$$
$$+ \nabla_x K \left(\tilde{x}, u^0 \right)' \nabla_x K \left(\tilde{x}, u^0 \right) = 0,$$
$$X(0) = N. \qquad (20)$$

and $\phi(t) \triangleq \frac{1}{2} \int_0^t [(y^0 - C(\tilde{x}))' R \left(y^0 - C(\tilde{x}) \right)$
$$- K \left(\tilde{x}, u^0 \right)' K \left(\tilde{x}, u^0 \right)] d\tau. \qquad (21)$$

The function $V(x,t)$ was approximated by a function of the form

$$\tilde{V}(x,t) = \frac{1}{2}\left(x - \tilde{x}(t)\right)' X(t)\left(x - \tilde{x}(t)\right) + \phi(t).$$

Hence, it follows from Theorem 1 that an approximate formula for the set $\mathcal{X}_s$ is given by

$$\tilde{\mathcal{X}}_s = \{x \in \mathbb{R}^n : \frac{1}{2}\left(x - \tilde{x}(s)\right)' X(s)\left(x - \tilde{x}(s)\right)$$
$$\leq d - \phi(s)\}$$

This amounts to the so called Robust Extended Kalman Filter(REKF) generalization presented in [18].

Experiences from the Empirical Evaluation of Two Physical Layers for Node Localization

Dimitrios Lymberopoulos[1] and Andreas Savvides[2]

[1] Department of Electrical Engineering, Yale University
    dimitrios.lymberopoulos,andreas.savvides@yale.edu
[2] Department of Electrical Engineering, Yale University
    andreas.savvides@yale.edu

1 Introduction

Node localization has attracted significant multidisciplinary research effort over the last few years. Despite these efforts, there is still no clear consensus on a particular technology or approach that can be used in a wide variety of environments. While this may be attributed to several reasons including cost, power and computation complexity, many researchers would agree that physical layer issues impose the largest barrier in the wider deployment and use of node localization services.

In this chapter, we present our results and experiences from the empirical evaluation of two very different physical layers, using our wireless sensor network testbed at Yale. The evaluation was conducted in two phases. The first phase evaluated the RF physical layer, through an extensive data collection of Received Signal Strength Indicator (RSSI). Our measurements were collected using a network of XYZ sensor nodes [15] featuring the popular IEEE 802.15.4 compliant, CC2420 radio from Chipcon coupled with a monopole wire antenna. As we will describe in section 2.4 after a preliminary characterization of the RSSI values, we decided to focus on the often overlooked antenna orientation and radio calibration effects. Our study revealed that radio calibration discrepancies across different radio chips are minor, and antenna orientation plays a significant role in the resulting RSSI measurements. The details of this RSSI evaluation are given in section 2.

The second phase of our empirical evaluation focused on the use of a more specialized measurement modality, small camera modules designed to mount on top of XYZ sensor nodes. This study was preceded by the development of two lightweight algorithms of node localization and camera calibration algorithms based on camera epipole's information. This process is overviewed in section 3.3 and it is described in detail in sections 3.4 and 3.5. The following sections in this book chapter will overview each sensing modality and highlight some prior research results. The results of our evaluation on the specific testbed and our conclusions regarding the weaknesses and limitations of each technology are also discussed.

Technique	Design Approach	Technology	Testbed	Error
Ecolocation [37]	Constrained based approach applied on the ordered sequence of raw RSSI data	MICA2	26x49(ft) (Indoor)	10ft
Probability Grid [30]	Probabilistic: RSSI values are used to estimate the one-hop distance in a grid	MICA2	410x410(ft) (Outdoor)	66ft
RADAR [1]	RSSI fingerprint map	802.11b	$141x72(ft)$	15ft
MoteTrack [14]	RSSI fingerprint map	MICA2	$18751\ ft^2$	13ft
LEASE [13]	Online Fingerprinting and signal propagation modeling	802.11b	225x144(ft)	15ft
Bayesian Indoor Positioning System [16]	Learning Based	802.11b	225x144(ft)	20ft
Stochastic Indoor Location System [26]	Optimal positioning of a given number of clusterheads	N/A	N/A	N/A
Monte Carlo Localization [2]	Learning-based with signal strength map	802.11b	N/A	7.2ft
Nibble [3]	Bayesian Networks	802.11b	224x96(ft) (Indoor)	20ft

Table 1. Characteristics of different RSSI techniques

2 Node Localization Using RSSI Measurements

Radio signal strength indicator (RSSI), a standard feature in most radios, has attracted a lot of attention in the recent literature for obvious reasons. RSSI eliminates the need for additional hardware in small wireless devices, and exhibits favorable properties with respect to power consumption, size and cost. As a result, several RSSI based algorithms have been proposed that either assume a complete profiling of the network deployment area (map based approaches) [1], [14] [3], [26], [2], [12], [23], [27], [32] or a specific signal attenuation model that can provide distance or area information directly or indirectly from the raw RSSI data (distance or area prediction approaches) [37], [7], [11], [30], [16], [21], [20], [6]. The characteristics of the most representative RSSI-based localization techniques can be seen in Table 1.

Some of the issues related with received signal strength ranging were presented by Whitehouse et. al. in an outdoor scenario characterization described in [36]. Three recent sensor network localization algorithms using low power sensor node radios are Ecolocation [37], MoteTrack [14] and Probability Grid [30]. Ecolocation determines the location of unknown nodes by examining the ordered sequence of received signal strength measurements taken at multiple reference nodes. The key idea of Ecolocation is that the distance-based rank order of reference nodes constitutes a unique signature for different regions in the localization space. Ecolocation reports a location error of $10ft$ for a very small outdoor network deployment area ($26ft$ x $49ft$) while Probability Grid reports a location error that is equal to the 70%-80% of the communication range for a $410ft$ x $410ft$ outdoor network deployment. In the case of Probability Grid it is assumed that the goal of the sensor network deployment is to form a grid topology. Given this a priori knowledge, Probability Grid attempts to compute in a probabilistic way the one-hop distance and the number of hops that an unknown node is far away from an anchor node. MoteTrack is very similar to RADAR [1] but it does not require a back-end server where all the data have to be transferred and processed. Conversely, in MoteTrack the location of each mobile node is computed using a received radio signal strength signature from numerous beacon nodes to a database of signatures that is replicated across the beacon nodes themselves. The location error reported by MoteTrack was reported to be approximately $13ft$ for an indoor network deployment area of $18751ft^2$.

Several schemes have also been presented using IEEE 802.11 radios. In [7] a comparative study of many RSSI based localization techniques using 802.11 cards is presented. According to the results of this study all the localization techniques produce approximately the same location error over a range of environments.

Despite the increasing interest in signal strength localization using IEEE 802.14.5 radios, there is still a lack of detailed characterization of the fundamental factors contributing to large signal strength variability. To investigate these factors, and to get a better understanding of the asymmetries that arise in 3-D scenarios, we present a detailed characterization of signal strength behaviors in an IEEE 802.15.4 sensor network with monopole antennas. Our evaluation focuses on showing and quantifying the sources of signal strength variability. We do so by collecting a large number of measurements from a 40-node testbed, both in an indoor and an open-field environment. This characterization differs from previous studies using IEEE 802.11 radios, since it examines a new radio technology with less powerful radio transmissions. Furthermore, a large fraction of the measurements are taken in a three-dimensional testbed deployment that emulates a realistic environment where sensor network deployments are likely to occur.

Our findings demonstrate that the relative antenna orientation between receiver-transmitter pairs is a major factor in signal strength variability, even in the absence of multipath effects. This suggests that many schemes using radio signal strength on similar radios should carefully consider these factors before going to actual deployments. The approximately 150,000 measurements collected for this study are available online at: $http://www.eng.yale.edu/enalab/rssidata/$.

2.1 Experimental Infrastructure

In the next sections we quantify the sources of RSSI variability using our Zigbee based infrastructure. A three dimensional, battery operated scalable testbed in our lab is used for indoor sensor network deployments. The testbed illustrated in Figure 1(a) is a 3-D structure measuring 4.5m(W) x 6m(L) x 3m(H) and it is designed to host a large number of static and mobile nodes to instrument a variety of application scenarios. The centerpiece of our infrastructure is the XYZ sensor node [15], an open-source general purpose sensing platform designed around the

OKI ML67Q500x ARM/THUMB microprocessor and the IEEE 802.15.4 compliant CC2420 radio from Chipcon [4](Figure 1(b)).

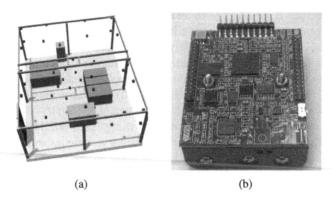

(a)   (b)

Fig. 1. a) Testbed node placement. b) The XYZ sensor node.

The Chipcon CC2420 IEEE 802.15.4 radio transceiver operates in the 2.4GHz ISM band and includes a digital direct sequence spread spectrum (DSSS) modem providing a spreading gain of $9dB$ and an effective data rate of $250Kbps$. It was specifically designed for low power wireless applications and supports 8 discrete power levels: $0dBm$, $-1dBm$, $-3dBm$, $-5dBm$, $-7dBm$, $-10dBm$, $-15dBm$ and $-25dBm$ at which its power consumption varies from $29mW$ to $52mW$ [15]. A built-in received signal strength indicator gives an 8-bit digital value: $RSSI_{VAL}$. The $RSSI_{VAL}$ is always averaged over 8 symbol periods($128\mu s$) and a status bit indicates when the $RSSI_{VAL}$ is valid (meaning that the receiver was enabled for at least 8 symbol periods). The power P at the RF pins can be obtained directly from $RSSI_{VAL}$ using the following equation:

$$P = RSSI_{VAL} + RSSI_{OFFSET}[dBm] \qquad (1)$$

where the $RSSI_{OFFSET}$ is found empirically from the front-end gain and it is approximately equal to $-45dBm$. In the next sections when we refer to the RSSI value we refer to the $RSSI_{VAL}$ and not the power P unless otherwise stated.

A straight piece of wire is used as a monopole antenna for our sensor node. The length of our antenna is equal to $1.1inch$, the optimal antenna length according to the CC2420's datasheet [4]. In all of the experiments described in the next sections, the length of the antenna on all the nodes was $1.1inch$ unless otherwise stated.

2.2 Sources of RSSI Variability

In addition to multipath, fading and shadowing of the RF channel, signal strength measurements are also affected by the following factors:

1. **Transmitter variability**: Different transmitters behave differently even when they are configured exactly in the same way. In practice, this means that when a transmitter is configured to send packets at a power level of d dBm then the transmitter will send these packets at a power level that is very close to d dBm but not necessarily exactly equal to d dBm. This can alter the received signal strength indication and thus it can lead to inaccurate distance estimation.

2. **Receiver variability**: The sensitivity of the receivers across different radio chips is different. In practice, this means that the RSSI value recorded at different receivers can be different even when all the other parameters that affect the received signal strength are kept constant.

3. **Antenna orientation**: Each antenna has its own radiation pattern that is not uniform. In practice, this means that the RSSI value recorded at the receiver for a given pair of communicating nodes and for a given distance between them varies as the pairwise antenna orientations of the transmitter and the receiver are changed.

Path Loss Prediction Model

The majority of RSSI localization algorithms that do not use full location profiling of the deployment environment make use of a signal propagation model that maps RSSI values to distance estimates [29]. The most widely used signal propagation model is the log-normal shadowing model:

$$RSSI(d) = P_T - PL(d_0) - 10\eta \log_{10} \frac{d}{d_0} + X_\sigma \qquad (2)$$

where, P_T is the transmit power, $PL(d_0)$ is the path loss for a reference distance d_0, η is the path loss exponent and X_σ is a gaussian random variable with zero mean and σ^2 variance, that models the random variation of the RSSI value.

Using the CC2420 radio we were able to verify the log-normal shadowing model in an obstacle-free environment(basketball court). The effects of orientation and calibration were isolated by taking measurements with a single pair of nodes, with the receiver and the transmitter on the same plane. Figure 5a shows the RSSI vs Distance plots. Based on our measurements in the basketball court, RSSI changes linearly with the log of the distance.

2.3 Variations Across Different Radios

In order to quantify the variability among different transmitter-receiver pairs we conducted 2 different experiments. To characterize transmitter variations we used a single receiver and 9 different transmitters. In all of our experiments the receiver was exactly at the same position and with the same antenna orientation. One transmitter at a time was placed at a specific location that was $1.31 ft$ far away from the receiver. Each transmitter was transmitting packets at $-15 dBm$ while in 4 different orientations (0, 90, 180, and 270 degrees). The nodes under test were placed in the middle

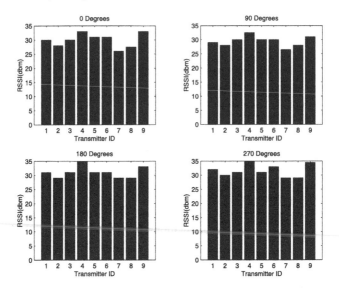

Fig. 2. Quantifying transmitter's variability.

of a room without furniture in order to minimize the effect of the reflections in our measurements.

Figure 2 shows the RSSI values recorded at the receiver for all the transmitter and for all 4 orientations. For each orientation the average RSSI value and its standard deviation are computed. Averaging over all the average standard deviations for all different orientations we find that the overall standard deviation of the received RSSI value is equal to: $2.24dBm$. Using the log-normal signal propagation model shown in Figure 5a we find that the $2.24dBm$ RSSI standard deviation corresponds to $0.4ft$ distance standard deviation.

To quantify the variability in the receiver we used a similar setup using 1 transmitter and 5 different receivers. The transmitter was transmitting packets at $-15dBm$ while in 4 different orientations (0, 90, 180, and 270 degrees). Figure 3 shows the RSSI values recorded at the different receivers for all 4 orientations of the transmitter. For each orientation the average RSSI value and its standard deviation are computed. Averaging over all the average standard deviations for all different orientations we find that the overall standard deviation of the received RSSI value is equal to: $1.86dBm$. Using the log-normal signal propagation model shown in Figure 5a we find that the $1.86dBm$ RSSI standard deviation corresponds to $0.33ft$ distance standard deviation. The same experiment was performed several times with different transmitters in order to make sure that we were measuring the receiver variability and not something else that had to do with the specific transmitter.

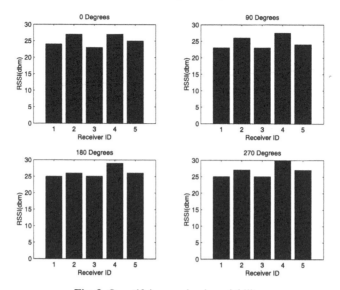

Fig. 3. Quantifying receiver's variability.

2.4 Antenna Characterization

The XYZ sensor node, as most of the generic sensor node platforms, uses a simple wire as a monopole antenna. Ideally, the radiation pattern of this antenna should be uniform and it should look like a circle (2-D space) or a sphere (3-D space). Of course, this does not hold in practice. However, without knowing our antenna's radiation pattern it would be impossible to attempt inferring distance or location information directly from RSSI measurements.

We characterized our antenna in a basketball court measuring $79ft$ in length and $46ft$ in width. The ceiling of the court was at a height of $30ft$. In order to avoid the interference of the floor we attached our transmitter node to a string running from the one side of the court to the other. The transmitter node was at a height of approximately 8ft from the ground at the center of the court. Its antenna was vertical to the PCB board pointing down towards the floor.

We measured RSSI with a receiver node at 3 different heights from the floor: $1.25ft$, $3.5ft$, and $6.5ft$. For each one of these heights we measured the RSSI values for 8 different angles of the receiver with respect to the transmitter: 0, 45, 90, 135, 180, 215, 270, and 315 degrees. For each of these orientations we recorded the RSSI values on the receiver at a distance resolution of $2ft$. We stopped taking measurements for a given height and orientation only when the receiver was not able to receive any packets.

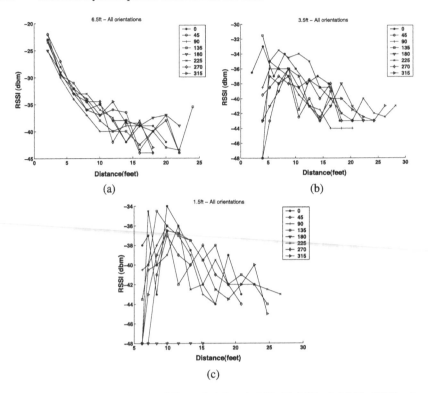

Fig. 4. RSSI vs. Distance plots at different heights a) 6.5ft, b) 3.5ft, c) 1.25ft. RSSI values equal to $-48dBm$ indicate absence of communication between receiver and transmitter.

Antenna Length: Using a suboptimal antenna

According to the Chipcon's Zigbee radio chip datasheet [4] the optimal length of the monopole antenna should be 1.1inch. However, we noticed that even at the lowest power level, transmitter and receiver could communicate for almost any position of the receiver in the basketball court. In addition, we noticed that the RSSI values recorded at the receiver were changing dramatically with very small changes in the distance between transmitter and receiver even when the orientation of the nodes was kept constant.By increasing the transmission power level of the radio we found out that even at slightly higher power levels two nodes could communicate over long distances even without line-of-sight. To reduce the effective communication range of the nodes we used a suboptimal antenna. Instead of using the recommended length (1.1inch) monopole antenna we used a 2.9inch wire as our monopole antenna. As it can be seen in Figure 5b, the communication range of the radio when using the suboptimal antenna is significantly reduced but the signal attenuation properties remain the same.

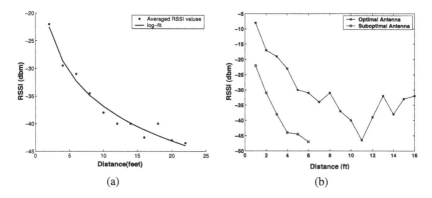

(a)   (b)

Fig. 5. a)RSSI vs Distance plots for an obstacle-free environment (basketball court). Each data point is the average RSSI value recorded for 20 packets, b) The effect of using a suboptimal antenna in an obstacle free indoor environment of size $24ft$ x $20ft$

Antenna Orientation in Basketball Court

After replacing the 1.1inch antennas, on both the receiver and the transmitter, with 2.9inch antennas, we were able to measure the antenna at the power level of $-15dBm$ using all the possible combinations described in the previous section.

Figures 4a, 4b, and 4c show the RSSI values versus distance for all the orientations and for the $6.5ft$, $3.5ft$, and $1.25ft$ receiver heights respectively. Note that when the receiver is at $1.25ft$ (Figure 4c) and $3.5ft$ (Figure 4b) height from the ground the raw RSSI data cannot be used to infer any distance information. The reason is that significantly different distances can produce the same or almost the same RSSI values. In addition, similar distances correspond to very different (even up to $11dBm$) RSSI values for different antenna orientations.

However, when the receiver is at $6.5ft$ height from the ground (Figure 4a) the RSSI versus distance plot can be easily fitted to the widely used log-normal signal propagational model. Note that as the distance between transmitter and receiver increases the variability in the RSSI value that corresponds to this distance also increases. In other words different ranges of RSSI values provide distance information with different levels of accuracy. This suggests that a probabilistic approach for translating RSSI values to distance information should be used. This can be easily implemented by computing the probability distribution of the raw RSSI values over the different distances. Using this probability distribution we can map an RSSI value to a specific distance with a given probability. The higher the probability the higher the accuracy of the distance estimation.

Figures 4a, 4b, and 4c clearly show that different antenna orientations can produce different sets of RSSI values for the same distances between receiver and transmitter. In practice, this implies that the raw RSSI values cannot be directly translated to distance information. Extra knowledge about the specific antenna orientation that corresponds to this set of RSSI values is needed. Furthermore, our results show that even if we are able to map a set of RSSI values to a specific antenna orientation this

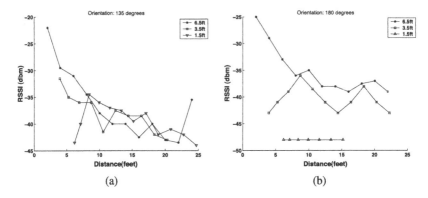

(a)   (b)

Fig. 6. a) The best angle between receiver and transmitter, b) One of the worst angles between the receiver and the transmitter. RSSI values equal to $-48dBm$ indicate absence of communication between receiver and transmitter.

does not necessarily mean that we can extract any useful distance information. The reason is that some antenna orientations do not provide a consistent signal attenuation.

Figures 6a and 6b provide some more insight to the antenna orientation effect. Figure 6 shows the best transmitter antenna orientation. Note that a single signal propagation model can be extracted that is independent of the height of the receiver.

On the other hand, Figure 6a shows the worst transmitter antenna orientation. It is obvious that any attempt to infer distance information directly from the RSSI values is impossible. Different heights of the receiver produce very different RSSI values. This implies that when the height difference between the transmitter and the receiver is small, antenna orientation does not affect the signal propagation model. But, as figure 6b shows, when the height difference between the transmitter and the receiver increases then the antenna orientation becomes a major factor that greatly affects the signal propagation model.

This can be seen in figure 7 where the radiation pattern of our monopole antenna is shown. The radiation pattern was constructed using all the measurements we collected in the basketball court. The shaded region of the antenna radiation pattern is the symmetric region for which a single signal propagation model can be extracted. Since the antenna orientation is not a major factor when the receiver and the transmitter are at the same height, the log-normal shadowing model is very accurate in the case of a 2-D sensor network deployment at an obstacle-free environment (outdoor deployment). However, the log-normal shadowing model is not able to capture the effect of the transmitter's antenna orientation in the case of a 3-D sensor network deployment even in an obstacle-free environment.

Consequently, a robust RSSI localization method should try to operate only in the shaded region of the antenna radiation pattern ,shown in Figure 7, where the log-normal shadowing model seems to hold. This requires isolating the shaded area from the rest of the communication region where the RSSI values are significantly affected by the antenna orientation of the transmitter and they cannot provide any

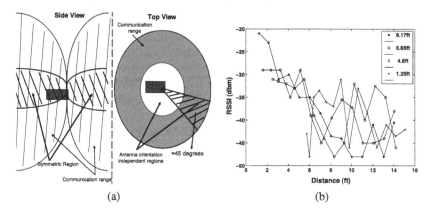

Fig. 7. a) Radiation Pattern of the monopole antenna, b) Indoors antenna characterization. RSSI values equal to $-48dBm$ indicate absence of communication between receiver and transmitter

reliable distance or location information. To demonstrate the difficulty of this task, consider the following case where a beacon is transmitting packets and a set of receivers listen the packets and record the RSSI values which are then send back to the beacon node. The beacon node is aware of a set of pairs of the following format: $\langle nodeID, RSSI \rangle$. How can the beacon identify the nodes that were in the shaded area of its communication range? The only case where the beacon is able to identify those nodes is the case where the RSSI values recorded on the nodes that were in the shaded area of the communication range of the beacon node are unique. In other words, there is a unique set of RSSI values that can be recorded on the receiver only when the receiver is in the symmetric region of the communication range of the transmitter. Unfortunately, Figures 4 and 6 show that this unique set of RSSI values is very small and covers only a small range of short distances.

Antenna Orientation in Indoor Environments

In this section we focus on the effect of the indoor environment on the received signal strength between a pair of communicating nodes. Our first indoors experiment focused on the effect of reflections on the antenna radiation pattern. We tried to replicate the experiment that we run in the basketball court in the 3-D testbed $(15ft(W) \times 20ft(L) \times 10ft(H))$ that is installed in our lab. Exactly the same transmitter that was used in the basketball court experiment was placed at a height of approximately $7ft$ from the ground. The same receiver that was used in the basketball court experiment was placed in four different heights from the ground: $1.25ft$, $4.6ft$, $5.65ft$, and $6.17ft$. For each one of these heights the receiver recorded the RSSI values for different distances from the transmitter with a distance resolution of $1ft$(the transmitter was transmitting packets at the same power level as in the basketball court, $-15dBm$). In this experiment we focused only on a single transmitter antenna orientation, the one that gave us a single signal propagation model that was independent of the height of the receiver(Figure 6) in the obstacle-free environment.

The RSSI values that were recorded on the receiver for all the different distances and for all the different heights of the receiver can be seen in Figure 7b. When the receiver is at $6.17ft$ from the ground a clear log-normal signal propagation model can be derived as in the case of the obstacle-free environment. However, for the other three heights of the receiver the RSSI values seem totally random and no actual distance information can be extracted from these sets of RSSI values. What is even more interesting, is the fact that the randomness that the reflections introduce in the RSSI values directly affects the symmetric region of the antenna and makes it significantly narrower. Note that every RSSI value that is equal or smaller than $-30dBm$ can actually correspond to any distance that is larger than $1.6ft$ and smaller than the communication range. The only RSSI values that can be used to accurately estimate the distance between the nodes are the RSSI values that are higher than $-30dBm$. This range of RSSI values can only be produced by the symmetric region of the antenna and it is not affected by the reflections in the room or the height of the receiver. In addition, this range of RSSI values can be fitted to a linear signal propagation model and not to a log-normal signal propagation model. Unfortunately, the maximum distance that this region of RSSI values can cover is very small and approximately $3ft$ to $4ft$. This suggests that even for small rooms a very large number of sensor nodes is required in order to perform accurate RSSI localization.

Indoor Testbed Experiment In order to verify the results of the previous section, we deployed 38^3 nodes with 2.9inch antennas on our 3-D testbed located inside our lab. The nodes were placed in 3-dimensions inside the testbed as shown in Figure 1b. The antennas of all the nodes on the floor were pointing to the ceiling and the antennas of the nodes on the testbed were pointing either towards the center of the testbed or towards the floor. In all cases, the antennas were vertical with respect to the PCB of the XYZ sensor node.

In our experiment, each node broadcasts 10 packets at each one of the eight available power levels. All the nodes that hear a packet record the RSSI value for this packet and the sender id. At every time instant only one node is broadcasting packets. After a node has finished transmitting packets, a gateway node connected to a PC polls the recorded data from each node in the testbed separately. This process continues until all nodes transmit 10 packets at each power level. The experiment took place during the night when no people were in the lab.

Received Signal Strength Data: Figures 8a, 8b, 8c show the recorded RSSI values versus the true distances that they correspond to for different power levels and for all 38 nodes. It is obvious that no actual distance information can be extracted directly from the RSSI values. This is due to the reflections and the random placement of the nodes which created communicating pairs of nodes with random pairwise antenna orientations. Note, that as the transmitting power level used decreases, the RSSI data starts looking less random. The reason is that as the power level increases the reflections in the testbed also increase. However, even at the low power level it is very difficult to fit the RSSI data to a signal propagation model.

---

[3] Initially we deployed 40 nodes. Unfortunately, as it can be seen in Figure 10, the batteries of nodes 20 and 21 were not full and therefore these nodes did not send/receive any packets

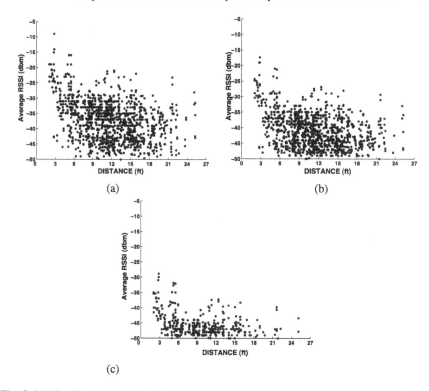

Fig. 8. RSSI vs Distance plots for the 38 node indoor sensor network deployment at different power levels: a) 0dBm, b) -5dBm, c) -15dBm.

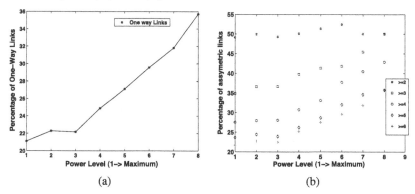

Fig. 9. Percentage of a) One-way links and b) Power-asymmetric links.

Connectivity and Link Symmetry: Our 38 node deployment also provided useful insight about the connectivity and the symmetry of the links in a real IEEE 802.15.4 sensor network. Figures 10a, 10b, and 10c show the connectivity achieved by the lowest, low and maximum power levels respectively.

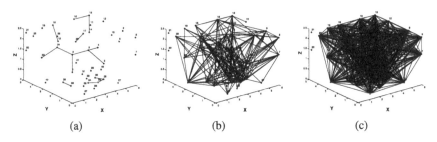

(a)	(b)	(c)

Fig. 10. Connectivity plots for the 38 node indoor sensor network a)-25dBm b) -15dBm, c) 0dBm. Nodes 20 and 21 were not equipped with fully charged batteries.

Figures 9a and 9b show the percentage of one-way links and the percentage of power-asymmetric[4] links respectively. As the power level of transmission decreases the percentage of power-asymmetric links and their absolute difference as well as the percentage of one-way links increase. What is even more important, is the fact that the power-asymmetry of the links does not depend on the actual RSSI values. Our experimental data shows that when node A transmits packets, node B might record a very high RSSI value that can be even equal to $-23dBm$ or $-25dBm$. However, when node B transmits packets, node A might record a very small RSSI value or it might not record any RSSI value at all because it is not able to receive any packets. Our experimental results, shown in Figure 9, show that the percentage of power-asymmetric links vary from 21% to 36% of the total number of links in the network depending on the power level used during transmission.

2.5 Discussion

Based on our detailed characterization we found that antenna orientation greatly impacts RSSI and link asymmetry in indoor and outdoor scenarios. This is especially the case in 3-D indoor deployments with random antenna orientations. These observations influence the assumptions of many node localization algorithms that utilize RSSI information. This includes RSSI distance prediction and profiling algorithms as well as other statistical approaches. Our results show that direct distance prediction from raw RSSI data in 3-D indoor environments is impossible. For profiling approaches, our measurements show that antenna orientation information should be included in the fingerprint. However, even if the antenna orientation that corresponds to a set of RSSI values is known it might be impossible to infer any distance information since some antenna orientations do not provide a consistent signal attenuation. This observation shows that modeling the antenna orientation effect as a random variable with Gaussian distribution, as it is modeled in equation 1, is not realistic.

---

[4] We call (A, B) link a power-asymmetric link if the RSSI value recorded at B when A is transmitting is different than the RSSI value recorded at A when B is transmitting. We call (A, B) link an one-way link if node A can reach node B but node B cannot reach node A. Power-asymmetric links include the one-way links.

Our experiments also show that the antenna orientation has a great impact on the ordering of the RSSI values. The ordering of RSSI values is meaningful only when the communication takes place in the symmetric region of the antenna as it is shown in Figure 7(a). According to our findings this region of the antenna is only a small fraction of the communication range and therefore the ranking of the RSSI values provides little or no information in the case of 3-D deployments where the antenna orientations of the communicating nodes are almost random.

The antenna orientation effect has also implications on the statistical RSSI localization algorithms. In most probabilistic algorithms a probability distribution, usually Gaussian, of the RSSI values is assumed. In general, such an assumption holds only in the symmetric region of the antenna. When the communication between two nodes takes place in the non-symmetric region of the antenna, which is generally the case in a 3-D network deployment, the variation in the RSSI values cannot be modeled by a Gaussian distribution since, according to our experiments, there is a huge variation in the RSSI values. Consequently, our observations suggest that new probabilistic models that better capture RSSI variations need to be developed for 3-D environments.

In the case of indoor environments, reflections become the main problem in performing RSSI distance prediction. Only a very small range of RSSI values can be used for extracting distance information for up to $3 - 4ft$. In this region, RSSI changes linearly with distance. In addition, our findings show that 3-D indoor sensor network deployments suffer of a high degree of link asymmetry. This link asymmetry is due to the multipath and fading effects as well as due to the random pairwise antenna orientations used during communication.

3 Sensor Assisted Camera Calibration and Localization

Cameras provide an alternative technology for localization. Although this technology is different and sometimes more expensive than other measurement systems, it still poses several advantages that justify its consideration. As a measurement modality, cameras measure the relative rotation (angle) between two objects they observe. This, combined with the camera calibration information, makes the problems associated with unique localizability much simpler. Our results show that a few simple deployment considerations and the use of imagers can result in lightweight, yet very accurate 3D localization that bypasses some of the challenges posed by rigidity constraints in the case of distance only localization [8]. Furthermore, as our experimental evaluation verified, the requirement of having distance measurements can be entirely eliminated by imposing a few distance measurements in the observed scene. For instance, one could require having a 3-LED triangle with known sides on some of the nodes. This distance information can provide sufficient information for scaling the coordinate system to Euclidean dimensions.

Our work builds up on concepts from computer vision but it also bears a few differences. We perform an in-situ evaluation in the context of sensor network applications based on a real camera enabled sensor network. We also examine the fea-

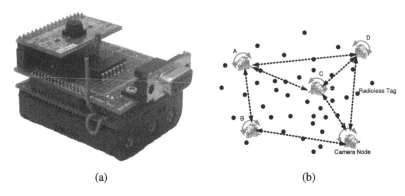

(a)   (b)

Fig. 11. a) XYZ sensor node with OV camera module, b)deployment scenario

sibility of solving the problem using small sensor nodes, and we propose an algorithm for reconstructing node coordinates that utilizes deployment information and is more lightweight than the stratification approaches used in computer vision. Our experiments bypass the vision correspondence problem by using modulated LED transmissions for communicating node IDs.

3.1 Problem Statement

Our presentation in this section deals with computing the relative translation and rotation matrices between cameras and localizing nodes, using information between camera views. A practical solution to this problem can be easily extended to multihop scenarios by using the translation and rotation information among adjacent cameras. This process is known as *transfer* in computer vision. With transfer, the coordinate system of any camera i, can be translated to the coordinate system of any other camera j if there is a path of cameras from i to j, on which the relative rotations and translations, among adjacent camera pairs are known. Assuming that transfer is possible because camera nodes can communicate with each other, we state our problem as follows:

Given a network of N sensor nodes $t_1, t_2, t_3...t_N$ where a subset $m \langle N$ nodes are equipped with cameras, compute all possible node locations and the relative camera translations and rotations with respect to a local coordinate system.. In some parts of our algorithms (measured epipoles), we also make minor assumptions about distances that will be explained in section 3.4.

For the rest of the chapter, our discussion focuses on how to exploit camera epipoles when two cameras have a shared field of view. Two algorithms are examined, measured epipoles and estimated epipoles. Camera coverage and mobility issues are discussed in section 3.8.

3.2 Related Work

The reconstruction of 3D imagery from images is a problem treated in computer vision for several years. In 1992 the work done by Tomasi and Kanade in [35] has proposed a way for reconstructing a scene and estimating camera parameters and feature point locations using matrix factorization. This initial work was performed under orthographic projection, and more recent work has treated the paraperspective case [22]. A more complete solution that compensates for the camera depth effects has been described in [31]. More recently, the works of Mantzel et. al. [17] and Devarajan et. al. [5] have begun to consider the problem of camera calibration in a networked setup. Both approaches have examined the problem from a vision perspective. The former has proposed an iterative approach for localizing cameras (position, relative translation and orientation up to a scale) using linear relationships and the DLT camera calibration developed by Faugeras in [18].The work of Devarajan et.al. builds upon, more computationally demanding factorization methods proposed by Sturm and Triggs in [31]. Cameras form microclusters with other nodes in the same coordinate system. The camera localization algorithm requires 4 cameras having at least 12 feature points in their common field of view. The algorithm also provides a scheme for aligning image frames.

Camera calibration using mobile entities such as people has also been considered in the computer vision community. Talor et. al. at MIT [34] considers camera calibration for cases of non-overlapping field of views. In sensor networks, the problem of ad-hoc node localization has been treated in great detail. Our work is more related to fine grained localization schemes such as the ones that have been demonstrated in MIT's Cricket system [24,25] and UCLA's AHLoS system [28]. In fact, many of the concepts presented here will be interoperable with the Cricket localization schemes recently presented in [19].

3.3 Background

The extrinsic camera calibration parameters of interest are the rotation matrix $R_{3\times3}$ and the translation vector $T_{3\times1}$, also referred to as the extrinsic calibration parameters. In the absolute sense, the two matrices R and T represent the camera coordinates with respect to a 3-D world origin O. In general, our discussion about cameras will deal with three types of coordinates. World coordinates w are the coordinates given with respect to a real 3-D world origin. The camera also has its own camera centric coordinates $\bar{w}$ that can also be expressed in world coordinates with (3).

$$\bar{w} = Rw + T \qquad (3)$$

The image coordinates u, v observed by the camera, can also be related to the camera centric coordinate system though the following equations.

$$\frac{u}{f} = \frac{\bar{w}_x}{\bar{w}_z}, \frac{v}{f} = \frac{\bar{w}_y}{\bar{w}_z} \qquad (4)$$

where f is the focal length of the camera. The above expressions can be obtained by considering the ratio of triangles as shown in figure 12.

Our work uses the epipoles between pairs of cameras. The epipoles between a pair of cameras are defined as the points where a straight line connecting the two camera centers intersect the image plane of each camera [9] as shown in Figure 12.

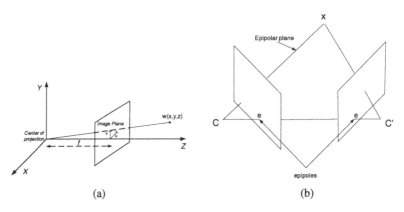

(a)   (b)

Fig. 12. a) Camera diagram. Camera coordinate system z-axis is piercing the normalized (u, v) image plain in the image plain origin, b)epipoles between a pair of cameras .

Throughout the paper we use the following notation:
l_{ab} - distance between nodes A and B
v_{ab} - unit vector from A to B
u_a, v_a - x, y coordinates of an object on the image plane of camera A.

3.4 Localization Based on Two Camera Views

We now examine two methods for extracting camera epipoles, direct observation, and fundamental matrix estimation.

Direct Epipole Observation - *Measured Epipoles* **(ME)**

If a tag C is observed by two cameras A and B that can also observe each other, the distances between A, B, C can be determined up to a scale. This was demonstrated in [33].

If cameras A and B can observe each other and a tag C, then we can derive the unit vectors v_{ab}, v_{ac}, v_{ba} (see Fig. 13), and v_{bc}. From these we can derive the normalized versions $n_a = v_{ab} \times v_{ac}$ and $n_b = v_{ba} \times v_{bc}$. Note that v_{ab}, v_{ba}, n_a and n_b are related by a rotation matrix R_{ab}, the relative orientation matrix between cameras A and B.

$$v_{ab} = -R_{ab}v_{ba}$$

$$n_a = -R_{ab}n_b$$

From the two perpendicular unit vectors v_{ab} and n_a, we can construct the orthonormal matrices R_a and R_b. $R_a = \begin{bmatrix} v_{ab} & n_a & (v_{ab} \times n_a) \end{bmatrix}$ and $R_b = \begin{bmatrix} v_{ba} & -n_b & (v_{ba} \times n_b) \end{bmatrix}$. Substituting we get the relative orientation as

$$R_{ab} = R_a(R_b)^T \tag{5}$$

from this we can extract the following linear system

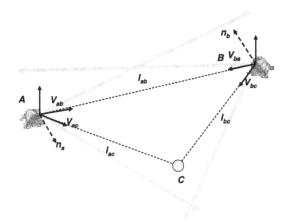

Fig. 13. Measured epipole: A tag can be observed by 2 cameras that can also observe each other

$$l_{ab}v_{ab} + l_{bc}(R_{ab}v_{bc}) - l_{ca}v_{ac} = 0 \qquad (6)$$

Solving these equations we can get l_{ab}, l_{ac} and l_{bc} up to a scale. Assuming, we also know the distance l_{ab} we can solve to Euclidian scale.

Extracting the Epipoles from the Fundamental Matrix - *Estimated Epipoles* **(EE)**

If two cameras are not facing each other, then the epipoles cannot be directly observed. Nonetheless, the camera epipoles can still be determined if the camera observations can provide enough information to compute the fundamental matrix between two cameras [10]. Although this can be done with a minimum of 5 nodes in the common field of view of two cameras, the resulting problem is highly non-linear and very difficult to solve. In our implementation we chose to use the widely used normalized eight-point algorithm proposed by Hartley in 1997 [9].

The eight-point algorithm uses eight or more points in the common field of view of a pair of cameras to estimate the fundamental matrix F between them. F is defined by the equation

$$u'^T F u = 0$$

u' and u are corresponding feature points in the images of the two cameras. The epipoles of the two cameras can be then extracted from F. For any point x, the epipolar line $l' = Fx$ contains the epipole e'. Thus $e'^T(Fx) = (e'^T F)x = 0$ for all x. From this it follows that $e'F = 0$ and $Fe = 0$, thus the epipoles of the two cameras, e' and e are the left and right null vectors of F respectively [10]. The estimation of both camera's epipoles allows us to compute the rotation between the

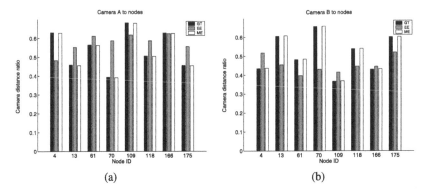

Fig. 14. Estimated distances between two camera nodes and other observed nodes. GT - Ground truth, EE - estimated epipoles, ME - measured epipoles

two cameras and the distances from both cameras to all the nodes up to a scale using the formulation described in the previous subsection. The only difference in this case is that instead of using the measured epipoles, as in ME, we use the estimated, from the fundamental matrix, epipoles.

3.5 Refining the Estimated Epipoles

The resulting epipole estimates are noisy and thus will provide noisy distance estimates. An illustrative example of the error based on our testbed measurements is shown in Fig. 14 . The figure shows the distance error ratio for ground truth (GT), ME and EE when the distances between the cameras A and B and the nodes are computed (by equations 4). The figure shows that ME produces accurate measurements comparable to ground truth while EE has noisy measurements. Furthermore note that the error from EE at the two cameras is complimentary. If a distance between camera A and a node is overestimated then the distance between the same node and camera B will be underestimated and vice-versa. This is of course an artifact of the lightweight algorithm we use. It does however suggest that refinement is possible by attempting to match the two camera views that have complimentary error as shown in Fig. 15.

To reduce this error, we formulate a more constrained optimization problem, where the distances between the observed nodes and the camera nodes are estimated simultaneously. To enforce matching views between the cameras, we enforce the constraint that all the pairwise distances between all the nodes observed by the two cameras should be equal from the view point of both cameras.

Consider the scenario in Figure 3.5 with cameras A and B and observed nodes i and j. The distance l_{ij} can be estimated by both cameras as

$$^{A}l_{ij} = ||l_{ai}v_{ai} - l_{aj}v_{aj}||$$
$$^{B}l_{ij} = ||l_{bi}v_{bi} - l_{bj}v_{bj}|| \tag{7}$$

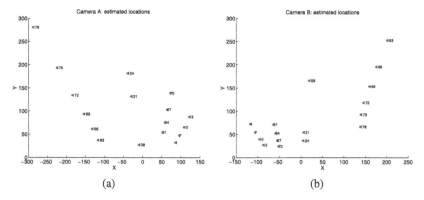

Fig. 15. Estimated locations using an optimized version of the distances in Fig. 14

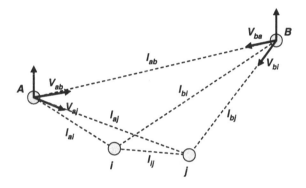

Fig. 16. Imposing the constraint that two cameras should agree on the l_{ij} estimate

The rest of the analysis is identical for both cameras, therefore we will focus on camera A. The required camera-to-node distances can be estimated up to a scale with respect to camera A's view point as:

$$l_{ab}v_{ab} + l_{bi}(R_{ab}v_{bi}) - l_{ai}v_{ai} = 0$$
$$l_{ab}v_{ab} + l_{bi}(R_{ab}v_{bj}) - l_{aj}v_{aj} = 0$$

(8)

As it was described in the previous section, in the case of the estimated epipoles method(EE), the computed distances from the camera to the nodes will be erroneous. These erroneous distances can be refined if a subset of distances between the nodes seen by the camera is known. Assuming that n distances $d_{i_x j_x}$, $x = 1, 2, \ldots, n$ are known, the following equations should hold:

$$^A l_{i_x j_x} = ||l_{ai_x}v_{ai_x} - l_{aj_x}v_{aj_x}|| = d_{i_x j_x}$$

(9)

where x is an index running over all n known edges and i_x, j_x are the nodes connected by the xth edge. In order to refine the distance estimates $l_a i_x$ and $l_a j_x$ based on the known n distances we would need to minimize the following function:

$$L = min \sum_x \left(d_{i_x j_x} - ||l_{a i_x} v_{a i_x} - l_{a j_x} v_{a j_x}|| \right)^2 \tag{10}$$

Since x is running over all n known distances, function L is basically a set of n equations where the number of unknowns depends on the known edges. If all the n known edges are independent[5] the number of variables is $2n$ because in that case each edge (i, j) would involve two unique unknowns: l_{ai} and l_{aj}. Since the number of equations is less than the number of unknowns the minimization of function L is impossible. However, if different known edges share common nodes then the minimization of L is possible. The simplest case, where the minimum number of edges is needed and function L can be minimized, is when all the distances among three nodes are known. In other words, when all the edges of a triangle that is formed by three nodes are known the set of equations represented by L can be solved. Each edge of the triangle provides an equation giving a total of 3 equations. For each node in the triangle there is only one unknown. Therefore the number of unknowns is equal to the number of equations and the distances from camera A to the nodes forming the triangle can be refined. This set of equations is a non-linear set of equations that can be solved with an Extended Kalman Filter(EKF) or another gradient descend method.

Note that any other closed geometric shape (square, pentagon, hexagon etc) could be used in the optimization process. We use triangles because they require the minimum number of known distances (only 3) and therefore they require the minimum number of local information. Therefore based on local triangles we can optimize the distances between the cameras and the non-camera nodes with respect to the camera's coordinate system. After the refinement of the distances between the camera nodes and the non-camera nodes, the rotation matrix between two camera nodes with overlapping fields of view can be refined using equation 8 and the already refined distances.

The preceding ME, EE and OEE algorithms provide us with a means of bootstrapping a coordinate system and also computes the relative rotation and translation between a pair of cameras. Next section shows how these two components can be applied on a distributed localization algorithm.

3.6 Multihop Localization Algorithm Overview

Although our focus in this paper is on how to perform calibration based on two camera views, we briefly describe an example of a distributed localization algorithm. This algorithm assumes the existence of a coordinate transformation service that runs in the background and distributes the rotations and transformations between pairs of cameras so that nodes can be localized on demand in multiple coordinate systems.

---

[5] Two edges are independent if they do not share any common nodes

Broadcast observed node IDs to radio neighbors
and wait for for radio neighbor broadcasts
Identify Vision neighbors
Transmit-Receive Observations {ID,image coordinates}
to vision neighbors that can execute ME,EE
if sufficient data
    Run ME or EE
    if for rotation from coordinates
        Transmit-Receive observations with other camera
        Compute rotation from camera coordinates
    endif
        Compute rotation from coordinates
    Update coordinate transformation service
endif

Fig. 17. Distributed Localization Algorithm

Immediately after deployment(or when triggered), each node broadcasts a list of IDs for all the nodes it can observe to its radio neighbors. The camera nodes take advantage of these broadcast to establish a vision graph that connects camera nodes that have one or more nodes in their shared field of view. The same information also allows each camera node to identify which other nodes can use its observations to run the ME or EE algorithm. Once a camera node identifies these nodes, it forwards them the observations (node id and image coordinates) for the observed nodes in their shared field of view.

After this phase, each camera node has the required observations to bootstrap its local coordinate system by running ME or EE. ME and EE provide a distance(magnitude) and a vector(direction) for each node they consider. This allows the camera node to estimate the coordinates of each of the considered nodes in its own coordinate system. The relative rotation between the two nodes is also computed, and its passed to the coordinate transformation service. At this point, each camera also examines if this bootstrapping phase produced any additional information that would allow computing a rotation from camera coordinates as described in section 3.4. If this is the case, it initiates an exchange of observations with the affected camera, and the relative rotation matrix is computed by both cameras.

If a new node appears in a camera's field of view, the camera repeats a similar process to discover if any of its neighbors can observe the same node. Depending on the number of nodes available in the shared field of view each camera node may execute ME, EE or simply compute the node's location based on previously computed rotation information. The multihop localization algorithm is outlined in Figure 17.

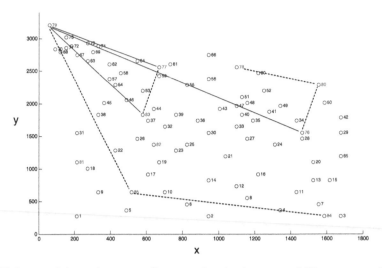

Fig. 18. Layout of the outdoor setup. Camera node pairs that can run ME are connected with solid lines and camera pairs that can run EE and REE are connected with dotted lines

3.7 Evaluation

The evaluation of the ME, EE and OEE algorithms is performed on real measurements obtained from an indoor and outdoor dataset collected using our camera enabled sensor nodes. Each node is equipped with an extension board carrying a COTS OV7649 camera module from OmniVision. The node processor can acquire frames from the camera and perform some basic feature extraction, such as sobel edge detection, differencing to detect motion and LED identification. All nodes also carry a Lumex CCl-CRS10SR omnidirectional LED with an axial intensity of 40 milli-candellas. In our lab setup, the cameras nodes can clearly observe these LEDs for distances up to 4 meters. In our tests, camera nodes can uniquely identify the nodes using an asynchronous protocol which can read the node id of each node by toggling the LED to send bits across.

Using this platform we acquired 2 datasets, one indoor and one outdoor. The indoor scenario was comprised of 2 camera nodes and 16 non-camera nodes identified with blinking LEDs. The two camera nodes were placed at different heights and angles with respect to the non-camera nodes yielding four different datasets. For the outdoor scenario, we placed 80 bright orange postit nodes in an outdoor plaza next to our building. In this outdoor test we were interested in the accuracy and range of our system, and we assumed that the correspondences for each post it node were known [6]. The layout of the outdoor scenario is shown in Fig 18.

---

[6] The image correspondences can be found by methods from computer vision not described here

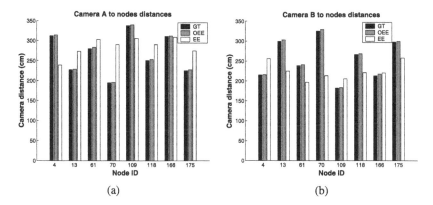

Fig. 19. Estimated distances between two camera nodes and other observed nodes for the indoor experiment. GT - Ground truth, OEE - optimized estimated epipoles, EE - estimated epipoles. All distances are in cm.

ME, EE, OEE Evaluation

The performance of the ME, EE and OEE algorithms in the case of one of the indoor experiments can be seen in Figure 19. The accuracy of the ME method shows that the camera is a very reliable measuring modality. On the other hand, the poor performance of the EE algorithm shows that the estimates of the epipoles are erroneous and therefore unreliable. However, as it can be seen in Figure 19 the OEE method proposed in this paper reduces drastically the error and performs almost equally well to the ME method.

The drastic reduction in the error in the case of the OEE method can be seen better in Figures 20 and 21 where the empirical CDF of the node-to-node distance error is shown. In the indoor setup, EE reports an error of $60cm$ with probability 90% while OEE reports an error of $7cm$ with the same probability. Given that the average node-to-node distance in the indoor setup is approximately $85cm$ the OEE algorithm performs fairly well. The ME algorithm has the best performance reporting an error of $2cm$ with 90% probability. Figures 20 and 21 show very similar results for the outdoor experiment. Again, ME has the best performance($20cm$ error with 90% probability) but OEE performs comparably well reporting an error of $60cm$ with 90% probability given that the average node-to node distance is approximately $297cm$. in the case of the outdoor setup a more detailed view of the performance of all the methods can be seen in Figure 22.

To make our results comparable to others we also evaluate the estimated internode distance error with respect to the following two metrics:

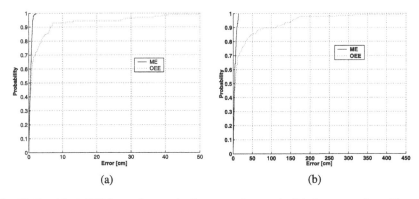

(a)   (b)

Fig. 20. Empirical CDF for node to node distance estimates a) all indoor scenarios with optimized EE, b) all outdoor scenarios with optimized EE

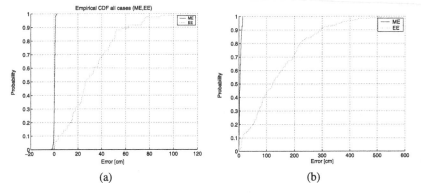

(a)   (b)

Fig. 21. Empirical CDF for node to node distance estimates a) all indoor scenarios with unoptimized EE, b) all outdoor scenarios with unoptimized EE

$$p = \sum_{i=1}^{N} \frac{|l_i - \hat{l}_i|}{N} \tag{11}$$

$$q = \sqrt{\sum_{i=1}^{N} \frac{(l_i - \hat{l}_i)^2}{N}}. \tag{12}$$

N is the number of pairwise distances, l_i are the measured distances and $\hat{l}_i$ are the estimated pairwise distances computed from the localization results of the algorithm. Metric (11) is the *mean error*. Metric (12) is the *square root of the mean square error* and it is also used in [19]. The standard deviation of the error, noted by σ, is also computed.

3.8 Discussion

Localization and camera calibration when sensors can generate stimuli that allows them to communicate with cameras is appealing in some applications since it implies

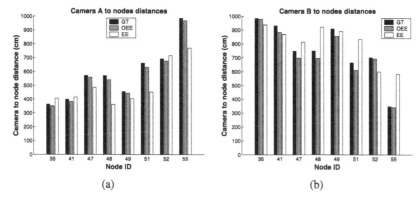

(a)   (b)

Fig. 22. Estimated distances between two camera nodes and other observed nodes for the outdoor experiment. GT - Ground truth, OEE - optimized estimated epipoles, EE - estimated epipoles. All distances are in cm.

Algorithm	Indoor		Outdoor	
	$p(cm)$	$q(cm)$	$p(cm)$	$q(cm)$
ME	4.0165	0.3779	17.9794	1.2493
OEE	0.9714	1.4159	29.6057	51.16818
Ultrasound	7.02	5.18	N/A	N/A

Table 2. The p and q distance metrics for the ME and OEE algorithms as well as for the ultrasound technology in the case of both indoor and outdoor experiments.

one could easily build very simple nodes that cost a few cents each with today's technologies. Imagine for instance, thousands of floatable chemical sensors dispersed to measure chemical concentrations in the sea, in water reservoirs or large tanks. Such nodes would consist of a small battery, a tiny 4-bit microcontroller and an LED serving both as a communication and localization device. A smaller set of camera nodes can observe them.

During our evaluation we also realized that camera placement is a main problem. To have adequate field of view, the cameras need to be mounted on the walls or the ceiling or need to be placed in positions higher than other nodes facing down. This together with the limited field of view of cameras suggest that articulation, and autonomous motion are necessary for many camera enabled sensor networks. the use of omni-directional cameras should also be considered. We anticipate that this would help mitigate the coverage problems and it would simplify calibration.

4 Conclusions

We have conducted an empirical evaluation of two different physical layers in the context of our wireless sensor network testbed at Yale. Our study of signal strength behavior using monopole antennas and the widely used Chipcon CC2420 radio in a large open space with minimal multipath effects, has shown that antenna orientation corrupts signal strength. This significantly alters the quality of information that RSSI can provide for deriving spatial relationships. Our results and experience from this work show that signal strength localization remains an extremely challenging task, especially in $3 - D$ deployments. Statistical techniques and specific deployment scenarios will mitigate some of these challenges. However, the large amount of characterization needed will make the use of signal strength approaches with low power radios practically impossible. Our study also provides valuable insight into link asymmetry in indoor $3D$ deployments.

Conversely to the RF physical layer, low resolution cameras can be used to perform accurate 3-D localization. We have compared using a real sensor network two basic computer vision algorithms for camera calibration and localization. Using the results of our evaluation we have developed a refinement algorithm that is more lightweight than traditional computer vision algorithms and therefore more suitable for resource constrained sensor nodes. Our evaluation has shown that low cost cameras are capable of producing very fine-grained three dimensional locations for hundreds or even thousands of nodes that could also be used to localize and track events, other than sensor nodes.

References

1. P. Bahl and V. N. Padmanabhan. RADAR: An in-building RF-based user location and tracking system. In *Proc. IEEE Infocom*, pages 775–784, Tel-Aviv, Israel, April 2000.
2. M. Berna, B. Lisien, B. Sellner, G. Gordon, F. Pfenning, and S. Thrun. A learning algorithm for localizing people based on wireless signal strength that uses labeled and unlabeled data. In *Proceedings of IJCAI 03, pages 1427-1428*, 2003.
3. P. Castro, P. Chiu, T. Kremenek, and R. Muntz. A Probabilistic Location Service for Wireless Network Environments. *Ubiquitous Computing 2001*, September 2001.
4. Chipcon: CC2420 802.15.4 compliant radio. http://www.chipcon.com.
5. D. Devarajan and R. Radke. Distributed metric calibration for large-scale camera networks. In *First Workshop on Broadband Advanced Sensor Networks (BASENETS) San Jose conjunction with BroadNets 2004*, October 25 2004.
6. D.Niculescu and B. Nath. Vor base stations for indoor 802.11 positioning. In *Proceedings of Mobicom*, 2004.
7. E. Elnahrawy, X. Li, and R. M. Martin. The limits of localization using signal strength: A comparative study. In *Proceedings of Sensor and Ad-Hoc Communications and Networks Conference (SECON), Santa Clara California*, October 2004.
8. D. Goldenberg, A. Krishnamurthy, W. Maness, Y. R. Yang, A. Young, A.S. Morse, A. Savvides, and B. Anderson. Network localization in partially localizable networks. In *To appear in the proceedings of INFOCOM 2005*, April 2005.
9. R. Hartley. In defense of the eight-point algorithm. In *IEEE Transactions on Pattern Analysis and Machine Intelligence*, June 1997.
10. R. Hartley and A. Zisserman. *Multiple View Geometry in Computer Vision*. Cambridge Press, 2003.
11. T. He, C. Huang, B. Blum, J. Stankovic, and T. Abdelzaher. Range-free localization schemes for large scale sensor networks. In *International Conference on Mobile Com-*

puting and Networking(Mobicom), September 14-19 San Diego California, September 2003.

12. Jeffrey Hightower, Roy Want, and Gaetano Borriello. SpotON: An indoor 3d location sensing technology based on RF signal strength. UW CSE 00-02-02, University of Washington, Department of Computer Science and Engineering, Seattle, WA, February 2000.

13. P. Krishnan, A. S. krishnakumar, W. Ju, C. Mallows, and S. Ganu. A system for lease: System location estimation assisted by stationary emitters for indoor rf wireless networks. In *Proceedings of IEEE Infocom 04*, 2004.

14. Konrad Lorincz and Matt Welsh. Motetrack: A robust, decentralized aproachto rf-based location tracking. In *Proceedings of the International Workshop on Location- and Context-Awareness (Loca 2005)*, 2005.

15. D. Lymberopoulos and A. Savvides. Xyz: A motion-enabled, power aware sensor node platform for distributed sensor network applications. In *Proceedings of Information Processing in Sensor Networks, IPSN/SPOTS, Los Angeles, CA*, April 2005.

16. D. Madigan, E. Elnahrawy, and R. Martin. Bayesian indoor positioning systems. In *Proceedings of INFOCOM 2005, Miami, Florida*, March 2005.

17. W.E. Mantzel, H. Choi, and R.G. Baraniuk. Distributed camera network localization. In *Proceedings of the 38th Asilomar Conference on Signals, Systems and Computers*, November 2004.

18. S.J Maybeck and O.D Faugeras. A theory of self-calibration of a moving camera. In *International Journal of Computer Vision 8(2):123-151*, 2004.

19. D. Moore, J. Leonard, D. Rus, and S. Teller. Robust distributed network localization with noisy range measurements. In *Proceedings of ACM SenSys, Baltimore, Maryland*, November 2004.

20. D. Niculescu and B. Nath. Localized positioning in ad hoc networks. In *Proceedings of the First IEEE International Workshop on Sensor Network Protocols and Applications*, San Diego, CA, May 2003.

21. N. Patwari and A. O. Hero III. Using proximity and quantized rss for sensor localization in wireless networks. In *WSNA03*, San Diego, CA, September 2003.

22. C. Poelman and T. Kanade. A paraperspective factorization method for shape and motion recovery. In *IEEE Transactions on Pattern Analysis and Machine Intelligence, Vol 19, NO 3*, March 1997.

23. Prasithsangaree, P. Krishnamurthy, and P. K. Chrysanthis. On indoor position location with wireless lans. In *The 13th IEEE International Symposium on Personal, Indoor, and Mobile Radio Communications (PIMRC 2002)*, 2002.

24. N. B. Priyantha, A. Chakraborty, and H. Balakrishnan. The cricket location-support system. In *Proceedings of 6th ACM Mobicom, Boston, MA*, August 2000.

25. N. B. Priyantha, A. Miu, H. Balakrishnan, and S. Teller. The cricket compass for context-aware mobile applications. In *Proceedings of the 7th ACM Mobicom, Rome, Italy*, July 2001.

26. S. Ray, W. Lai, and I. Pascalidis. Deployment optimization of sensornet-based stochastic location-detection systems. In *Proceedings of INFOCOM 2005, Miami, Florida*, March 2005.

27. S. Saha, K. Chaudhuri, D. Sanghi, and P. Bhagwat. Location determination of a mobile device using ieee 802.11 access point signals. In *IEEE Wireless Communications and Networking Conference (WCNC)*, 2003.

28. A. Savvides, C.C. Han, and M. B. Srivastava. Dynamic fine grained localization in ad-hoc sensor networks. In *Proceedings of the Fifth International Conference on Mobile Computing and Networking, Mobicom 2001, Rome, Italy*, pages pp. 166–179, July 2001.

29. Scott Y. Seidel and Theodore S. Rapport. 914 MHz path loss prediction model for indoor wireless communications in multifloored buildings. *IEEE Transactions on Antennas and Propagation*, 40(2):207–217, February 1992.

30. R. Stoleru and J. Stankovic. Probability grid: A location estimation scheme for wireless sensor networks. In *Proceedings of Sensor and Ad-Hoc Communications and Networks Conference (SECON), Santa Clara California*, October 4-7 2004.

31. P. Strum and B. Triggs. A factorization based algorithm for multi-image projective structure and motion. In *In Proceedings of European Conference on Computer Vision*, pages 709–720, 1996.

32. P. Myllymaki T. Roos and H. Tirri. A statistical modeling approach to location estimation. In *IEEE Trnsactions on Mobile Computing*, pages 1:59–69, 2002.

33. C. Taylor. A scheme for calibrating smart camera networks using active lights. In *SenSys04 Demostration Session, Baltimore, MD*, Novemeber 2004.

34. C. Taylor, A. Rahimi, J. Bachrach, and H. Shrobe. Simultaneous localization and tracking in an ad hoc sensor network. 2005.

35. C. Tomassi and T. Kanade. Shape and motion from image streams: a factorization method. In *Carnegie Mellon Technical Report CMU-CS-92-104*, 1992.

36. K. Whitehouse, A. Woo, C. Karlof, and D. Culler F. Jiang. The effects of ranging noise on multi-hop localization: An empirical study. In *Proceedings of Information Processing in Sensor Networks (IPSN), Los Angeles, CA*, April 2005.

37. K. Yedavalli, B. Krishnamachari, S. Ravula, and B. Srinivasan. Ecolocation: A technique for rf based localization in wireless sensor networks. In *Proceedings of Information Processing in Sensor Networks (IPSN), Los Angeles, CA*, April 2005.

Part II

Secure Localization

Robust Wireless Localization: Attacks and Defenses

Yanyong Zhang[1], Wade Trappe[1], Zang Li[1], Manali Joglekar[1], and Badri Nath[2]

[1] WINLAB, Rutgers University
    {yyzhang,trappe,zang,manali}@winlab.rutgers.edu
[2] Computer Science Department, Rutgers University badri@cs.rutgers.edu

Summary. Many sensor applications are being developed that require the location of wireless devices, and localization schemes have been developed to meet this need. However, as location-based services become more prevalent, the localization infrastructure will become the target of malicious attacks. These attacks will not be conventional security threats, but rather threats that adversely affect the ability of localization schemes to provide trustworthy location information. This paper identifies a list of attacks that are unique to localization algorithms. Since these attacks are diverse in nature, and there may be many unforseen attacks that can bypass traditional security countermeasures, it is desirable to incorporate an additional layer in the data path to classify/clean the corrupted location data. To address these attacks, we outline a general framework for validating location information through data classification and data cleansing techniques. Consistency checking methods can be used to verify that physical measurements are consistent with each other and with physical reality. We then explore more powerful techniques that employ robust statistical methods to make localization schemes attack-tolerant. We examine two broad classes of localization: triangulation and RF-based fingerprinting methods. For triangulation-based localization, we propose an adaptive least squares and least median squares position estimator that has the computational advantages of least squares in the absence of attacks and is capable of switching to a robust mode when being attacked. We introduce robustness to fingerprinting localization through the use of a median-based distance metric. We evaluate our robust localization schemes under different threat conditions.

1 Introduction

The infrastructure provided by wireless networks promises to have a significant impact on the way computing is performed. Not only will information be available while we are on the go, but new location-aware computing paradigms along with location-sensitive security policies will emerge. Already, many techniques have emerged to provide the ability to localize a communicating device [1–5].

Enforcement of location-aware security policies (e.g., this laptop should not be taken out of this building, or this file should not be opened outside of a secure room) requires trusted location information. As more of these location-dependent services get deployed, the very mechanisms that provide location information will become the target of misuse and attacks. Therefore, as we move forward with location services, it is prudent to integrate mechanisms that protect localization techniques from attacks.

The purpose of this paper is to examine the problem of secure localization from a viewpoint different from traditional network security services. In addition to presenting different attacks faced by wireless localization mechanisms, we present the viewpoint that these vulnerabilities can be mitigated by exploiting the redundancy present in typical wireless deployments– either the redundancy arising from the measurement of possibly several physical properties or the redundancy associated with multiple devices located at different positions. Rather than introduce countermeasures for every possible attack, our approach is to provide *localization-specific, attack-tolerant* mechanisms that shield the localization infrastructure from threats that bypass traditional security defenses. These attack-tolerant methods involve methods that identify the presence of suspicious data, as well as more robust localization algorithms capable of operating in the presence of adversarial attacks. The idea is to live with bad nodes rather than eliminate all possible bad nodes.

We begin in Section 7 by presenting an overview of several techniques used in wireless localization, as well as discuss efforts that have been made to provide security to localization. Following the review of localization, we explore attacks that can be mounted against a variety of different wireless localization algorithms in Section 2. We also report a case study on the effectiveness of various ways of altering the signal strength readings, and discuss the impact of signal strength errors on localization results associated with a typical RF fingerprinting localization algorithm. To classify corrupted location data, as well as correct them, we propose a generic location data validation/cleansing framework in Section 4. After introducing the generic framework, we next focus on robust statistical methods that can classify and clean corrupted location data at the same time. In Section 5 and Section 6 we turn our discussion to applying robust mechanisms to two broad classes of localization: range-based triangulation and fingerprinting methods. We present conclusions in Section 7.

2 Related Work

Broadly speaking, there are two main categories of localization techniques: those that involve range estimation, and those that do not [1]. Range-based localization algorithms involve measuring physical properties that are then used to calculate the distances or angles, such as between a sensor node and an anchor point whose location is known. Time of arrival, or time of flight, is an important property that can be used to measure ranges, such as GPS [6]. Here, the propagation speed of the medium is used to directly convert time measurements into distance. Time difference of arrival is also widely used, such as MIT Cricket [2], and [7, 8]. In addition, APS [3] pointed out that the angle of arrival can be used to calculate the relative angle between two nodes, which can be further used to calculate the distance. It is not necessary to use real-world, physical properties to arrive at distance or range estimates. For example, one can count the number of hops between a sensor node and an anchor point, and further convert the hop counts to physical distances, such as [9–11]. The received signal strength (RSS) value, together with the signal propagation model, is also a good indicator of the distance between two nodes [12], and may be used to perform multilateration.

Range-free localization algorithms, on the other hand, do not explicitly calculate distances from measured properties when conducting localization. A prime example of such a scheme is the Centroid method [13], whereby a sensor node can estimate its location using the centroid of those anchor nodes that are within its radio range. Another class of range-free localization algorithms is the class of RF fingerprinting algorithms. In RF fingerprinting, a node measures a set of radio properties (often just the RSS of a set of landmarks), called the *fingerprint*, and attempts to match these to known location(s) on a radio map that was previously measured. These approaches are often used in indoor environments, where reflection, diffraction and scattering can significantly affect the reliability of ranging estimates. Matching fingerprints to locations can be cast in statistical terms [14,15], or using machine-learning [16], or as a classification problem [17]. A primary example of an RF fingerprinting algorithm is the RADAR scheme [17]. Several variations of RADAR have been proposed [18]. Another range-free method is APIT [19], which employs an area-based estimation scheme to determine a node's location. Compared to range-based localization algorithms, these schemes do not require special hardware, and their accuracies are thus lower as well. Finally, some other properties of arriving signals can be exploited as well.

Secure localization has received attention only recently. In [4], the authors listed a few attacks that might affect the correctness of localization algorithms. One such attack, which we will explore later, are wormhole attacks, and packet leashes [20] have been proposed to defend against wormhole attacks. One strategy for addressing secure localization has been to use variations of a distance bounding protocol [21,22] to place an upperbound on the distance between two nodes. Such a strategy has been used in [5,23,24]. Location estimation can be verified against these bounds and any inconsistency will then indicate attack. For example, in SeRLoc [5], the authors propose a secure localization algorithm that makes use of the property that two sensor nodes that can hear from each other must be within the distance $2r$ assuming r is fixed. [25] uses hidden and mobile base stations to localize and verify location estimate. Since such base station locations are hard for attackers to infer, it is hard to launch an attack, thereby providing extra security. [24] uses both directional antenna and distance bounding to achieve security.

A different strategy to coping with localization, which the authors of this paper proposed in [26], takes the viewpoint that *we should learn how to live with bad guys rather than defeating each of them*. In particular, it might be difficult to identify the presence of an attack, or to build a defense for every possible attack that might be launched against a localization algorithm. Instead, it is desirable to build robust localization algorithms capable of automatically withstanding potential attacks. We will explore this strategy in further detail later in this paper.

Property	Example Algorithms	Attack Threats
Time of Flight	Cricket	Remove direct path and force radio transmission to employ a multipath; Delay transmission of a response message; Exploit difference in propagation speeds (speedup attack, transmission through a different medium).
Signal Strength	RADAR, SpoTON, Nibble	Remove direct path and force radio transmission to employ a multipath; Introduce different microwave or acoustic propagation loss model; Transmit at a different power than specified by protocol; Locally elevate ambient channel noise.
Angle of Arrival	APS	Remove direct path and force radio transmission to employ a multipath; Change the signal arrival angel by using reflective objects, e.g., mirrors; Alter clockwise/counter-clockwise orientation of receiver (up-down attack).
Region Inclusion	APIT, SeRLoc	Enlarge neighborhood by wormholes; Manipulate the one-hop distance measurements; Alter neighborhood by jamming along certain directions.
Hop Count	DV-Hop	Shorten the routing path between two nodes through wormholes; Lengthen the routing path between two nodes by jamming; Alter the hop count by manipulating the radio range; Vary per-hop distance by physically removing/displacing nodes.
Neighbor Location	Centroid Method, SeRLoc	Shrink radio region (jamming); Enlarge radio region (transmit at higher power, wormhole); Replay; Modify the message; Physically move locators; Change antenna receive pattern.

Table 1. Properties employed by different localization algorithms and attacks that may be launched against these properties.

3 Attacks Unique to Localization

The first step to tackling a security problem is to put oneself in the role of the adversary and attempt to understand the attacks. Broadly speaking, there are two main categories of localization techniques: those that involve range estimation, and those that do not [1]. These different localization methods are built upon the measurement of some basic properties. In Table 1, we enumerate several properties that are used by localization algorithms, along with different threats that may be employed against these properties. The threats that we describe are primarily non-cryptographic threats, though it should be recognized that some attacks may be difficult to classify exclusively as a cryptographic or non-cryptographic threat.

We start by looking at methods that employ time of flight. The basic concept behind time of flight methods is that there is a direct relationship between the distance between two points, the propagation speed, and the duration needed for a signal to propagate between these two points. For time of flight methods, an attacker may try to bias the estimation of distance to a larger value by forcing the observed signal to come from a multipath. This may be accomplished by placing a barrier sufficiently close to the transmitter and effectively removing the line-of-sight signal. Another technique that may be used to falsely increase the distance estimate occurs in techniques employing round-trip time of flight. Here, an adversarial target that does not wish to be located by the network, receives a transmission and holds it for a short time before retransmitting. An attack that skews the distances to smaller values can be accomplished by exploiting the propagation speed of different media. For example, in CRICKET [2], the combination of an RF signal and an ultrasound signal allows for the estimation of distance since the acoustic signal travels at a slower propagation velocity. An adversary located near the target may therefore hear the RF signal and

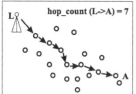 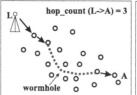 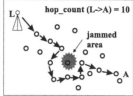

Fig. 1. (Left) Operation of localization using hop count, (Middle) Wormhole attack on hop count methods, and (Right) Jamming attack on hop count methods.

then transmit an ultrasound signal that would arrive before the original ultrasound signal can reach the receiver [4].

As another example, consider a range-based location system that uses signal strength as the basis for location. Such a system is very closely tied to the underlying physical-layer path loss model that is employed. In order to attack such a system, an adversary could introduce an absorbing barrier between the transmitter and the target, changing the underlying propagation physics. As the signal propagates through the barrier, it is attenuated, and hence the target would observe a significantly lower received signal strength. Consequently, the receiver would conclude that it is further from the transmitter than it actually is.

Hop count based localization schemes [10] usually consist of two phases. In the first phase, per-hop distance is measured. In the second phase, anchor points flood beacons to individual sensor nodes, which count the number of hops between them, and these hop counts are translated into physical distances. As a result, adversaries can initiate attacks as follows: (1) manipulate the hop count measurement, and (2) manipulate the translation from hop count to physical distance. A number of tricks can be played to tweak hop count measurements, ranging from PHY-layer attacks, such as increasing/decreasing transmission power, to network layer attacks that tamper with the routing path. Since PHY-layer attacks have been discussed earlier, we now focus on discussing the possible network layer attacks, namely, jamming and wormholes [20]. By jamming a certain area between two nodes, beacons may take a longer route to reach the other end (as shown in Figure 1), which increases the measured hop count. While jamming may not always increase the hop count, for it may not block the shortest path between the two nodes, the other type of attacks, which involve wormhole links, are more harmful because they can often significantly shorten the shortest path and result in a much smaller hop count. Figure 1 illustrates such a scenario: the shortest path between anchor L and node A has 7 hops, while the illustrated wormhole brings the hop count down to 3. Consequently, these attacks can also affect the translation from hop count to physical distance. In addition, if adversaries can manage to physically remove or displace some sensor nodes, even correct hop counts are not useful for obtaining accurate location calculations.

Localization methods that use neighbor location are built upon the implicit assumption that neighbors are uniformly distributed around the wireless device. These localization methods, such as the Centroid method, can be attacked by altering the shape of the received radio region. For example, an attacker can shrink the effective radio region through blocking some neighbors by introducing a strong absorbing barrier around several neighbors. Another approach to shrinking the radio region is for an adversary to employ a set of strategically located jammers. Since these neighbors are not heard by the wireless device, the location estimate will be biased toward the unblocked side.

3.1 Case Study I: Signal Strength Attacks

In order to support the claim that such measurement-oriented attacks are feasible and capable of significantly affecting the results of a localization algorithm, we will examine signal strength attacks. In this section, we report results of actual experiments to quantify the effectiveness of various ways of attenuating signal strength.

Our experiments were performed on the third floor of the CoRE building at Rutgers University, as in Figure 3. Six access points (APs) are available on the third floor, and we measured the strength of their signals from a Dell laptop with an Orinoco Silver wireless card. The laptop runs Linux, and we used `iwlist` to sample the signal strength. In order to avoid fluctuations, we collected samples once every second for 10 minutes, and averaged the signal strength over 600 samples.

An adversary may attack the signal strength by attenuating the RSS readings. This can be done either at the receiver or at the transmitter. The symmetry of RF propagation allows us to perform the measurements at the laptop and infer that similar affects would have been witnessed if the measurements were performed at an AP. Therefore, in the experiments, we placed various obstruction materials close to the laptop's wireless card and measured the RSS values from each AP at the laptop. The following obstructions were used: a thin book (referred to as With Thin Book), a thick book (referred to as With Thick Book), a layer of metal foil (referred to as With Foil), three layers of foil (referred to as With More Foil), a mug filled with water (referred to as With Water), a glass mug (referred to as With Mug), a metal cabinet (referred to as Inside Cabinet), and a human hand (referred to as With Hand). The original signal strength values, together with the signal strength measurements

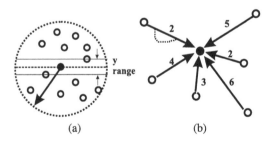

(a)   (b)

Fig. 2. (a) This figure depicts a scenario where use of a sectored antenna allows one to narrow the y-value region. (b) A wormhole attack (indicated by the dotted line) introduces a false hop value in DV-hop. A long distance is perceived as shorter than it really is due to the wormhole.

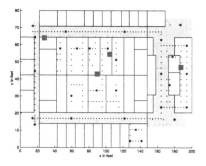

Fig. 3. The third floor of CoRE and the experimental deployment (squares are APs, stars and dots are training and testing points).

AP	Normal Readings	With Thin Book	With Thick Book
1	-85.73	-87.67	-88.26
2	-59.07	-62.03	-64.43
3	-69.29	-71.03	-73.01
4	-65.26	-67.85	-67.46
5	-86.89	-87.40	-89.04
6	-80.86	-79.34	-81.94
AP	With Foil	With More Foil	With Water
1	-87.77	-85.85	-87.64
2	-64.75	-62.68	-62.35
3	-75.05	-78.40	-73.59
4	-64.59	-69.90	-69.67
5	-86.99	-86.75	-87.02
6	-83.11	-85.42	-83.13
AP	With Mug	Inside Cabinet	With Hand
1	-89.29	-88.45	N/A
2	-54.24	-74.05	-76.31
3	-73.28	-83.50	-85.92
4	-64.13	-84.35	-76.22
5	-89.82	-86.00	N/A
6	-84.82	-86.73	-92.33

Table 2. Signal strength readings (RSS) from six APs without and with different obstructing materials. The units of the RSS readings are dBm.

in the presence of these objects, are recorded in Table 2. From these results, we have the following observations:

- Placing a blocking object between the transmitter and receiver can attenuate the signal strength.
- Across different APs, weaker signals attenuate more as compared to stronger ones. For example, APs 1 and 5, which are closer to the laptop, have stronger signals than others, and as a result, get affected less by the blocking objects.
- Across different blocking objects, both metal cabinets and human hands are more effective than others. (We would like to point out that in the case of With Hand, we couldn't get the RSS readings for two APs because they were lower than the threshold.) This may be due to more absorbent materials, or the moisture of

size	case 1 (%)		case 2 (%)		case 3 (%)		case 4 (%)		case 5 (%)		case 6 (%)		case 7 (%)		case 8 (%)	
	nnss	avg	nnss	avg	nnss	avg	nnss	avg	nnss	avg	nnss	avg	nnss	avg	nnss	avg
145	100	100	73.33	73.33	100	100	6.67	66.67	100	100	6.67	53.33	100	100	6.67	33.33
212	100	100	93.33	93.33	100	100	20	6.67	100	100	26.67	26.67	100	100	26.67	6.67

Table 3. The localization errors due to signal strength attacks.

human hands, or the fact that both metal cabinets and human hands can cover the laptop card better than others.

3.2 Case Study II: Impact of Signal Strength Errors on Localization Results

RADAR is a popular in-door localization algorithm [17, 27], involving a training phase and localization phase. In the training phase, the APs collect RSS readings from known locations, forming the *training data sets*. The training set thus contains a set of RSS vectors and the corresponding locations, i.e. $\{x_i, y_i, RSS_{i,1}, RSS_{i,2}, \ldots, RSS_{i,m}\}$ for $0 \leq i \langle n$, where m is the number of APs, and n is the size of training set. In the localization phase, to localize a wireless device, the APs need to measure the RSS from the device and search the training set to find one location whose RSS vector is closest to the device's RSS vector, and use this location as the device's location.

We conducted a simulation study to evaluate the effect that signal strength attacks would have on RADAR. In our study, we used training sets of size 145 and 212 sampled points on the third floor of CoRE. We took a random RSS vector from the training set, varied one or more RSS values, obtained the corresponding localization results using the RADAR algorithm, and determined whether the attacked result corresponded to the same room in CoRE as the non-attacked result. If the original RSS vector is $\{RSS_1, RSS_2, RSS_3, RSS_4\}$, then the following attack cases were studied: (case 1) $\{RSS_1, RSS_2, RSS_3, RSS_4 + 15\}$, (case 2) $\{RSS_1, RSS_2, RSS_3, RSS_4 - 15\}$, (case 3) $\{RSS_1, RSS_2, RSS_3 + 15, RSS_4 + 15\}$, (case 4) $\{RSS_1, RSS_2, RSS_3 - 15, RSS_4 - 15\}$, (case 5) $\{RSS_1, RSS_2 + 15, RSS_3 + 15, RSS_4 + 15\}$, (case 6) $\{RSS_1, RSS_2 - 15, RSS_3 - 15, RSS_4 - 15\}$, (case 7) $\{RSS_1 + 15, RSS_2 + 15, RSS_3 + 15, RSS_4 + 15\}$, and (case 8) $\{RSS_1 - 15, RSS_2 - 15, RSS_3 - 15, RSS_4 - 15\}$. Table 3 presents the likelihood that RADAR identified the correct room when the RSS values are attacked. For each case, we looked at two variations of RADAR: the nearest neighbor in signal space (NNSS), and the average of the two closes matches (AVG). From these results, we can observe that signal strength errors will lead to localization errors. More specifically, when the RSS values decrease by the adversary attenuating the signal, the resulting localization errors are more pronounced than the cases in which the RSS values increase.

4 Framework for Validating Location Information in Sensor Networks

Location information is of critical importance to many sensor applications, but as shown in last section, localization processes can be easily tampered, and lead to erroneous location data. The important problem is thus to assure that these data are correct, i.e. they reflect the true location of each sensor node. In this paper, we propose a location validation framework that consists of the following two parts: (1) location data classification, which is intended to verify whether each individual location measurement is valid or not, and (2) location data cleansing, which is intended to estimate the correct location value from a group of reported readings.

4.1 Location Data Classification

The first stage to coping with false measurement data is the detection or classification of such data as suspicious. We propose to base the diagnosis of localization error on a series of consistency checks. It is to be noted that different consistency checking policies have varying levels of resource requirements, such as CPU power, memory capacity, or battery usage. Since practical sensor network deployments consist of nodes with heterogeneous capabilities, possibly in the form of hierarchies, one can envision that each sensor node can adopt suitable consistency checking policies.

Constraint-based Consistency Checks: Consistency checking has been an important technique for information assurance in the database community, and the most widely adopted approach is through the usage of pre-defined constraints. In the most common form, these constraints specify the range for a node's location. For example, the statement "*check* (x *between* 0 *and* 100)" specifies that the x coordinate must be between 0 and 100. Such constraints are feasible because most sensor networks are deployed within a pre-determined area. A more careful deployment can even lead to constraints with finer granularity. In addition to the constraints on each individual sensor node's location values, we can also specify constraints that govern the relationship of the relative locations of multiple sensor nodes. For example, nodes i and j are within a certain distance. An example of such constraints is "two nodes who can hear each other should not be more than $2r$ apart where r is the radio range". In fact, this constraint has been used in [5].

Setting up these constraints requires certain knowledge about the network deployment, and such knowledge is usually available as long as the deployment is not completely random. This type of consistency checking is very simple, and only consumes a small amount of resources.

Multimodal Consistency Checks: Most current localization techniques employ only a single property at a time, thereby facilitating attacks by an adversary that target a single property. It should be realized, however, that there are correlations between the different properties that might not be maintained by attacking a single property. It is possible to exploit several properties simultaneously to corroborate each other and improve the robustness of localization.

Attacks on localization methods utilizing neighbor locations, such as the centroid method, generally involve modifying the neighbor list of the sensor. One way to combat these attacks is to deploy both neighbor location *and* a two-sector antenna on each sensor. The sensor knows the direction of the global axis so that it can align its antenna section border to the x-axis or y-axis. To get its coordinates, the sensor first aligns its antenna to the x-axis, as shown in Figure 2 (a). Then every neighbor heard in the upper sector should have a larger y-coordinate than that of the sensor, while every neighbor heard in the lower sector should have a smaller y-coordinate. The sensor can estimate its own y-coordinate by simply averaging the smallest y-coordinate in the upper neighbors and the largest y-coordinate in the lower neighbors. If no neighbor is heard in a sector, say the upper sector, the sensor estimates its own y-coordinate as the largest y-coordinate in the lower neighbors. Similarly, the sensor's x-coordinate can be estimated. In this scheme, only the neighbors that are closest to the sensor in the x-coordinate or y-coordinate will affect the estimation. When wrong neighbor information is injected by an attacker, if the forged neighbor is far away from the sensor in both coordinate, it has no effect at all. If it's close in a coordinate, it won't hurt the estimation much. Similar statement holds for jamming attack. On the other hand, if the attacker tries to harm the sensor's orientation capability, it can be easily detected since the neighbor coordinate rule will not hold when the antenna orientation is not aligned to the global axis. Therefore, the extra information provided by the sectored antenna enhances the robustness of the localization significantly.

Challenge-Response Consistency Checks: There are two broad types of classifier schemes: one type makes conclusions based on available measurements and predefined rules, which we may refer to as *static classification*, while the other type may dynamically require some new information with which the system can refine its classification. We refer to this second type of classifier as *dynamic classification*. There are many different strategies for dynamic classification, one type of strategy is motivated by challenge-response protocols from network security [28–30]. To explain the challenge-response consistency checking method, let us look at an example. Consider an indoor wireless localization infrastructure [2, 17, 31] that enables a device to report its position to a neighboring basestation. A localization attack will skew the reported position. To verify position claims, we may place a low-power transponder in each room of the building, where each transponder has its own sequence of pseudo-random numbers that it transmits periodically. If a device A reports that it is located in Room 523 then, assuming the basestation is synchronized with the transponder, the basestation might challenge A to respond with the random number announced by the transponder in Room 523. If A responds correctly, then the basestation will authenticate the fact that A is indeed in Room 523.

4.2 Robust Location Data Cleansing

Localization attacks will behave dramatically different from measurement noise. Although measurement noise can degrade the performance of a localization scheme, such errors are not intentional, and hence not likely to have a persistent bias to any specific localization scheme. However, the attacks mentioned in Section 2 will be intelligent and coordinated, causing significant bias to the localization results. Although it is desirable to detect and classify such anomalies, it should be realized that it is unlikely that any single technique will be able to identify all possible threats and many adversarial attacks will bypass classification. Based on this observation, we take the viewpoint that instead of coming up with a security solution for each attack, it is essential to be robust to unforseen attacks.

The second component to our framework involves robust data cleansing algorithms that automatically cleanse potentially attacked measurements by deemphasizing such data in the position estimation process. To accomplish this, we take advantage of the redundancy in the deployment of the localization infrastructure to provide stability to contaminated measurements.

Data cleansing focuses on estimating the correct value from a set of readings (including false ones) by using robust statistical methods. Unlike data classification schemes that can be implemented on resource-constrained sensor nodes, data cleansing strategies are more appropriate for more powerful nodes, such as aggregators, for two reasons. First, these nodes are usually powerful enough to run robust statistical methods. Second, these nodes will have access to a set of multiple readings, upon which the robust statistical methods can be executed.

In general, measurement errors will lead to false values that deviate enough to cause significant damage to actuation. For example, a decision based upon the average is easily thrown off by a single, corrupted reading. Consequently robust estimates of the values should be used. The application of the median as a robust estimator for the mean is a simple example of robust statistical methods that have found application in data aggregation in sensor networks [32, 33]. For the more general case where $\mathbf{Z}$ (the actual value) has a functional or statistical relationship with $\mathbf{X} = [X_1, \cdots, X_N]$ (measurement vector), it is necessary to employ more complex robust methods.

One promising family of robust estimators are those that improve upon least squares (LS) estimators. LS estimators are susceptible to outliers, and to address this disadvantage, Rousseeuw proposed the method of least median of squares (LMS) [34], which can tolerate up to 50% measurement outliers. Unfortunately, the exact solution for LMS is computationally prohibitive even for aggregators. Hence, the challenge lies in carefully modifying and designing such algorithms to make them suitable for deployment in light-weight computing platforms.

In the remainder of this paper, we shall focus on robust algorithms for two classes of localization schemes: range-based triangulation, and the range-free method of RF fingerprinting. We have chosen these two methods since they represent a broad survey of the methods used. For example, triangulation methods constitute a large class of localization algorithms that exploit some measurement to estimate distances to anchors, and from these distances an optimization procedure is used to determine the optimal position. The robust methods that we describe can be easily extended to other localization techniques, such as the centroid method.

5 Robust Methods for Range-based Triangulation

The range-based triangulation methods involve gathering a collection of $\{(x, y, d)\}$ values, where d represents an estimated distance from the wireless device to an anchor at (x, y). These distances d may stem from different types of measurements, such as hop counts (as in the case of DV-hop [10]), time of flight (as in the case of CRICKET), or signal strength. For example, in a hop-based scheme like DV-hop, following the flooding of beacons by anchor nodes, hop counts are measured between anchor points and the wireless device, which are then transformed into distance estimates.

In the ideal case, where the distances are not subjected to any measurement noise, these $\{(x, y, d)\}$ values map out a parabolic surface

$$d^2(x, y) = (x - x_0)^2 + (y - y_0)^2, \tag{1}$$

whose minimum value (x_0, y_0) is the wireless device location. Gathering several $\{(x_j, y_j, d_j)\}$ values and solving for (x_0, y_0) is a simple least squares problem that accounts for overdetermination of the system and the presence of measurement noise.

However, such an approach is not suitable in the presence of malicious perturbations to the $\{(x, y, d)\}$ values. For example, if an adversary alters the hop count, perhaps through a wormhole attack (as depicted in Figure 2 (b)) or jamming attack, the altered hop count may result in significant deviation of the distance measurement d from its true value. The use of a single, significantly incorrect $\{(x, y, d)\}$ value will drive the location estimate significantly away from the true location in spite of the presence of other, correct $\{(x, y, d)\}$ values. This exposes the vulnerability of least squares, and we would like to find a robust alternative, as discussed below, to reduce the impact of attacks on localization.

5.1 Robust Fitting: Least Median of Squares

The vulnerability of the least squares (LS) algorithm to attacks is essentially due to its non-robustness to "outliers". The general formulation for the LS algorithm minimizes the cost function

$$J(\theta) = \sum_{i=1}^{N} [u_i - f(v_i, \theta)]^2, \tag{2}$$

where θ is the parameter to be estimated, u_i corresponds to the i-th measured data sample, v_i corresponds to the abscissas for the parameterized surface $f(v_i, \theta)$, $|y_i - f(x_i, \theta)|$ is the residue for the i-th sample, and N is the total number of samples. Due to the summation in the cost function, a single influential outlier may ruin the estimation.

To increase robustness to outliers, a robust cost function is needed. For example, the method of least median of squares (LMS), proposed by Rousseeuw and described in detail in [34], is one of the most commonly used robust fitting algorithms. Instead of minimizing the summation of the residue squares, LMS fitting minimizes the median of the residue squares $J(\theta) = \text{med}_i [y_i - f(x_i, \theta)]^2$. Now a single outlier has little effect on the cost function, and won't bias the estimate significantly. Actually it was shown in [34] that in absence of noise, LMS can tolerate up to 50 percent outliers among N total measurements, and still give the correct estimate.

In general, there is no analytical form for LMS, and an exhaustive search over the parameter space is computationally prohibitive. An efficient and statistically robust alternative [34] is to solve random subsets of $\{(x_i, y_i)\}$ values to get several candidate $\hat{\theta}$. The median of the residue squares for each candidate is then computed, and the one with the least median of residue squares is chosen as a tentative estimate. However, this tentative estimate is obtained only from a small subset of samples. It is desirable to include more samples that are not outliers for a better estimation. So, the samples are reweighted based on their residues for the tentative estimate, followed by a reweighted least squares fitting to get the final estimate.

The samples can be reweighted in various ways. A simple thresholding method given by [34] is

$$w_i = \begin{cases} 1, & |\frac{r_i}{s_0}| \leq \gamma \\ 0, & \text{otherwise} \end{cases} \tag{3}$$

where γ is a predetermined threshold, r_i is the residue of the i-th sample for the least median subset estimate $\hat{\theta}$, and s_0 is the scale estimate given by [34]

$$s_0 = 1.4826(1 + \frac{5}{N-p})\sqrt{\text{med}_i r_i^2(\hat{\theta})}, \tag{4}$$

where p is the dimension of the estimated variable. The term $(1 + \frac{5}{N-p})$ is used to compensate the tendency for a small scale estimate when there are few samples.

Assume we are given a set of N samples, and that we aim to estimate a p-dimensional variable θ from this ensemble. The procedure of implementing the robust LMS algorithm is summarized as follows:

1. Choose an appropriate subset size n, the total number of subsets randomly drawn M, and a threshold γ.
2. Randomly draw M subsets of size n from the data ensemble. Find the estimate $\hat{\theta}_j$ for each subset. Calculate the median of residues r_{ij}^2 for every $\hat{\theta}_j$. Here $i = 1, 2, \cdots, N$ is the index for samples, while $j = 1, 2, \cdots, M$ is the index for the subsets.
3. Define $m = \arg\min_j \text{med}_i \{r_{ij}^2\}$, then $\hat{\theta}_m$ is the subset estimate with the least median of residues, and $\{r_{im}\}$ is the corresponding residues.
4. Calculate $s_0 = 1.4826(1 + \frac{5}{N-p})\sqrt{\text{med}_i r_{im}^2}$.
5. Assign weight w_i to each sample according to Equation (3).
6. Do a weighted least squares fitting to all data with weights $\{w_i\}$ to get the final estimate $\hat{\theta}$.

5.2 Robust Localization with LMS

In the absence of attacks, the device location estimate $(\hat{x}_0, \hat{y}_0)$ can be found by least squares, i.e.

$$(\hat{x}_0, \hat{y}_0) = \arg \min_{(x_0, y_0)} \sum_{i=1}^{N} [\sqrt{(x_i - x_0)^2 + (y_i - y_0)^2} - d_i]^2. \tag{5}$$

In presence of attacks, however, the adversary produces "outliers" in the measurements. Instead of identifying this misinformation, we would like to live with them and still get a reasonable location estimate (identification of misinformation will come out as a byproduct naturally). To achieve this goal, we use LMS instead of least squares to estimate the location. That is, we can find $(\hat{x}_0, \hat{y}_0)$ such that

$$(\hat{x}_0, \hat{y}_0) = \arg \min_{(x_0, y_0)} \operatorname{med}_i [\sqrt{(x_i - x_0)^2 + (y_i - y_0)^2} - d_i]^2. \tag{6}$$

Then the above LMS procedure can be used.

However, before using the algorithm, we need to consider two issues: First, how to choose the appropriate n and M for LMS-based localization? Second, how to get an estimate from the samples efficiently? The answers depend on the required performance and the allowed computational complexity. Considering that power is limited for sensor networks, and that the computational complexity of LMS depends on both the parameters and algorithmic implementation, we would like to gain the robustness of LMS with minimal additional computation compared to least squares, while exhibiting only negligible performance degradation. These two issues are addressed below.

1) How to choose the appropriate n and M?

The basic idea of LMS is that, hopefully, at least one subset among all subsets does not contain any contaminated samples, and the estimate from this good subset will thus fit the inlier data well. Since the inlier data are the majority ($>50\%$) of the data, the median of residues corresponding to this estimate will be smaller than that from the bad subsets.

We now calculate the probability P to get at least one good subset without contamination. Assume the contamination ratio is ϵ, i.e, ϵN samples are outliers, it is easy to get that $P = 1 - (1 - (1 - \epsilon)^n)^M$. For a fixed M and ϵ, the larger n, the smaller is P. So the size of a subset n is often chosen such that it's just enough to get an estimate. In our case, although the minimum number of samples needed to decide a location is 3, we have chosen $n = 4$ to reduce the chance that the samples are too close to each other to produce a numerically stable position estimate.

Once n is chosen, we can decide the value of P for a given pair of M and ϵ. For larger ϵ, a larger M is needed to obtain a satisfactory probability of at least one good subset. Depending on how much contamination the network localization system is required to tolerate and how much computation the system can afford, M can be chosen correspondingly. Because the power of the sensors are limited, and the functionality of the sensor network may be ruined when the contamination ratio is high, we chose $M = 20$ in our simulations, so that the system is resistant up to 30 percent contamination with $P = 0.99$.

2) How to get a location estimate from the samples efficiently?

To estimate (x_0, y_0) from the measurements $\{x_i, y_i, d_i\}$, we can use the least squares solution specified by equation 5. This is a nonlinear least square problem, and usually involves some iterative searching technique, such as gradient descent or Newton method, to get the solution. Moreover, to avoid local minimum, it is necessary to rerun the algorithm using several initial starting points, and as a result the computation is relatively expensive. Considering that sensors have limited power, and LMS finds estimates for M subsets, we may want to have a suboptimal but more computationally efficient algorithm.

Recall that equation (5) is equivalent to solving the following equations when $N \geq 2$:

$$(x_1 - x_0)^2 + (y_1 - y_0)^2 = d_1^2$$
$$(x_2 - x_0)^2 + (y_2 - y_0)^2 = d_2^2 \qquad (7)$$
$$\vdots$$
$$(x_N - x_0)^2 + (y_N - y_0)^2 = d_N^2$$

Averaging all the left parts and right parts respectively, we get

$$\frac{1}{N}\sum_{i=1}^{N}[(x_i - x_0)^2 + (y_i - y_0)^2] = \frac{1}{N}\sum_{i=1}^{N}d_i^2. \qquad (8)$$

Subtracting each side of the equation above from equation (7), we get the new equations

$$(x_1 - \frac{1}{N}\sum_{i=1}^{N}x_i)x_0 + (y_1 - \frac{1}{N}\sum_{i=1}^{N}y_i)y_0 =$$
$$\frac{1}{2}[(x_1^2 - \frac{1}{N}\sum_{i=1}^{N}x_i^2) + (y_1^2 - \frac{1}{N}\sum_{i=1}^{N}y_i^2) - (d_1^2 - \frac{1}{N}\sum_{i=1}^{N}d_i^2)]$$
$$\vdots \qquad (9)$$
$$(x_N - \frac{1}{N}\sum_{i=1}^{N}x_i)x_0 + (y_N - \frac{1}{N}\sum_{i=1}^{N}y_i)y_0 =$$
$$\frac{1}{2}[(x_N^2 - \frac{1}{N}\sum_{i=1}^{N}x_i^2) + (y_N^2 - \frac{1}{N}\sum_{i=1}^{N}y_i^2) - (d_N^2 - \frac{1}{N}\sum_{i=1}^{N}d_i^2)],$$

which can be easily solved using linear least squares.

The optimum solution of the linear equations (9) is not exactly the same as the optimum solution of the nonlinear equations (7), or equivalently equation (5). However, it can save computation and also serve as the starting point for the nonlinear LS problem. We noticed that there is a non-negligible probability of falling into a local minimum of the error surface when a random initial value is used with Matlab's *fminsearch* function to find the solution to equation 5. However, we observed that initiating the nonlinear LS from the linear estimate does not get trapped in a local minimum. In other words, the linear estimate is always close to the global minimum of the error surface. Nonlinear searching from linear estimate performs better than the linear method at the price of a higher computational complexity.

5.3 Simulation

To test the performance of range-based localization using LMS, we need to build a threat model first. In this work, we assume that the adversary successfully gains the ability to arbitrarily modify the distance measurements for a fraction ϵ of the total anchor nodes. The contamination ratio ϵ should be less than 50 percent, the highest contamination ratio LMS can tolerate. The goal of the adversary is to drive the location estimate as far away from the true location as possible. Rather than randomly perturbing the measurements of these contaminated devices, the adversary should coordinate his corruption of the measurements so that they will push the localization toward the same wrong direction. The adversary will thus tamper measurements so they lie on the parabolic surface $d_a^2(x, y)$ with a minimum at (x_a, y_a). As a result the localization estimate will be pushed toward (x_a, y_a) from the true position (x_0, y_0) in the absence of robust countermeasures. The larger distance between (x_a, y_a) and (x_0, y_0), the larger the estimate deviates from (x_0, y_0). So we use the distance $d_a = \sqrt{(x_a - x_0)^2 + (y_a - y_0)^2}$ as a measurement of the strength of the attack.

In our simulation, we had $N = 30$ anchor nodes and one sensor that is to be localized. These devices were randomly deployed in a 500×500 region. We assume that the sensor gets a set of $\{x_i, y_i, d_i\}$ observations by either DV-hop or another distance measurement scheme, and that the measurement noise obeys a Gaussian distribution with mean 0 and variance σ_n^2. The adversary tampers $N\epsilon$ measurements such that they all "vote" for (x_a, y_a). Least squares and LMS localization algorithms are applied to the data to obtain the location estimates $(\hat{x}_0, \hat{y}_0)$. For computational simplicity, we use linear least squares to get location estimate, being aware that a nonlinear least squares approach will improve the performance a little, but won't change the other features of the algorithms. The distance between the estimate and the true location is the corresponding estimation error.

For each contamination ratio ϵ and measurement noise level σ_n, we observed the change of the square root of mean square error (MSE) with the distance $d_a = \sqrt{(x_a - x_0)^2 + (y_a - y_0)^2}$. As an example, the performances at two different pairs of σ_n and ϵ are presented in Figure 4, where the value at each point is the average over 2000 trials. As expected, the estimation error of ordinary least squares increases as d_a increases due to the non-robustness of the least squares to outliers. The estimation error of LMS increases first until it reaches the maximum at some critical value of d_a. After this critical value, the error decreases slightly and then stabilizes. In other words, if LMS is used in localization, it's useless or even harmful for the adversary to attempt to conduct too powerful of an attack.

The performance of the LS and LMS algorithms are affected by both the contamination ratio and the noise level. Figure 5 (a) illustrates the degradation of the performance as ϵ increases at a fixed $\sigma_n = 15$, while Figure 5 (b) illustrates the impact of measurement noise σ_n on the performance at a fixed $\epsilon = 0.2$. Not surprisingly, the higher the contamination ratio, the larger the measurement noise, the larger is the estimation error. Also, since we designed the system to be robust up to

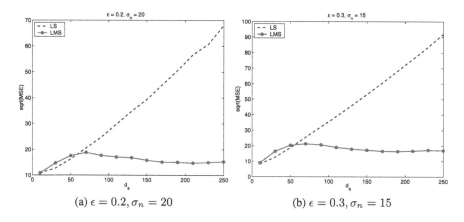

(a) $\epsilon = 0.2, \sigma_n = 20$ (b) $\epsilon = 0.3, \sigma_n = 15$

Fig. 4. Performance comparison between LS and LMS for localization in presence of attack.

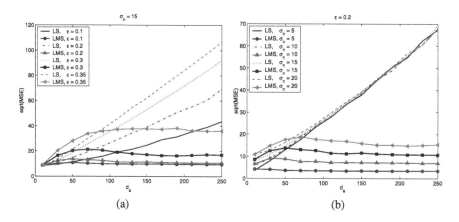

(a)   (b)

Fig. 5. (a) The impact of ϵ on the performance of LS and LMS algorithms at $\sigma_n = 15$. (b) The impact of σ_n on the performance of LS and LMS algorithms at $\epsilon = 0.2$.

30 percent contamination, 35 percent contamination results in severe performance degradation.

We also noticed from Figure 4 and Figure 5 (b) that at low attacking strength, the performance of LS is actually better than LMS. In order to elucidate the reason for this behavior, let us look the simpler problem of fitting a line through data. In Figure 6, we present the line-fitting scenario using an artificial data set with 40 percent contamination. We generated 50 samples, among which 20 samples with $x = 31, \cdots, 50$ are the contaminated outliers. When the outlier data are well separated from the inlier data, LMS can detect this and fit the inlier data only, which gives a better fitting than LS. However, when the outlier data are close to the inlier data, it's hard for LMS

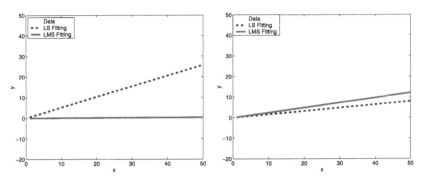

Fig. 6. LMS performs worse than LS when the inlier and outlier data are too close.

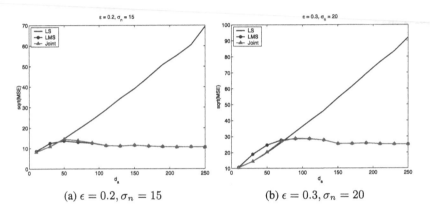

(a) $\epsilon = 0.2, \sigma_n = 15$ (b) $\epsilon = 0.3, \sigma_n = 20$

Fig. 7. The performance of the joint algorithm comparing to LS and LMS algorithms.

to tell the difference, so it may fit part of the inlier data and part of the outlier data, thus giving a worse estimate than LS.

Therefore, when the attack strength is low, LS performs better than LMS. Further, in this case, LS also has a lower computational cost. Since power consumption is an important concern for sensor networks, we do not want to use LMS when not necessary. We next present a method by which we may switch between LS and LMS estimation and achieve the performance advantages of each.

5.4 An Efficient Switched LS-LMS Localization Scheme

We use the observation that when outliers exist, the variance of the data will be larger than that when no outlier exists. Moreover, the farther outliers are from the inliers, the larger the variance. This suggests that the variance of the data can be used to indicate the distance between inliers and outliers. Therefore, we can do a LS estimate over the data first, and use the residues to estimate the data variance $\hat{\sigma}_n$ from the residuals

r_i, i.e. $\hat{\sigma}_n = \sqrt{\frac{\sum_{i=1}^{N} r_i^2}{N-2}}$. Then the ratio $\frac{\hat{\sigma}_n}{\sigma_n}$ represents the variance expansion due to possible outliers. If the normal measurement noise level σ_n is known, which is a reasonable assumption in practice, we can compare the $\frac{\hat{\sigma}_n}{\sigma_n}$ to some threshold T. If $\frac{\hat{\sigma}_n}{\sigma_n} > T$, we choose to apply the LMS algorithm; otherwise, we just use the LS estimate we have calculated. We call this a joint algorithm. In our simulation, we found that $T = 1.5$ gives quite good results over all tested ϵ and σ_n pairs. Two examples with different ϵ and σ_n are shown in Figure 7. After the selection strategy is deployed, the performance curves (the triangle in Figure 7) are very close to the lower envelop of the performance of LS and LMS algorithms.

6 Robust Methods for RF-Based Fingerprinting

A different approach to localization is based upon radio-frequency fingerprinting. One of the first implementations was the RADAR system [17, 27]. The system was shown to have good performance in an office building. In this section, we will show how robustness can be applied to such a RF-based system to obtain attack-tolerant localization.

In RADAR, multiple base stations are deployed to provide overlapping coverage of an area, such as a floor in an office building. During set up, a mobile host with known position broadcasts beacons periodically. The signal strengths at each base station are measured and stored. Each record has the format of $\{x, y, ss_1, \cdots, ss_N\}$, where (x, y) is the mobile position, and ss_i is the received signal strength in dBm at the i-th base station. N, the total number of base stations, should be at least 3 to provide good localization performance. To reduce the noise effect, each ss_i is usually the average of multiple measurements collected over a time period. The collection of all measurements forms a radio map that consists of the featured signal strengths, or fingerprints, at each sampled position.

Following setup, a mobile may be localized by broadcasting beacons and using the signal strengths measured at each base station. To localize the mobile user, we search the radio map collected in the setup phase, and find the fingerprint that best matches the signal strengths observed. That is, the central base station compares the observed signal energy $\{ss_1', \cdots, ss_N'\}$ with the recorded $\{x, y, ss_1, \cdots, ss_N\}$, and pick the location (x, y) that minimizes the Euclidean distance $\sqrt{\sum_{i=1}^{N} (ss_i - ss_i')^2}$ as the location estimate of the mobile user. This technique is called *nearest neighbor in signal space (NNSS)*. A slight variant of the technique involves finding the k nearest neighbors in signal space, and averaging their coordinates to get the location estimate. It was shown in [17] that averaging 2 to 4 nearest neighbors improves the location accuracy significantly.

The location estimation method described above is not robust to possible attacks. If the reading of signal strength at one base station is corrupted, the estimate can be dramatically different from the true location. Such an attack can be easily launched by inserting an absorbing barrier between the mobile host and the base station. Sudden change of local environment, such as turning on a microwave near one base

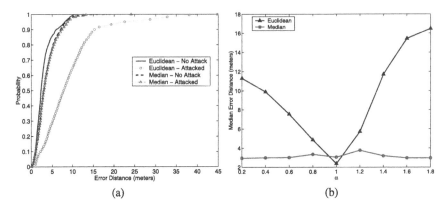

(a)   (b)

Fig. 8. (a) The CDF of the error distance for the NNSS method in Euclidean distance and in median distance, with and without an attack (one reading is modified to $\alpha \cdot ss_i$, where $\alpha = 0.6$). (b) Median of the error distance vs. the attacking strength α (one reading is modified to $\alpha \cdot ss_i$).

station, can also cause incorrect signal strength readings. To obtain reasonable location estimates, in spite of attacks or sudden environmental changes, we propose to deploy more base stations and use a robust estimation method to utilize the redundancy introduced. In particular, instead of minimizing the Euclidean distance $\sqrt{\sum_{i=1}^{N}(ss_i - ss_i')^2}$ to find nearest neighbors in signal space, we can minimize the median of the distances in all dimensions, i.e. minimize $med_{i=1}^{N}(ss_i - ss_i')^2$ to get the "nearest" neighbor. In this way, a corrupted estimate won't bias the neighbor searching significantly.

We tested the proposed method through simulations. As pointed out in [17], the radio map can be generated either by empirical measurements, or by signal propagation modeling. Although the modeling method is less accurate than the empirical method, it still captures the data fairly well and provides good localization. In [17] a *wall attenuation factor* model was used to fit the collected empirical data and, after compensating for attenuation due to intervening walls, it was found that the signal strength varies with the distance in a trend similar to the generic exponential path loss [35]. In our simulation, we use the model, which we adopted from [17], $P(d)[dBm] = P(d_0)[dBm] - 10\gamma log(\frac{d}{d_0})$, to generate signal strength data. We used the parameter $d_0 = 1m$, $P(d_0) = 58.48$ and $\gamma = 1.523$, which were obtained in [17] when fitting the model with the empirical data. We emphasize that the trends shown in our results are not affected by the selection of the parameters. We also added random zero-mean Gaussian noise with variance 1dBm so that the received signal strengths at a distance have a similar amount of variation as was observed in [17].

The rectangular area we simulated was similar to the region used in [17], and had a size $45m \times 25m$, which is a reasonable size for a large indoor environment. Instead of three base stations, we employed six to provide redundancy for robust localization.

We collected samples on a grid of 50 regularly spaced positions in order to form the radio map. During localization, a mobile sends beacons, and the signal strengths at the base stations are recorded. The nearest neighbors in signal space in terms of Euclidean distance and median distance are each found. The coordinates of the four nearest neighbors are averaged to get the final location estimate of the mobile user.

To simulate the attack, we randomly choose one reading ss_i and modify it to $\alpha \cdot ss_i$, where α indicates the attacking strength. $\alpha = 1$ means no attack. Figure 8 (a) shows the cumulative distribution function (CDF) of the error distance for the NNSS method in Euclidean distance and in median distance, with and without an attack. In presence of an attack with $\alpha = 0.6$, which is very easy to launch from a practical point of view, the Euclidean-based NNSS method shows significantly larger error than when there is no attack, while for the median-based NSSS approach there is little change (the curves with and without attack almost completely overlap in Figure 8 (a)). Although its performance is slightly worse than Euclidean-NNSS in the absence of attacks, median-NNSS is much more robust to possible attacks. In Figure 8 (b), we plot the 50th percentile value of the error distance for a series of α from 0.2 to 1.8. NNSS in median distance shows good performance across all α's.

With six base stations, the system can tolerate attacks on up to two readings. For simplicity, we assume the adversary randomly selects two readings and modifies them to $\alpha \cdot ss_i$. We note that such an approach is not a coordinated attack, and there may be better attack strategies able to produce larger localization error. Figure 9 (a) shows the CDF of the error distance at $\alpha = 0.6$, and Figure 9 (b) shows the change of median error distance with α. Again, the median-NNSS exhibits better resistance to attacks. We observed the same phenomenon as that in the triangulation method: it is better for the adversary to not be too greedy when attacking the localization scheme. Finally, we note that the computational requirements for Euclidean-NNSS and median-NNSS are comparable. The fact that there is only marginal performance improvement for Euclidean-NNSS when there are no attacks suggests that a switched algorithm is not critical for fingerprinting-based localization.

7 Conclusions

As wireless networks are increasingly deployed for location-based services, these networks are becoming more vulnerable to misuses and attacks that can lead to false location calculation. Towards the goal of securing localization, this paper has made three main contributions. It first enumerates a list of novel attacks that are unique to wireless localization algorithms. It proposes a generic framework that can be used to classify and clean corrupted/faked location data. Further, this paper proposes the idea of tolerating attacks, instead of eliminating them, by exploiting redundancies at various levels within wireless networks. We explored robust statistical methods to make localization attack-tolerant. We examined two broad classes of localization: triangulation and RF-based fingerprinting methods. For triangulation-based localization, we examined the use of a least median squares estimator for estimating position. We provided analysis for selecting system parameters. We then proposed an adaptive

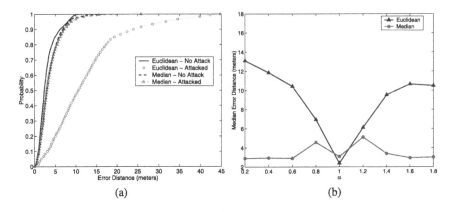

Fig. 9. (a) The CDF of the error distance for the NNSS method in Euclidean distance and in median distance, with and without an attack (two readings are modified to $\alpha \cdot ss_i$, where $\alpha = 0.6$). (b) Median of the error distance vs. the attacking strength α (two readings are modified to $\alpha \cdot ss_i$).

least squares and least median squares position estimator that has the computational advantages of least squares in the absence of attacks and switches to a robust mode when being attacked. For fingerprinting-based localization, we introduced robustness through the use of a median-based distance metric.

References

1. K. Langendoen and N. Reijers, "Distributed localization in wireless sensor networks: a quantitative comparison," *Comput. Networks*, vol. 43, no. 4, pp. 499–518, 2003.
2. N. Priyantha, A. Chakraborty, and H. Balakrishnan, "The CRICKET location-support system," in *Proceedings of the 6th annual international conference on Mobile computing and networking (Mobicom 2000)*, 2000, pp. 32–43.
3. D. Nicelescu and B. Nath, "Ad hoc positioning (APS) using AOA," in *Proceedings of IEEE Infocom 2003*, 2003, pp. 1734 – 1743.
4. S. Capkun and J.P. Hubaux, "Secure positioning in sensor networks," Technical report EPFL/IC/200444, May 2004.
5. L. Lazos and R. Poovendran, "SeRLoc: Secure range-independent localization for wireless sensor networks," in *Proceedings of the 2004 ACM Workshop on Wireless Security*, 2004, pp. 21–30.
6. B. H. Wellenhoff, H. Lichtenegger, and J. Collins, *Global Positions System: Theory and Practice, Fourth Edition*, Springer Verlag, 1997.
7. A. Harter, A. Hopper, P. Steggles, A. Ward, and P.Webster, "The anatomy of a context-aware application," in *Proceedings of the MOBICOM 99*, 1999.
8. A. Savvides, C. C. Han, and M. B. Srivastava, "Dynamic fine-grained localization in ad-hoc networks of sensors," in *Proceedings of the MOBICOM 01*, 2001.
9. C. Savarese, K. Langendoen, and J. Rabaey, "Robust positioning algorithms for distributed ad-hoc wireless sensor networks," in *Proceedings of USENIX Technical Annual Conference*, 2002.

10. D. Nicelescu and B. Nath, "DV based positioning in ad hoc networks," *Telecommunication Systems*, vol. 22, no. 1-4, pp. 267–280, 2003.

11. A. Savvides, H. Park, and M. Srivastava, "The bits and flops of the n-hop multilateration primitive for node localization problems," in *Proceedings of First ACM International Workshop on Wireless Sensor Networks and Application (WSNA)*, 2002, pp. 112–121.

12. J. Hightower, G. Boriello, and R. Want, "SpotON: An indoor 3D Location Sensing Technology Based on RF Signal Strength," Tech. Rep. Technical Report 2000-02-02, University of Washington, February 2000.

13. N. Bulusu, J. Heidemann, and D. Estrin, "Gps-less low cost outdoor localization for very small devices," *IEEE Personal Communications Magazine*, vol. 7, no. 5, pp. 28–34, 2000.

14. M. Youssef, A. Agrawal, and A. U. Shankar, "WLAN location determination via clustering and probability distributions," in *Proceedings of IEEE PerCom'03*, Fort Worth, TX, Mar. 2003.

15. T. Roos, P. Myllymaki, and H.Tirri, "A statistical modeling approach to location estimation," *IEEE Transactions on Mobile Computing*, vol. 1, pp. 59–69, 2002.

16. R. Battiti, M. Brunato, and A. Villani, "Statistical learning theory for location fingerprinting in wireless lans," Technical Report DIT-02-086, University of Trento, Informatica e Telecomunicazioni, Oct. 2002.

17. P. Bahl and V.N. Padmanabhan, "RADAR: An in-building RF-based user location and tracking system," in *Proceedings of IEEE Infocom 2000*, 2000, pp. 775–784.

18. E. Elnahrawy, X. Li, and R. P. Martin, "The limits of localization using signal strength: A comparative study," in *Proceedings of the First IEEE International Conference on Sensor and Ad hoc Communcations and Networks (SECON 2004)*, Oct. 2004.

19. T. He, C. Huang, B. Blum, J. Stankovic, and T. Abdelzaher, "Range-free localization schemes for large scale sensor networks," in *Proceedings of the 9th annual international conference on Mobile computing and networking (Mobicom 2003)*, 2003, pp. 81 – 95.

20. Y.C. Hu, A. Perrig, and D. Johnson, "Packet leashes: a defense against wormhole attacks in wireless networks," in *Proceedings of IEEE Infocom 2003*, 2003, pp. 1976–1986.

21. S. Brands and D. Chaum, "Distance-bounding protocols," in *Proceedings of the Workshop on the Theory and Application of Cryptographic Techniques on Advances in Cryptology*, 1994, pp. 344–359.

22. N. Sastry, U. Shankar, and D. Wagner, "Secure verification of location claims," in *Proceedings of the 2003 ACM workshop on Wireless security*, 2003, pp. 1–10.

23. S. Capkun and J. P. Hubaux, "Secure positioning of wireless devices with application to sensor networks," in *Proceedings of the IEEE International Conference on Computer Communications (INFOCOM)*, 2005.

24. L. Lazos, R. Poovendran, and S. Capkun, "Rope: robust position estimation in wireless sensor networks," in *Proceedings of the Fourth International Symposium on Information Processing in Sensor Networks (IPSN 2005)*, 2005, pp. 324–331.

25. S. Capkun and J.P. Hubaux, "Securing localization with hidden and mobile base stations," to appear in Proceedings of IEEE Infocom 2006.

26. Z. Li, W. Trappe, Y. Zhang, and B. Nath, "Robust statistical methods for securing wireless localization in sensor networks," in *Proceedings of the Fourth International Symposium on Information Processing in Sensor Networks (IPSN 2005)*, 2005.

27. P. Bahl, V.N. Padmanabhan, and A. Balachandran, "Enhancements to the RADAR User Location and Tracking System," Tech. Rep. Technical Report MSR-TR-2000-12, Microsoft Research, February 2000.

28. W. Stallings, *Network Security Essentials, Applications and Standards, 2nd Edition*, Prentice Hall, 2003.

29. C. Kaufman, R. Perlman, and M. Speciner, *Network Security: Private Communication in a Public World*, Prentice Hall, 1995.

30. W. Trappe and L.C. Washington, *Introduction to Cryptography with Coding Theory*, Prentice Hall, 2002.

31. D. Niculescu and B. Nath, "VOR base stations for indoor 802.11 positioning," in *Proceedings of the Mobicom 2004, Philadelphia, PA*, September 2004.

32. B. Przydatek, D. Song, and A. Perrig, "SIA: secure information aggregation in sensor networks," in *SenSys '03: Proceedings of the 1st International Conference on Embedded Networked Sensor Systems*, 2003, pp. 255–265.

33. D. Wagner, "Resilient aggregation in sensor networks," in *SASN '04: Proceedings of the 2nd ACM workshop on Security of ad hoc and sensor networks*, 2004, pp. 78–87.

34. P. Rousseeuw and A. Leroy, "Robust regression and outlier detection," Wiley-Interscience, September 2003.

35. A. Goldsmith, *Wireless Communications*, Cambridge University Press, to appear 2005.

Secure and Resilient Localization in Wireless Sensor Networks

Peng Ning[1], Donggang Liu[2], and Wenliang Du[3]

[1] Department of Computer Science, North Carolina State University pning@ncsu.edu
[2] Department of Computer Science and Engineering, University of Texas at Arlington dliu@cse.uta.edu
[3] Department of Electrical Engineering and Computer Science, Syracuse University wedu@ecs.syr.edu

1 Introduction

Recent technological advances have made it possible to develop distributed sensor networks consisting of a large number of low-cost, low-power, and multi-functional sensor nodes that communicate in short distances through wireless links [1]. Such sensor networks are ideal candidates for a wide range of applications such as health monitoring, data acquisition in hazardous environments, and military operations. The desirable features of distributed sensor networks have attracted many researchers to develop protocols and algorithms that can fulfill the requirements of these applications (e.g., [1, 12, 14, 17, 28, 29, 32]).

Sensors' locations play a critical role in many sensor network applications. Not only do applications such as environment monitoring and target tracking require sensors' location information to fulfill their tasks, but several fundamental techniques developed for wireless sensor networks also require sensor nodes' locations. For example, in geographical routing (e.g., FACE [2, 3], GPSR [3, 4, 18]), sensor nodes make routing decisions at least partially based on their own and their neighbors' locations. As another example, in some data-centric storage applications such as GHT [33, 36], storage and retrieval of sensor data highly depend on sensors' locations. Indeed, many sensor network applications will not work without sensors' location information.

The Global Positioning System (GPS) is a popular outdoor localization system for mobile devices. However, due to the cost reasons, it is highly undesirable to have a GPS receiver on every sensor node. Moreover, in some situations such as indoor sensor network applications, GPS cannot be used for localization because of interferences and obstacles. This creates a demand for efficient and cost-effective location discovery algorithms in sensor networks. In the past several years, a number of location discovery protocols have been proposed to reduce or completely remove the dependence on GPS in wireless sensor networks [4, 8, 13, 26, 27, 30, 31, 34, 35].

These protocols share a common feature: They all (with one exception [11]) use some special nodes, called *beacon (or anchor) nodes*, which are assumed to know their own locations (e.g., through GPS receivers or manual configuration). (For this reason, we call these location discovery techniques *beacon-based location discovery*.) These protocols work in two stages. In the first stage, non-beacon nodes receive radio signals called *beacon signals* from the beacon nodes. The packet carried by a beacon signal, which we call a *beacon packet*, usually includes the location of the beacon node. The non-beacon nodes then estimate certain measurements (e.g., distance between the beacon and the non-beacon nodes) based on features of the beacon signals. Features that may be used for location determination include Received Signal Strength Indicator (RSSI), Time of Arrival (ToA), Time Difference of Arrival (TDoA), and Angle of Arrival (AoA). We refer to such a measurement (e.g., the distance) and the location of the corresponding beacon node collectively as a *location reference*. In the second stage, when a sensor has enough number of location references from different beacon nodes, it determines its own location in the network field. A typical approach is to consider the location references as constraints that a sensor node's location must satisfy, and estimate it by finding a mathematical solution that satisfy these constraints with minimum estimation error. Existing approaches either employ *range-based* methods [8, 27, 30, 34, 35], which use the exact measurements obtained in stage one, or *range-free* ones [4, 13, 26, 31], which only need the existences of beacon signals in stage one.

As a fundamental service critical to many sensor network applications, location discovery in *hostile environments* is certainly subject to attacks. As illustrated in Figure 1, an attacker may provide incorrect location reference information by pretending to be valid beacon nodes (Figure 1(a)), compromising beacon nodes (Figure 1(b)), or replaying the beacon packets that he/she intercepted in different locations (Figure 1(c)). In either of the above cases, non-beacon nodes will determine their locations incorrectly. As a result, the attacker can effectively disable or mislead the localization service, and none of the sensor network applications that require sensors' locations will work as expected.

Both security and sensor network researchers have realized this problem. In the past two years, a number of techniques have been proposed for secure and resilient localization, from target localization (e.g., [5]), where a number of trusted nodes determine the location of a potentially malicious node, to node localization (e.g., [21, 24]), where a non-beacon node determines its own location based on information provided by potential attackers. Some of these techniques target at range-based localization (e.g., [21, 24]), while some others focus on range-free localization (e.g., [19, 20]).

In this chapter, we would like to discuss two recent efforts for achieving secure and resilient localization in sensor networks: a suite of attack-resistant location estimation techniques [24], and a mechanism for detecting malicious beacon nodes [25]. Other related techniques can be found in the other chapters of this edited volume.

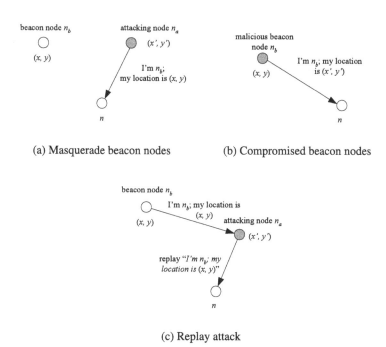

(a) Masquerade beacon nodes   (b) Compromised beacon nodes

(c) Replay attack

Fig. 1. Attacks against location discovery schemes

2 Attack-Resistant Location Estimation

In this section, we present two attack-resistant estimation approaches to dealing with malicious attacks against location discovery in wireless sensor networks. The first approach is extended from the minimum mean square estimation (MMSE). It uses the mean square error as an indicator to identify and remove malicious location references. The second one adopts an iteratively refined voting scheme to tolerate malicious location references introduced by attackers. More details of these approaches can be found in [23, 24].

Our techniques are purely based on a set of location references. The location references may come from beacon nodes that are either single hop or multiple hops away, or from those non-beacon nodes that already estimated their locations. We do not distinguish these location references, though the effect of "error propagation" may affect the performance of our techniques due to the estimation errors at non-beacon nodes. Since our techniques only utilize the location references from beacon nodes, there is no extra communication overhead involved when compared to the previous localization schemes.

We assume all beacon nodes are uniquely identified. In other words, a non-beacon node can identify the original sender of each beacon packet based on the cryptographic key used to authenticate the packet. This can be easily achieved with a pairwise key establishment scheme [6, 9, 10] or a broadcast authentication scheme [32].

We assume each non-beacon node uses at most one location reference derived from the beacon signals sent by each beacon node. As a result, even if a beacon node is compromised, the attacker that has access to the compromised key can only introduce at most one malicious location reference to a given non-beacon node by impersonating the compromised node.

For simplicity, we assume the distances measured from beacon signals (e.g., with RSSI or TDoA [34]) are used for location estimation. For the sake of presentation, we denote a location reference obtained from a beacon signal as a triple $\langle x, y, \delta \rangle$, where (x, y) is the location of the beacon declared in the beacon packet, and δ is the distance measured from its beacon signal.

We assume an attacker may change any field in a location reference. In other words, it may declare a wrong location in its beacon packets, or carefully manipulate the beacon signals to affect the distance measurement by, for example, adjusting the signal strength when RSSI is used for distance measurement. We also assume multiple malicious beacon nodes may collude together to make the malicious location references appear to be "consistent". Our techniques can still defeat such colluding attacks as long as the majority of location references are benign.

2.1 Attack-Resistant Minimum Mean Square Estimation

Intuitively, a location reference introduced by a malicious attack is aimed at misleading a sensor node about its location. Thus, it is usually "different" from benign location references. When there are redundant location references, there must be some "inconsistency" between the malicious location references and the benign ones. (An attacker may still have a location reference consistent with the benign ones after changing both the location and the distance values. However, such a location reference will not generate significantly negative impact on location determination.) To take advantage of this observation, we propose to use the "inconsistency" among the location references to identify the malicious ones, and discard them before finally estimating the locations at sensor nodes.

We assume a sensor node uses a MMSE-based method (e.g., [8,27,30,31,34,35]) to estimate its own location. Thus, most current range-based localization methods can be used with this technique. To harness this observation, we first estimate the sensor's location with the MMSE-based method and then assess if the estimated location could be derived from a set of consistent location references. If yes, we accept the estimation result; otherwise, we identify and remove the most "inconsistent" location reference, and repeat the above process. This process may continue until we find a set of consistent location references or it is not possible to find such a set.

Checking the Consistency of Location References

We use the mean square error ς^2 of the distance measurements based on the estimated location as an indicator of the degree of inconsistency, since all the MMSE-based

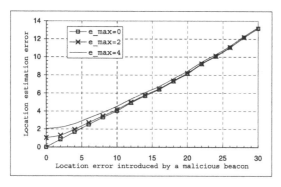

Fig. 2. Location estimation error. Unit of measurement for x and y axes: meter

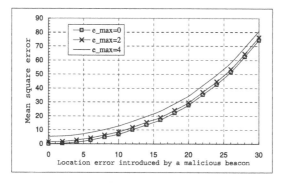

Fig. 3. Mean square error ς^2. Unit of measurement for x-axis: meter

methods estimate a sensor node's location by (approximately) minimizing this mean square error. Other indicators are possible but need further investigation.

Definition 1. *Given a set of location references* $\mathcal{L} = \{\langle x_1, y_1, \delta_1 \rangle, \langle x_2, y_2, \delta_2 \rangle, ..., \langle x_m, y_m, \delta_m \rangle\}$ *and a location* $(\tilde{x}_0, \tilde{y}_0)$ *estimated based on* $\mathcal{L}$, *the mean square error of this location estimation is*

$$\varsigma^2 = \sum_{i=1}^{m} \frac{(\delta_i - \sqrt{(\tilde{x}_0 - x_i)^2 + (\tilde{y}_0 - y_i)^2})^2}{m}.$$

Intuitively, the more inconsistent a set of location references is, the greater the corresponding mean square error should be. To gain further understanding, we performed an experiment through simulation with the MMSE-based method in [34]. We assume the distance measurement error is uniformly distributed between $-e_{max}$ and e_{max}. We used 9 honest beacon nodes and 1 malicious beacon node evenly deployed in a $30m \times 30m$ field. The node that estimates location is positioned at the center of the field. The malicious beacon node always declares a false location that is x meters away from its real location, where x is a parameter in our experiment.

Figures 2 and 3 show the location estimation error (i.e., the distance between a sensor's real location and the estimated location) and the mean square error ς^2

when x increases. As these figures show, if a malicious beacon node increases the location estimation error by introducing greater errors, it also increases the mean square error ς^2 at the same time. This further demonstrates that the mean square error ς^2 is potentially a good indicator of inconsistent location references.

We choose a simple, threshold-based method to determine if a set of location references is consistent. Specifically, a set of location references $\mathcal{L} = \{\langle x_1, y_1, \delta_1 \rangle, \langle x_2, y_2, \delta_2 \rangle, ..., \langle x_m, y_m, \delta_m \rangle\}$ obtained at a sensor node is τ-consistent w.r.t. a MMSE-based method if the method gives an estimated location $(\tilde{x}_0, \tilde{y}_0)$ such that the mean square error of this location estimation

$$\varsigma^2 = \sum_{i=1}^{m} \frac{(\delta_i - \sqrt{(\tilde{x}_0 - x_i)^2 + (\tilde{y}_0 - y_i)^2})^2}{m} \leq \tau^2.$$

Determining Threshold τ

The determination of threshold τ depends on the measurement error model, which is assumed to be available for us to perform simulation off-line and determine an appropriate τ. The threshold is stored on each sensor node. Usually, the movement of sensor nodes (beacon or non-beacon nodes) does not have significant impact on this threshold, since the measurement error model will not change significantly in most cases. However, when the error model changes frequently and significantly, the performance of our techniques may be affected. In this work, we assume the measurement error model will not change.

Note that the malicious beacon signals usually increase the variance of estimation. Thus, having a lower bound (e.g., Cramer-Rao bound) is not enough for us to filter malicious beacon signals. In fact, the upper bound or the distribution of the mean square error are more desirable. In this work, we study the distribution of the mean square error ς^2 when there are no malicious attacks, and use this information to help determine the threshold τ.

Since there is no other error besides the distance measurement error, a benign location reference $\langle x, y, \delta \rangle$ obtained by a sensor node at (x_0, y_0) must satisfy:

$$|\delta - \sqrt{(x - x_0)^2 + (y - y_0)^2}| \leq \epsilon,$$

where ϵ is the maximum distance measurement error.

All the localization techniques are aimed at estimating a location as close to the sensor's real location as possible. Thus, we may assume the estimated location is very close to the real location when there are no attacks. Next, we derive the distribution of the mean square error ς^2 using the real location as the estimated location, and compare it with the distribution obtained through simulation when there are location estimation errors.

The measurement error of a benign location reference $\langle x_i, y_i, \delta_i \rangle$ can be computed as $e_i = \delta_i - \sqrt{(x_0 - x_i)^2 + (y_0 - y_i)^2}$, where (x_0, y_0) is the real location of the sensor node. Assuming the measurement errors introduced by different benign location references are independent, we can get the distribution of the mean square error through the following Lemma. (The proof of the Lemma can be found in [24].)

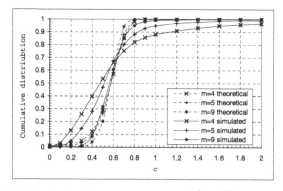

Fig. 4. Cumulative distribution function $F(\varsigma^2 \leq \varsigma_0^2)$. Let $c = \frac{\varsigma_0}{\epsilon}$.

Lemma 1. *Let $\{e_1, ..., e_m\}$ be a set of independent random variables, and μ_i, σ_i^2 be the mean and the variance of e_i^2, respectively. If the estimated location of a sensor node is its real location, the probability distribution of ς^2 is*

$$\lim_{m \longrightarrow \infty} F[\varsigma^2 \leq \varsigma_0^2] = \Phi(\frac{m\varsigma_0^2 - \mu'}{\sigma'}),$$

where $\mu' = \sum_{i=1}^{m} \mu_i$, $\sigma' = \sqrt{\sum_{i=0}^{m} \sigma_i^2}$, and $\Phi(x)$ is the probability of a standard normal random variable being less than x.

Lemma 1 describes the probability distribution of ς^2 based on a sensor's real location. Though it is different from the probability distribution of ς^2 based on a sensor's estimated location, it can be used to approximate such distribution in most cases.

Let us further assume a simple model for measurement errors, where the measurement error is evenly distributed between $-\epsilon$ and ϵ. Then the mean and the variance for e_i are 0 and $\frac{\epsilon^2}{3}$, respectively, and the mean and the variance for any e_i^2 are $\frac{\epsilon^2}{3}$ and $\frac{4\epsilon^4}{45}$, respectively. Let $c = \frac{\varsigma_0}{\epsilon}$, we have

$$F(\varsigma^2 \leq (c \times \epsilon)^2) = \Phi(\frac{\sqrt{5m}(3c^2 - 1)}{2}).$$

Figure 4 shows the probability distribution of ς^2 derived from Lemma 1 and the simulated results using sensors' estimated locations. We can see that when the number of location references m is large (e.g., $m = 9$) the theoretical result derived from Lemma 1 is very close to the simulation results. However, when m is small (e.g., $m = 4$), there are observable differences between the theoretical results and the simulation. The reasons are twofold. First, our theoretical analysis is based on the central limit theorem, which is only an approximation of the distribution when m is a large number. Second, we used the MMSE-based method proposed in [34] in the simulation, which estimates a node's location by only *approximately* minimizing the mean square error. (Otherwise, the value of ς^2 for benign location references should never exceed ϵ^2.)

Figure 4 gives three hints about the choice of the threshold τ. First, when there are enough number of benign location references, a threshold less than the maximum measurement error is enough. For example, when $m = 9$, $\tau = 0.8\epsilon$ can guarantee the nine benign location references are considered consistent with high probability. Besides, a large threshold may lead to the failure to filter out malicious location references. Second, when m is small (e.g. 4), the cumulative probability becomes flatter and flatter when $c\rangle 0.8$. This means that setting a large threshold τ for small m may not help much to guarantee the consistency test for benign location references; instead, it may give an attacker high chance to survive the detection. Third, the threshold cannot be too small; otherwise, a set of benign location references has high probability to be determined as a non-consistent reference set.

Based on the above observations, we propose to choose the value for τ with a hybrid method. Specifically, when the number of location references is large (e.g., more than 8), we determine the value of τ based on Lemma 1. Specifically, we choose a value of τ corresponding to a high cumulative probability (e.g., 0.9). When the number location references is small, we perform simulation to derive the actual distribution of the mean square error, and then determine the value of τ accordingly. Since there are only a small number of simulations to run, we believe this approach is practical.

Identifying the Largest Consistent Set

Since the MMSE-based methods can deal with measurement errors better if there are more benign location references, we should keep as many benign location references as possible when the malicious ones are removed. This implies we should get the largest set of consistent location references.

Given a set $\mathcal{L}$ of n location references and a threshold τ, a simple approach to computing the largest set of τ-consistent location references is to check all subsets of $\mathcal{L}$ with i location references about τ-consistency, where i starts from n and drops until a subset of $\mathcal{L}$ is found to be τ-consistent or it is not possible to find such a set. Thus, if the largest set of consistent location references consists of m elements, a sensor node has to use the MMSE method at least $1 + \binom{n}{m+1} + \cdots + \binom{n}{n}$ times to find out the right one. If $n = 10$ and $m = 5$, a node needs to perform the MMSE method for at least 387 times. It is certainly not desirable to do such expensive operations on resource constrained sensor nodes.

To reduce the computation on sensor nodes, we may use a greedy algorithm, which is simple but suboptimal. This greedy algorithm works in rounds. It starts with the set of all location references in the first round. In each round, it first verifies if the current set of location references is τ-consistent. If yes, the algorithm outputs the estimated location and stops. Optionally, it may also output the set of location references. Otherwise, it considers all subsets of location references with one fewer location reference, and chooses the subset with the least mean square error as the input to the next round. This algorithm continues until it finds a set of τ-consistent location references or when it is not possible to find such a set (i.e., there are only 3 remaining location references).

The greedy algorithm significantly reduces the computational overhead in sensor nodes. To continue the earlier example, a sensor node only needs to perform MMSE operations for about 50 times (instead of 387 times) using this algorithm. In general, a sensor node needs to use a MMSE-based method for at most $1 + n + (n - 1) + \cdots + 4 = 1 + \frac{(n-3)(n+4)}{2}$ times.

However, as we mentioned, the greedy algorithm cannot guarantee that it can always identify the largest consistent set. It is possible that benign location references are removed. In [24], we note that this generates a big impact on the accuracy of location estimation – especially when there are multiple malicious location references. To deal with this problem, we develop an enhanced greedy algorithm in the following. The new algorithm is based on an efficient approach to identifying the most suspicious location reference from a set of location references.

In the previous discussion, we only consider the consistency of 3 or more location references. A further investigation also reveals that two benign location references are usually "consistent" with each other in the sense that there exists at least one location in the deployment field on which both location references agree. Hence, when the majority of location references are benign, we can usually find many location references that are consistent with a benign location reference. In addition, when a malicious location reference tries to create a larger location error, the number of location references that are consistent with the malicious one will decrease quickly.

According to the above discussion, for each location reference, we simply count the number of location references that are consistent with this location reference. We call this number the *degree of consistency* and use it to rank the suspiciousness of the location references received at a particular non-beacon node. The smaller the degree is, the more likely that the corresponding location reference is malicious.

The consistency between two location references can be verified as follows. For any location reference $\langle x, y, \delta \rangle$, the non-beacon node derives the area that it may reside based on this location reference. This area can be represented by a ring centered at (x, y), with the inner radius $\max\{\delta - \epsilon, 0\}$ and the outer radius $\delta + \epsilon$, where ϵ is the maximum distance error. For the sake of presentation, we refer to such a ring a *candidate ring (centered) at location* (x, y). The non-beacon node then check whether the candidate rings of two location references overlap each other. If yes, they are consistent; otherwise, they are not consistent.

The algorithm to check whether the candidate rings of two location references $a = \langle x_a, y_a, \delta_a \rangle$ and $b = \langle x_b, y_b, \delta_b \rangle$ overlap can be done efficiently in the following way. Let d_{ab} denote the distance between (x_a, y_a) and (x_b, y_b). Let $rmax(x)$ and $rmin(x)$ denote the outer radius and the inner radius of the candidate ring of location reference x respectively. We can easily figure out that the candidate rings of location references a and b will not overlap when either of the following three conditions is true: (1) $d_{ab} \rangle rmax(a) + rmax(b)$, (2) $d_{ab} + rmax(a) \langle rmin(b)$ and (3) $d_{ab} + rmax(b) \langle rmin(a)$.

Similar to the greedy algorithm, the enhanced algorithm to identify the largest consistent set starts with the set of all location references in the first round. In each round, it verifies whether the current set of location references is τ-consistent. If yes, the algorithm outputs the estimated location and stops. Optionally, it may also output the set of location references. If not, it removes the location reference corresponding to the smallest degree and use the remaining location references as the input to the next round. This algorithm continues until it finds a set of τ-consistent location references or when it is not possible to find such a set (i.e., there are only 3 remaining location references).

The enhanced algorithm not only improves the accuracy of location estimation in the presence of malicious attacks, but also reduces the computation overhead significantly since it can identify the most suspicious location reference efficiently and effectively. To continue the earlier example, a non-beacon node only needs to perform MMSE operations for 5 times. In general, a non-beacon node needs to use a MMSE-based method for at most $n - 3$ times.

Performance of Attack-Resistant MMSE Method

We performed simulation experiments to evaluate the attack-resistant MMSE method. Three attack scenarios are considered. The first scenario considers a single malicious location reference that declares a wrong location e meters away from the beacon node's real location. (An attacker may also modify the distance component δ in a location reference, which will generate a similar impact.) In the second scenario, there are multiple non-colluding malicious location references, and each of them independently declares a wrong location that is e meters away from the beacon node's real location. In the third scenario, multiple colluding malicious location references are considered. In this case, the malicious location references declare false locations by coordinating with each other to create a virtual location e meters away from the sensor's real location. Thus, the malicious location references may appear to be consistent to a victim node.

In all simulations, a set of benign beacon nodes and a few malicious beacon nodes are evenly deployed in a $30m \times 30m$ target field. The non-beacon sensor node is located at the center of this target field. We assume the maximum transmission range of beacon signals is $R_b = 22m$, so that the non-beacon node can receive the beacon signal from every beacon node located in the target field. We assume the entire field is much larger than this target field so that an attacker can create a very large location estimation error inside the deployment field. Each malicious beacon node declares a false location according to the three attack scenarios discussed above. We assume a simple distance measurement error model. That is, the distance measurement error is uniformly distributed between $-\epsilon$ and ϵ, where the maximum distance measurement error ϵ is set to $\epsilon = 4m$.

In the simulation, we use the MMSE-based method proposed in [34], which we call the *basic MMSE method*, to perform the basic location estimation. Our attack-resistant MMSE method is then implemented on the basis of this method. We set $\tau = 0.8\epsilon$ according to Figure 4, which guarantees 9 benign location references are considered consistent with probability of 0.999.

Figure 5 shows the performance of the attack resistant MMSE method when the enhanced greedy algorithm is used to identify the largest consistent set. We can see that that the attack-resistant MMSE with the enhanced greedy algorithm reduces

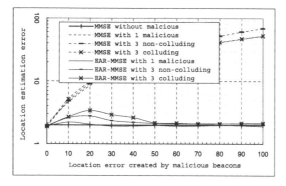

Fig. 5. Performance of attack-resistant MMSE using the enhanced greedy algorithm.

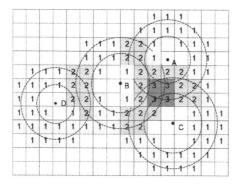

Fig. 6. The voting-based location estimation

the location estimation error significantly even when there are malicious location references.

2.2 Voting-Based Location Estimation

In this voting-based approach, we have each location reference "vote" on the locations at which the node of concern may reside. To facilitate the voting process, we quantize the target field into a grid of cells, and have each sensor node determine how likely it is in each cell based on each location reference. We then select the cell(s) with the highest vote and use the "center" of the cell(s) as the estimated location. To deal with the resource constraints on sensor nodes, we further develop an iterative refinement scheme to reduce the storage overhead, improve the accuracy of estimation, and make the voting scheme efficient on resource constrained sensor nodes.

The Basic Scheme

After collecting a set of location references, a sensor node should determine the target field. The node does so by first identifying the minimum rectangle that covers all the locations declared in the location references, and then extending this rectangle by R_b, where R_b is the maximum transmission range of a beacon signal. This extended rectangle forms the target field, which contains all possible locations for the sensor node. The sensor node then divides this rectangle into M small squares (cells) with the same side length L, as illustrated in Figure 6. (The node may further extend the target field to have square cells.) The node then keeps a voting state variable for each cell, initially set to 0.

At the beginning of this algorithm, the non-beacon node needs to identify the candidate ring of each location reference. For example, in Figure 6, the ring centered at point A is a candidate ring at A, which is derived from the location reference with the declared location at A.

For each location reference $\langle x, y, \delta \rangle$, the sensor node identifies the cells that overlap with the corresponding candidate ring, and increments the voting variables for these cells by 1. After the node processes all the location references, it chooses the cell(s) with the highest vote, and uses its (their) geometric centroid as the estimated location of the sensor node.

A critical problem in the voting-based approach is to determine if a candidate ring overlaps with a cell. Suppose we need to check if the candidate ring at A overlaps with the cell shown in Figure 7(a). Let $d_{min}(A)$ and $d_{max}(A)$ denote the minimum and maximum distances from a point in the cell to point A, respectively. We can see that the candidate ring does not overlap with the cell only when $d_{min}(A) \rangle r_o$ or $d_{max}(A) \langle r_i$, where $r_i = \max\{0, \delta - \epsilon\}$ and $r_o = \delta + \epsilon$ are the inner and the outer radius of the candidate ring, respectively.

To compute d_{min} and d_{max}, we divide the target field into 9 regions based on the cell, as shown in Figure 7(b). It is easy to see that given the center of any candidate ring, we can determine the region in which it falls with at most 6 comparisons between the coordinates of the center and those of the corners of the cell. When the center of a candidate ring is in region 1 (e.g., point A in Figure 7(b)), it can be shown that the closest point in the cell to A is the upper left corner, and the farthest point in the cell from A is the lower right corner. Thus, $d_{min}(A)$ and $d_{max}(A)$ can be calculated accordingly. These two distances can be computed similarly when the center of a candidate ring falls into regions 3, 7, and 9.

Consider point B in region 2. Assume the coordinate of point B is (x_B, y_B). We can see that $d_{min}(B) = y_B - y_2$. Computing $d_{max}(B)$ is a little more complex. We first need to check if $x_B - x_1 \rangle x_2 - x_B$. If yes, the farthest point in the cell from B must be the lower left corner of the cell. Otherwise, the farthest point in the cell from B should be the lower right corner of the cell. Thus, we have

$$d_{max}(B) = \sqrt{(\max\{x_B - x_1, x_2 - x_B\})^2 + (y_B - y_1)^2}.$$

These two distances can be computed similarly when the center of a candidate ring falls into regions 4, 6, and 8.

Consider a point C in region 5. Obviously, $d_{min}(C) = 0$ since point C itself is in the cell. Assume the coordinate of point C is (x_c, y_c). The farthest point in the cell from C must be one of its corners. Similarly to the above case for point B, we

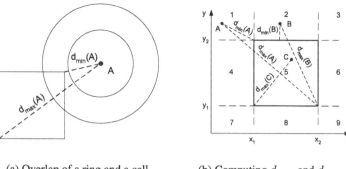

(a) Overlap of a ring and a cell   (b) Computing d_{min} and d_{max}

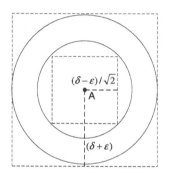

(c) Limiting cell examination

Fig. 7. Determine whether a ring overlaps with a cell

may check which point is farther away from C by checking $x_c - x_1 \rangle x_2 - x_c$ and $y_c - y_1 \rangle y_2 - y_c$. As a result, we get

$$d_{max}(C) = \sqrt{(\max\{x_c - x_1, x_2 - x_c\})^2 + (\max\{t_c - y_1, y_2 - y_c\})^2}.$$

Based on the above discussion, we can determine if a cell and a candidate ring overlap with at most 10 comparisons and a few arithmetic operations. To prove the correctness of the above approach only involves elementary geometry, and thus is omitted.

For a given candidate ring, a sensor node does not have to check all the cells for which it maintains voting states. As shown in Figure 7(c), with simple computation, the node can get the outer bounding box centered at A with side length $2(\delta + \epsilon)$. The node only needs to consider the cells that intersect with or fall inside this box. Moreover, the node can get the inside bounding box with simple computation, which is centered at A with side length $\sqrt{2}(\delta - \epsilon)$, and all the cells that fall into this box need not be checked.

Iterative Refinement

The number of cells M (or equivalently, the quantization step L) is a critical parameter for the voting-based algorithm. It has several implications to the performance of our approach. First, the larger M is, the more state variables a sensor node has to keep, and thus the more storage is required. Second, the value of M (or L) determines the precision of location estimation. The larger M is, the smaller each cell will be. As a result, a sensor node can determine its location more precisely based on the overlap of the cells and the candidate rings.

However, due to the resource constraints on sensor nodes, the granularity of the partition is usually limited by the memory available for the voting state variables on the nodes. This puts a hard limit on the accuracy of location estimation. To address this problem, we propose an *iterative refinement* of the above basic algorithm to achieve fine accuracy with reduced storage overhead.

In this version, the number of cells M is chosen according to the memory constraint in a sensor node. After the first round of the algorithm, the node may find one or more cells having the largest vote. To improve the accuracy of location estimation, the sensor node then identifies the smallest rectangle that contains all the cells having the largest vote, and performs the voting process again. For example, in Figure 6, the same algorithm will be performed in a rectangle which exactly includes the 4 cells having 3 votes. Note that in a later iteration of the basic voting-based algorithm, a location reference does not have to be used if it does not contribute to any of the cells with the highest vote in the current iteration.

Due to a smaller rectangle to quantize in a later iteration, the size of cells can be reduced, resulting in a higher precision. Moreover, a malicious location reference will most likely be discarded, since its candidate ring usually does not overlap with those derived from benign location references. For example, in Figure 6, the candidate ring centered at point D will not be used in the second iteration.

The iterative refinement process should terminate when a desired precision is reached or the estimation cannot be refined. The former condition can be tested by checking if the side length L of each cell is less than a predefined threshold S, while the latter condition can be determined by checking whether L remains the same in two consecutive iterations. The algorithm then stops and outputs the estimated location obtained in the last iteration. It is easy to see that the algorithm will fall into either of these two cases, and thus will alway terminate. In practice, we may set the desired precision to 0 in order to get the best precision.

Performance of Voting-Based Method

We also studied the performance of the voting-based scheme under malicious attacks through simulation. In the simulation, we set $S = 0$ to get the minimum location estimation error achievable by this method. We set $M = 100$ for a reasonable storage cost. The other parameters are the same as in the evaluation of the attack-resistant MMSE method. Figure 8 compares the accuracy of the basic MMSE method and our voting-based scheme under different types of attacks. Similar to the attack-resistant

MMSE method, we can clearly see that the location estimation error is bounded even if there are malicious attacks.

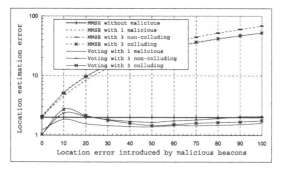

Fig. 8. Performance of the voting-based scheme ($M = 100, S = 0$).

3 Detecting Malicious Beacon Nodes

In hostile environments, a compromised beacon node or an attacking node that has access to compromised cryptographic keys may send out malicious beacon signals that include incorrect locations, or manipulate the beacon signals so that a receiving node obtains, for example, incorrect distance measurements. Sensor nodes that use such beacon signals for location determination may estimate incorrect locations. In this section, we first describe a simple but effective method to detect malicious beacon signals. With this method, we then develop techniques to filter out replayed beacon signals and thus detect malicious beacon nodes.

We assume that two communicating nodes share a unique pairwise key. A number of random key pre-distribution schemes (e.g., [6, 9, 22]) can be used for this purpose. We assume that a beacon node cannot tell if it is communicating with a beacon or non-beacon node simply from the radio signal or the key used to authenticate the packet. We also assume that communication is two way; that is, if node A can reach node B, then node B can reach node A as well. Moreover, we assume beacon signals are unicasted to non-beacon nodes, and every beacon packet is authenticated (and potentially encrypted) with the pairwise key shared between two communicating nodes. Hence, beacon packets forged by external attackers that do not have the right keys can be easily filtered out.

We assume location estimation is based on the distances measured from beacon signals (through, e.g., RSSI). Nevertheless, our approach can be easily revised to deal with location estimation based on other measurements.

3.1 Detecting Malicious Beacon Signals

The technique to detect malicious beacon signals is the basis of detecting malicious beacon nodes. The basic idea is to take advantage of the (known) locations of beacon nodes and the constraints that these locations and the measurements (e.g., distance, angle) derived from their beacon signals must satisfy to detect malicious beacon signals.

A beacon node can perform detection on the beacon signals it hears from other beacon nodes. For the sake of presentation, we call the node making this detection the *detecting (beacon) node*, and the node being detected the *target (beacon) node*. Note that if a malicious beacon node knows that a detecting beacon node is requesting for its beacon signal, it can send out a normal beacon signal that does not lead to incorrect location estimation, and thus pass the detection mechanism without being noticed. To deal with this problem, the detecting node uses a different node ID, called *detecting ID*, during the detection. This ID should be recognized as a non-beacon node ID. The detecting node also has all keying materials related to this ID so that it can communicate securely with other beacon nodes as a non-beacon node. To increase the probability of detecting a malicious beacon node, we may allocate multiple detecting IDs as well as the related keying materials to each beacon node. With the help of these detecting IDs, it is very difficult for an attacker to distinguish the requests from detecting beacon nodes and those from non-beacon nodes when sensor nodes are densely deployed. If sensor nodes have certain mobility and/or the detecting node can carefully craft its request message (e.g., adjust the transmission power in RSSI technique), it will become even more difficult for the attacker to determine the source of a request message. For simplicity, we assume that the attacker cannot tell if a request message is from a beacon node or a non-beacon node.

The proposed method works as follows. The detecting node n first sends a request message to the target node n_a as a non-beacon node. Once the target node n_a receives this message, it sends back a beacon packet (beacon signal) that includes its own location (x', y'). The detecting node n then estimates the distance between them from the beacon signal upon receiving it. Since the detecting node n knows its own location, it can also calculate the distance between them based on its own location (x, y) and the target node's location (x', y'). The detecting node n then compares the estimated distance and the calculated one. If the difference between them is larger than the maximum distance error, the detecting node can infer that the received beacon signal must be malicious. Figure 9 illustrates this idea.

A potential problem in the above method is that even if the calculated distance is consistent with the estimated distance, it is still possible that the beacon signal comes from a compromised beacon node or is replayed by an attacking node. However, a further investigation reveals that this will not generate impact on location estimation. Consider a malicious beacon node that declares a location (x', y'). If the estimated distance from its beacon signal is consistent with the calculated one, it is equivalent to the situation where a benign beacon node located at (x', y') sends a benign beacon signal to the requesting node. In fact, to mislead the location estimation at a non-beacon node, the attacker has to manipulate its beacon signal and/or beacon packet to

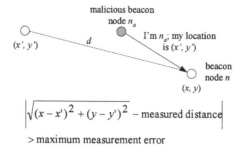

$$\left| \sqrt{(x-x')^2 + (y-y')^2} - \text{measured distance} \right|$$

$$> \text{maximum measurement error}$$

Fig. 9. Detect malicious beacon signals

make the estimated distance inconsistent with the calculated one. This manipulation will certainly be detected if the requesting node happens to be a detecting node.

3.2 Filtering Replayed Beacon Signals

Suppose a beacon signal from a target node is detected to be malicious, it is still not clear if this node is malicious, since an attacker may replay a previously captured beacon signal. However, if we can determine that a malicious beacon signal indeed comes directly from this target node, this target node must be malicious. Thus, it is necessary to filter out as many replayed beacon signals between benign beacon nodes as possible in the detection.

A beacon signal may be replayed through a *wormhole attack* [16], where an attacker tunnels packets received in one part of the network over a low latency link and replays them in a different part [16]. Wormhole attacks generate big impacts on the security of many protocols (e.g., localization, routing). A number of techniques have been proposed recently to detect such attacks, including geographical leashes [16], temporal leashes [16], and directional antenna [15]. These techniques can be used to filter out beacon signals replayed through a wormhole.

A beacon signal received from a neighbor beacon node may also be replayed by an attacking node. We call such replayed beacon signals *locally replayed beacon signals*. Most of wormhole detectors cannot deal with such attacks, since they can only tell if two nodes are neighbor nodes. It is possible to use temporal leashes [16] to filter out locally replayed beacon signal, since replaying a beacon signal may introduce delay that is detectable with temporal leashes. However, this technique requires a secure and tight time synchronization, and large memory space to store authentication keys. Instead, we study the effectiveness of using round trip time to filter out locally replayed beacon signals, and demonstrate that using round trip time does not require time synchronization method but can detect locally replayed beacon signals effectively.

Replayed Beacon Signals from Wormholes

We assume that there is a wormhole detector installed on every beacon and non-beacon node. This wormhole detector can tell whether two communicating nodes are neighbor nodes or not with certain accuracy. The purpose of the following method is to filter out the replayed beacon signals due to the wormhole between two benign beacon nodes that are far away from each other. An observation regarding such replayed beacon signals is that the distance between the location of the detecting node and the location contained in the beacon packet is larger than the communication range of the target node. Thus, we combine the wormhole detector with the location information in the following algorithm.

Once a beacon signal is detected to be a malicious beacon signal, the detecting node begins to verify if it is replayed through a wormhole with the help of the wormhole detector. The detecting node first calculates the distance to the target beacon node based on its own location and the location declared in the beacon packet. If the calculated distance is larger than the radio communication range of the target node and the wormhole detector determines that there is a wormhole attack, the beacon signal is considered as a replayed beacon signal and is ignored by the detecting node. Otherwise, the beacon signal will go through the process to filter locally replayed signals in Section 3.2.

Let us briefly study the effectiveness of this method. Since a malicious target node can always manipulate its beacon signals to convince the detecting node that there is a wormhole attack and they are far from each other even if they are neighbor nodes, it is possible that the beacon signal from a malicious target node is removed. Fortunately, non-beacon nodes in the network are also equipped with this wormhole detector. This means that a malicious target node cannot convince all detecting nodes that there are wormhole attacks, and at the same time convince all non-beacon nodes that there are no wormhole attacks so that its beacon signals are not removed by non-beacon nodes. This is because a malicious beacon node does not know if a requesting node is a detecting beacon node.

It is also possible that a replayed beacon signal through a wormhole from a benign target node is not removed. The reason is that the wormhole detector cannot guarantee that it can always detect wormhole attacks.

Locally Replayed Beacon Signals

The method to filter out locally replayed beacon signals is based on the observation that the replay of a beacon signal introduces extra delay. In most cases, this delay is large enough to detect whether there is a locally replayed beacon signal through the round trip time (RTT) between two neighbor nodes. In the following, we first investigate the characteristics of RTT between two neighbor sensor nodes in a typical sensor network, and then use this result to filter out locally replayed signals between benign beacon nodes.

To remove the uncertainty introduced by the MAC layer protocol and the processing delay, we measure the RTT in the following way. As shown in Figure 10, the sender sends a request message to the receiver, and the receiver responds with a reply message. t_1 is the time of finishing sending the first byte of the request from a sender, t_2 is the time of finishing receiving the first byte of this request at a receiver, t_3 is the time of finishing sending the first byte of the reply from the receiver, and t_4 is the time of finishing receiving the first byte of this reply at the original sender. The sender estimates RTT by computing $RTT = (t_4 - t_1) - (t_3 - t_2)$, where t_4 and t_1 are available at the sender, and $t_3 - t_2$ can be obtained from the receiver by exchanging messages.

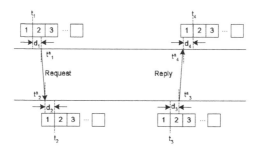

Fig. 10. Round trip time

Characteristics of RTT between neighbor nodes: We may perform experiments on actual sensor platform to obtain the characteristics of RTT. To gain further insights and examine our approach, we performed experiments on MICA2 motes [7] running TinyOS [14]. For simplicity, we assume the same type of sensor nodes in the sensor network. Nevertheless, our technique can be easily extended to deal with different types of nodes in the network.

In the experiment, t_1 is measured by recording the time right after the communication module (CC1000) moves the second byte of the request message to the SPDR register, which is used to store the byte being transmitted over the radio channel. In other words, t_1 is the time of finishing shifting the first byte of the request message out of this register. Assume the absolute time of finishing sending the first byte of the request message is t_1^a. We have $t_1 + d_1 = t_1^a$, where d_1 is the delay between shifting the data byte out of the SPDR register and finishing sending this byte over the radio channel. Similarly, we have $t_3 + d_3 = t_3^a$. Similarly, t_2 is measured by recording the time right after the first byte of the request message is ready at the SPDR register. Assume the absolute time of finishing receiving this byte from the radio channel is t_2^a. We have $t_2 = t_2^a + d_2$, where d_2 is the delay between receiving this byte from the radio channel and reading this byte from the SPDR register. Similarly, we have $t_4 = t_4^a + d_4$. Since the radio signal travels at the speed of light, we have $t_4^a - t_1^a - (t_3^a - t_2^a) = \frac{2D}{c}$, where D is the distance between two neighbor nodes and c is the speed of light. Thus, we have $RTT = d_1 + d_2 + d_3 + d_4 + \frac{2D}{c}$.

Note that d_1, d_2, d_3 and d_4 are mainly affected by the underlying radio communication hardware. Since two neighbor nodes are usually close to each other, the value of $\frac{2D}{c}$ in the above equation is negligible. Hence, the RTT measured by computing

$RTT = (t_4 - t_1) - (t_3 - t_2)$ is not affected by the MAC protocol or any processing delay. This means that the distribution of RTT should be within a narrow range. Let F denotes the cumulative distribution function of RTT when there are no replay attacks, x_{min} denotes the maximum value of x such that $F(x) = 0$, and x_{max} denotes the minimum value of x such that $F(x) = 1$.

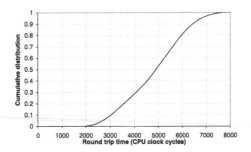

Fig. 11. Cumulative distribution of round trip time

Figure 11 shows the cumulative distribution of RTT when there are no replay attacks. We use one CPU clock cycle as the basic unit to measure the time. This figure is derived by measuring RTT $100,000$ times. The result shows that $x_{min} = 1,951$ and $x_{max} = 7,506$. Since the transmission time of one bit is about 384 clock cycles, we can detect any replayed signal if the delay introduced by this replay is longer than the transmission time of $\frac{7506-1951}{384} \approx 14.5$ bits.

The detector for locally replayed beacon signals: With RTT's cumulative distribution, we can detect locally replayed signals between benign beacon nodes effectively. The basic idea is to check if there is any significant difference between the observed RTT and the range of RTT derived during our experiments. For example, if the observed RTT at the requesting node is larger than the maximum RTT in Figure 11, it is very likely that the reply signal is replayed. The following local replay detector will be installed on every beacon and non-beacon node.

The requesting node u communicates with a beacon node v following the request-reply protocol shown in Figure 10. As a result, node u can compute $RTT = (t_4 - t_1) - (t_3 - t_2)$. There are two cases: (1) When $RTT \leq x_{max}$, the beacon signal is considered as not locally replayed. If the requesting node is a detecting node, it will report an alert when the beacon signal is detected to be malicious. If the requesting node is a non-beacon node, this beacon signal will be used in its location estimation. (2) When $RTT \rangle x_{max}$, this beacon signal is considered as locally replayed, and will be ignored by the requesting node.

When the target node is a benign beacon node and is a neighbor of the detecting node, but the beacon signal is replayed by a malicious node, the detecting node will report an alert if the delay introduced by the locally replayed signal is less than the transmission time of 14.5 bits data. However, it is very difficult for the attacker to achieve, since the attacker has to replay the beacon signal to the detecting node when the target node is still sending its beacon signal. This implies that the attacker has to physically shield every signal to the detecting node and replay the intercepted packet at the same time. When the target benign beacon node is not a neighbor node of the detecting node, the detecting node will report an alert if the delay introduced by the undetected wormhole attack is less than the transmission time of 14.5 bits data. Note that this implies this replayed beacon signal has bypassed the wormhole detector.

Note that the purpose of the above method is to filter the replayed beacon signals between benign beacon nodes to avoid false positives. This method becomes trivial when the target node is a malicious beacon node, since it can easily convince a detecting node that the beacon signal is locally replayed and thus prevent the detecting node from reporting an alert. However, the malicious target node cannot convince all detecting nodes that the beacon signals are locally replayed, and at the same time convince all non-beacon nodes that its beacon signals are not locally replayed so that its beacon signals are accepted by non-beacon nodes.

3.3 Experimental Results

We have implemented the detection and filtering techniques on TinyOS [14], an operation system for networked sensors. In the following, we present the simulation results obtained through the TinyOS simulator, with a focus on the detection rate (i.e., $\frac{\#detected\ malicious\ beacon\ nodes}{\#total\ malicious\ beacon\ nodes}$) and the false positive rate (i.e., $\frac{\#incorrect\ detected\ beacons}{\#total\ benign\ beacons}$).

We assume 1,000 sensor nodes ($N = 1000$) are randomly deployed in a sensing field of 1000×1000 square feet. Among these sensor nodes, there are 100 beacon nodes ($N_b = 100$) with 10 compromised beacon nodes ($N_a = 10$). We assume the maximum communication range of a beacon or non-beacon node is 150 feet, and a malicious beacon node only contacts the nodes within its communication range. We assume there is a wormhole between location A (100,200) and location B (800,700), which forwards every message received at one side immediately to the other side. We assume malicious beacon nodes collude together to report alerts against benign beacon nodes. We also assume that there is a technique (e.g., RSSI) used to estimate the distance to the beacon node that has the maximum error of 10 feet.

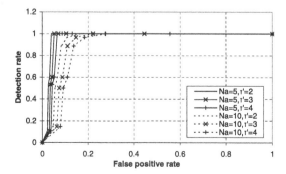

Fig. 12. ROC curves of the proposed detection method.

Figure 12 shows the ROC (Receiver Operating Characteristic) curves for the proposed techniques with different parameters. We can see that our technique can detect most of malicious beacon nodes with small false positive rate (e.g., 5%) when there are a small number of compromised beacon nodes. However, when the number of compromised beacon nodes increases, the performance decreases accordingly. For example, when there are 10 malicious beacon nodes, the false positive rate will reach 20% in order to detect most of malicious beacon nodes. Nevertheless, the figure still shows that our techniques are practical and effective in detecting malicious beacon nodes. More details about our experiments can be found in [25].

4 Conclusions

Sensor network security is a challenging problem, particularly due to the resource constraints on sensor nodes, the threat of node compromises resulting from unattended deployment, and the imbalance between the threat and the defense in sensor networks. Foundational cryptographic services such as broadcast authentication and key management are definitely a necessary condition to ensure secure and resilient sensor network applications. Other foundational services (such as localization and clock synchronization) in sensor network applications also require protection.

In this chapter, we discussed two recent efforts to provide secure and resilient localization in wireless sensor networks: a suite of attack-resistant MMSE methods, and a mechanism to detect malicious beacon nodes. This chapter and other chapters in this edited volume offer a snapshot of recent advances in protecting localization and clock synchronization services in sensor networks. Other secure foundational services such as secure aggregation and in-network processing, cluster formation, and cluster head election also deserve careful investigation. Moreover, intrusion detection in sensor networks is highly desirable, particularly due to the fact that unattended sensor nodes may be easily captured and compromised. We expect to see more advances in sensor network security in the next several years.

References

1. I.F. Akyildiz, W. Su, Y. Sankarasubramaniam, and E. Cayirci. Wireless sensor networks: A survey. *Computer Networks*, 38(4):393–422, 2002.
2. P. Bose, P. Morin, I. Stojmenovic, and J. Urrutia. Routing with guaranteed delivery in ad hoc wireless networks. In *Proceedings of 3rd ACM International Workshop on Discrete Algorithms and Methods for Mobile Computing and Communications*, pages 48–55, 1999.
3. P. Bose, P. Morin, I. Stojmenovic, and J. Urrutia. Routing with guaranteed delivery in ad hoc wireless networks. *ACM Wireless Networks*, 7(6):609–616, 2001.
4. N. Bulusu, J. Heidemann, and D. Estrin. GPS-less low cost outdoor localization for very small devices. In *IEEE Personal Communications Magazine*, pages 28–34, October 2000.
5. S. Capkun and J.P. Hubaux. Secure positioning of wireless devices with application to sensor networks. In *Proceedings of IEEE InfoCom'05 (to appear)*, 2005.

6. H. Chan, A. Perrig, and D. Song. Random key predistribution schemes for sensor net-works. In *IEEE Symposium on Research in Security and Privacy*, pages 197–213, 2003.

7. Crossbow Technology Inc. Wireless sensor networks. http://www.xbow.com/Products/Wireless_Sensor_Networks.htm. Accessed in May 2005.

8. L. Doherty, K. S. Pister, and L. E. Ghaoui. Convex optimization methods for sensor node position estimation. In *Proceedings of INFOCOM'01*, 2001.

9. W. Du, J. Deng, Y. S. Han, and P. Varshney. A pairwise key pre-distribution scheme for wireless sensor networks. In *Proceedings of 10th ACM Conference on Computer and Communications Security (CCS'03)*, pages 42–51, October 2003.

10. L. Eschenauer and V. D. Gligor. A key-management scheme for distributed sensor net-works. In *Proceedings of the 9th ACM Conference on Computer and Communications Security*, pages 41–47, November 2002.

11. L. Fang, W. Du, and P. Ning. A beacon-less location discovery scheme for wireless sensor networks. In *Proceedings of IEEE InfoCom'05 (to appear)*, 2005.

12. D. Gay, P. Levis, R. von Behren, M. Welsh, E. Brewer, and D. Culler. The nesC language: A holistic approach to networked embedded systems. In *Proceedings of Programming Language Design and Implementation (PLDI 2003)*, June 2003.

13. T. He, C. Huang, B. M. Blum, J. A. Stankovic, and T. F. Abdelzaher. Range-free local-ization schemes in large scale sensor networks. In *Proceedings of ACM MobiCom 2003*, 2003.

14. J. Hill, R. Szewczyk, A. Woo, S. Hollar, D.E. Culler, and K. S. J. Pister. System ar-chitecture directions for networked sensors. In *Architectural Support for Programming Languages and Operating Systems*, pages 93–104, 2000.

15. L. Hu and D. Evans. Using directional antennas to prevent wormhole attacks. In *Proceed-ings of the 11th Network and Distributed System Security Symposium*, pages 131–141, February 2003.

16. Y.C. Hu, A. Perrig, and D.B. Johnson. Packet leashes: A defense against wormhole attacks in wireless ad hoc networks. In *Proceedings of INFOCOM 2003*, April 2003.

17. C. Intanagonwiwat, R. Govindan, and D. Estrin. Directed diffusion: A scalable and robust communication paradigm for sensor networks. In *Proceedings of the sixth annual inter-national conference on Mobile computing and networking (Mobicom '00)*, pages 56–67, Nov 2003.

18. B. Karp and H. T. Kung. GPSR: Greedy perimeter stateless routing for wireless networks. In *Proceedings of ACM MobiCom 2000*, 2000.

19. L. Lazos and R. Poovendran. Serloc: Secure range-independent localization for wireless sensor networks. In *ACM workshop on Wireless security (ACM WiSe 2004)*, Philadelphia, PA, October 1 2004.

20. L. Lazos and R. Poovendran. Serloc: Robust localization for wireless sensor networks. *ACM Transactions on Sensor Networks*, 1(1):73–100, August 2005.

21. Z. Li, W. Trappe, Y. Zhang, and B. Nath. Robust statistical methods for securing wireless localization in sensor networks. In *Proceedings of the Fourth International Conference on Information Processing in Sensor Networks (IPSN '05)*, April 2005.

22. D. Liu and P. Ning. Establishing pairwise keys in distributed sensor networks. In *Pro-ceedings of 10th ACM Conference on Computer and Communications Security (CCS'03)*, pages 52–61, October 2003.

23. D. Liu, P. Ning, and W.K. Du. Attack-resistant location estimation in wireless sensor networks. Technical Report TR-2004-29, North Carolina State University, Department of Computer Science, 2004.

24. D. Liu, P. Ning, and W.K. Du. Attack-resistant location estimation in wireless sensor networks. In *Proceedings of the Fourth International Conference on Information Processing in Sensor Networks (IPSN '05)*, April 2005.

25. D. Liu, P. Ning, and W.K. Du. Detecting malicious beacon nodes for secure location discovery in wireless sensor networks. In *Proceedings of the 25th International Conference on Distributed Computing Systems (ICDCS '05)*, pages 609–619, June 2005.

26. R. Nagpal, H. Shrobe, and J. Bachrach. Organizing a global coordinate system from local information on an ad hoc sensor network. In *IPSN'03*, 2003.

27. A. Nasipuri and K. Li. A directionality based location discovery scheme for wireless sensor networks. In *Proceedings of ACM WSNA '02*, September 2002.

28. J. Newsome and D. Song. GEM: graph embedding for routing and data-centric storage in sensor networks without geographic information. In *Proceedings of the First ACM Conference on Embedded Networked Sensor Systems (SenSys '03)*, pages 76–88, Nov 2003.

29. D. Niculescu and B. Nath. Ad hoc positioning system (APS). In *Proceedings of IEEE GLOBECOM '01*, 2001.

30. D. Niculescu and B. Nath. Ad hoc positioning system (APS) using AoA. In *Proceedings of IEEE INFOCOM 2003*, pages 1734–1743, April 2003.

31. D. Niculescu and B. Nath. DV based positioning in ad hoc networks. In *Journal of Telecommunication Systems*, 2003.

32. A. Perrig, R. Szewczyk, V. Wen, D. Culler, and D. Tygar. SPINS: Security protocols for sensor networks. In *Proceedings of Seventh Annual International Conference on Mobile Computing and Networks*, July 2001.

33. S. Ratnasamy, B. Karp, L. Yin, F. Yu, D. Estrin, R. Govindan, and S. Shenker. GHT: A geographic hash table for data-centric storage. In *Proceedings of 1st ACM International Workshop on Wireless Sensor Networks and Applications*, Sep 2002.

34. A. Savvides, C. Han, and M. Srivastava. Dynamic fine-grained localization in ad-hoc networks of sensors. In *Proceedings of ACM MobiCom '01*, pages 166–179, July 2001.

35. A. Savvides, H. Park, and M. Srivastava. The bits and flops of the n-hop multilateration primitive for node localization problems. In *Proceedings of ACM WSNA '02*, September 2002.

36. S. Shenker, S. Ratnasamy, B. Karp, R. Govindan, and D. Estrin. Data-centric storage in sensornets. In *Proceedings of the First ACM Workshop on Hot Topics in Networks*, October 2002.

Secure Localization for Wireless Sensor Networks using Range-Independent Methods

Loukas Lazos and Radha Poovendran

Network Security Lab, Electrical Engineering Department, University of Washington, Seattle, WA, 98195
llazos@u.washington.edu, rp3@u.washington.edu

1 Introduction

Wireless Sensor Networks (WSNs) are envisioned to be integrated into our everyday lives, enabling a wealth of commercial applications such as environmental and habitat monitoring, disaster relief and emergency rescue operations, patient monitoring, as well as military applications such as target detection and tracking. These applications are facilitated by the collaborative processing of the physical properties monitored by the sensors, such as temperature, light, sound, humidity, vibration, acceleration, or air quality.

For most applications of WSNs, knowledge of the origin of the sensed information is critical for taking appropriate action based on the observations. As an example, if a smoke detector reports the break out of a fire, this information, while useful, is not sufficient to initiate proper action. On the other hand, associating the report from the smoke detector in space, enables the timely response to the reported event. Hence, the association of the observations reported by sensors in space increases the quality of the information aggregated via the sensor network. Furthermore, location is assumed to be known in many network operations such as routing protocols where a family of geographically-aided algorithms have been proposed [2], or security protocols where location information is used to prevent threats against network services [13, 16]. In WSNs, enabling sensors to associate their reports with space is achieved via the location estimation process also known as *localization*.

The majority of the localization techniques that are proposed for WSNs, [4, 12, 25, 27, 31, 34] are designed to operate in a benign environments with no security threats. However, WSNs may be deployed in hostile environments and operating unsupervised, and hence, are vulnerable to conventional and novel attacks [11, 30] aimed at interrupting the functionality of location-aware applications by exploiting the vulnerabilities of the localization scheme.

In this chapter, we study the problem of *enabling nodes of a WSN to determine their location even in the presence of malicious adversaries*. This problem will be referred to as *Secure Localization*. We consider secure localization in the context of the following design goals: (a) decentralized implementation, (b) resource efficiency, and (c) robustness against security threats.

We illustrate a series of attacks against localization schemes for WSNs [11, 13, 26, 28] and propose *SeRLoc*, a robust location estimation scheme for WSNs that achieves decentralized, resource-efficient sensor localization even in the presence of adversaries. We also propose a high resolution localization algorithm called *HiRLoc*, that improves the localization accuracy at the expense of more complicated hardware. Since sensors are hardware and power limited, SeRLoc and HiRLoc rely on a two-tier network architecture. The network consists of a small number of nodes equipped with known coordinates and orientation we call *locators* and a large number of resource-constrained sensor devices with unknown location.

Moreover, since distance measurements are susceptible to distance enlargement/reduction [5], we do not use any such measurements to compute the sensor location. Instead sensors rely on beacon broadcasts from the locator containing localization information to infer their location. We refer to methods that are not using distance measurements as range-independent localization schemes [4, 12, 25]. Methods for securing range-dependent localization schemes are presented in [5, 7].

Since range independent schemes do not rely on any distance measurements to estimate location, they are not vulnerable to range-alteration attacks. However an adversary may launch relay type of attacks such as the wormhole attack [13, 28], impersonation attacks such as the Sybil attack [11, 26], or compromise network entities. First, we describe the impact of these attacks on the location estimation process, and then, we provide mechanisms that allow each sensor to determine its location *even* in the presence of these threats. Furthermore, we analytically evaluate the probability of success for each type of attack using *spatial statistics* theory [9].

The remainder of the chapter is organized as follows. In Section 2 we illustrate different attacks against range-independent location estimation schemes. In Section 3, we state our network model. In Section 4 we describe two algorithms for robustly estimating the position of sensors. In Section 5, we present a threat analysis. In Section 6, we evaluate the performance of SeRLoc and HiRLoc. In Section 7, we present related work and open problems. Section 8 presents our conclusions.

2 Attacks on Range-independent Localization Schemes

In this section we first define the adversarial model considered for WSNs. We then illustrate different types of attacks against range-independent localization schemes.

2.1 Adversarial Model

We assume that the adversary's goal is to mislead sensors to falsely estimate their location. We also assume that in its effort to mislead the sensors, the adversary must remain undetected. We do not consider Denial-of-Service (DoS) attacks against the localization scheme. Such attacks can be easily detected, since sensors will not be

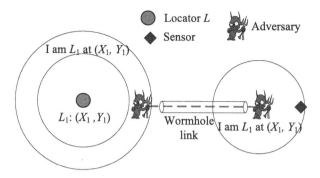

Fig. 1. The adversary records the broadcast of reference point L_1, tunnels it to the region of the sensor under attack and replays it. The sensors believes it is within range of L_1.

able to compute their position. We also do not address attacks against the physical medium such as frequency jamming. Spread spectrum [38] and coding [39] are known to be efficient mechanisms to shield the physical layer against jamming attacks. Also, we do not consider any attack against the Medium Access Control (MAC) protocol that may lead to a denial-of-service (DoS). Secure location estimation schemes that take into account jamming are presented in [5, 20].

2.2 Attack Models

In range-independent location estimation methods, nodes rely on localization information included in beacons transmitted from reference points in order to estimate their position. In order to bias the location estimation process, the adversary attempts to inject bogus localization information into the network. This can be achieved by performing a wormhole (relay) attack [13, 28, 30], an impersonation (Sybil) attack [11, 26], or compromise of reference points. In any of those attacks we assume that at least some valid information not altered by the adversary is present, that allows the node to estimate its position. We now discuss the different attacks against range-independent localization schemes in more detail.

The Wormhole (Relay) Attack

The wormhole attack is a relay type of attack where an adversary relays information transmitted at one part of the network to some distant part of the network, thus violating the geometry of the network and the communication range constraint. To mount a wormhole attack, the adversary initially establishes a direct link referred to as a *wormhole link* between two points in the network. Once the wormhole link is established, the adversary eavesdrops (records) messages at one end of the link, referred to as the *origin point*, tunnels them through the wormhole link and replays them at the other end, referred to as the *destination point*. The wormhole attack is very difficult to detect, since it is launched without compromising any host, or the integrity and authenticity of the communication [13, 28].

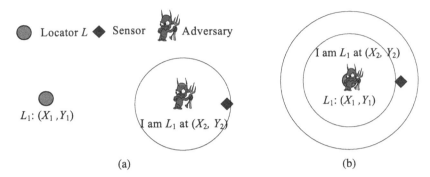

(a)   (b)

Fig. 2. (a) The adversary impersonates reference point L_1 to a sensor under attack. The sensor is misled to believe it is within range of L_1 with L_1 located at (X_2, Y_2), (b) reference point L_1 is compromised and falsely reports its location.

When an adversary launches a wormhole attack against the location estimation process, sensors located at the destination point of the attack hear beacons transmitted from reference points located at the origin point of the attack. Hence, sensors are misled to believe they are within proximity of reference points at the origin point of the attack. The bogus localization information is properly authenticated by the sensors (since beacons are indeed authentic) and can significantly bias the location estimation at each sensor under attack.

One mechanism for detecting relay type of attacks, is synchronizing the nodes of the network and timestamping each message [13]. Every recipient of a message compares the timestamp with the time when the message is received to determine whether the message has traveled a distance longer than the communication range of the sender. However, when the RF medium is used for transmitting beacons synchronization has to be achieved with nanosecond accuracy [13]. Using a slower medium such as the acoustic medium to transmit beacons to avoid the tight synchronization requirement, leaves the system vulnerable to wormholes when the adversary uses an RF wormhole link to relay the localization information in a timely manner.

2.3 The Impersonation (Sybil) Attack

In the impersonation attack, the adversary assumes one or multiple identities from network nodes and impersonates those nodes to other entities within the network [11, 26]. With respect to the localization process, the adversary impersonates reference points and injects bogus localization information into the network. Unlike the wormhole attack, in the Sybil attack model, the adversary must compromise cryptographic quantities necessary to prove its impersonated IDs to the nodes under attack. Hence, nodes properly authenticate an adversary as a trustable source.

In Figure 2, the adversary impersonates locator L_1 to a sensor that is not within the range of L_1. The sensor under attack is misled to believe that it can hear locator L_1 located at coordinates (X_2, Y_2). The adversary can modify the coordinates contained within the beacon to any arbitrary position within the network.

2.4 Compromise of Network Nodes

The adversary may be also able to compromise network nodes used in the location estimation process and force them to misbehave. For example the adversary may compromise reference points and force them to falsely report their positions. Under node compromise, we assume that the adversary gains full control over the behavior of the entity that has been compromised. This assumption is significantly stronger than the assumption made for launching an impersonation attack where the adversary can only impersonate a node and not alter its behavior (controls only the impersonators).

We assume that the sensors have to receive at least some localization information from uncompromised reference points in order to perform any kind of robust location estimation. In Figure 2(b), we show the compromise of locator L_1 and the broadcast of bogus localization information. The sensor is misled to believe that locator L_1 is located at position (X_2, Y_2).

3 Network Model

In this section, we state our network model assumptions for building our secure location estimation algorithm.

Network Setup: We assume a two-tier network architecture where a set of sensors S of unknown location is randomly deployed with a density ρ_s within an area $\mathcal{A}$, and a set of reference points L we call *locators*, with known location[1] and orientation, also randomly deployed with a density ρ_L.

Antenna Model: We assume that sensors are equipped with omnidirectional antennas and transmit with power P_s, while locators are equipped with M directional antennas with directivity gain $G\rangle 1$, and transmit with power $P_L\rangle P_s$. Since the locator transmission power is higher than the sensor transmission power, the locator-sensor communication channel is asymmetric. For the rest of the chapter, we denote the sensor-to-locator communication range as r, and the locator-to-sensor communication range as R.

System Parameters: Since both locators and sensors are randomly and independently deployed, it is essential to select the system parameters so that sufficient number of locators can communicate with sensors. The random deployment of the locators with a density $\rho_L = \frac{|L|}{\mathcal{A}}$ ($|\cdot|$ denotes the cardinality of a set) is equivalent to a sequence of events following a *homogeneous Poisson point process* of rate ρ_L [9]. The random deployment of sensors with a density $\rho_s = \frac{|S|}{\mathcal{A}}$, is equivalent to a random sampling of the area $\mathcal{A}$ with rate ρ_s [9]. Making use of *Spatial Statistics*

---

[1] We presume that locators acquire their position either through manual insertion or through GPS receivers [36]. Though GPS signals can be spoofed, knowledge of the coordinates of several nodes is essential to achieve any kind of node localization for any localization scheme.

theory [9], if LH_s denotes the set of locators heard by a sensor s, that is, within range R from s, the probability that s hears exactly k locators, given that the locators are randomly and independently deployed, is given by the Poisson distribution:

$$P(|LH_s| = k) = \frac{(\rho_L \pi R^2)^k}{k!} e^{-\rho_L \pi R^2}. \tag{1}$$

Based on (1) and the independent deployment of sensors, the probability for *every* sensor to hear at least k locators $P(|LH_s|\rangle k)$:

$$P(|LH_s| \geq k, \forall s \in S) = (1 - \sum_{i=0}^{k-1} \frac{(\rho_L \pi R^2)^i}{i!} e^{-\rho_L \pi R^2})^{|S|}. \tag{2}$$

Equation (2) allows the choice of ρ_L, R so that a sensor hears at least k locators with any desired probability.

4 Secure Location Estimation in WSN

In this section we describe two location estimation schemes. We first present the SEcure Range-independent LOCalization scheme (*SeRLoc*) that enables sensors to determine their location based on beacon information transmitted by the locators, even in the presence of security threats. We then present the HIgh-resolution LOCalizaion scheme (*HiRLoc*) that improves the location resolution.

4.1 Location Determination in SeRLoc

In SeRLoc, sensors determine their location based on the localization information included in beacons transmitted by the locators. Figure 3(a) illustrates the idea behind SeRLoc. Each locator transmits beacons at each antenna sector containing (a) the locator's coordinates and, (b) the angles of the antenna boundary lines with respect to a common global axis.

For each locator L_i heard at a sensor s, sensor s defines the sector S_i corresponding to the transmission of that locator where s has to be included. Combining information from multiple locators it defines the *Region Of Intersection* (ROI), as the region where the maximum number of sectors overlap:

$$ROI = \bigcap S_i. \tag{3}$$

The sensor s determines its location as the center of gravity (CoG) of the ROI. The CoG is the least square error solution given that a sensor can lie with equal probability at any point of the ROI. In Figure 3(a), the sensor hears beacons from locators $L_1 \sim L_4$ and determines its position as the CoG of the ROI. We now present the algorithmic details of SeRLoc.

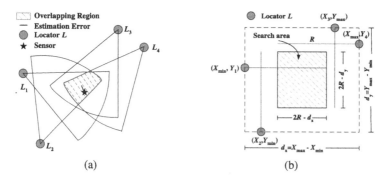

Fig. 3. (a) The sensor hears locators $L_1 \sim L_4$ and estimates its location as the Center of Gravity (CoG) of the region of intersection. (b) Determination of the search area.

- *Step 1:* **Collection of localization information:** In Step 1, the sensor collects information from all the locators that it can hear. A sensor s can hear all locators $L_i \in L$ that lie within a circle of radius R, centered at s.

$$LH_s = \{L_i : \|s - L_i\| \le R, \quad L_i \in L\}. \tag{4}$$

- *Step 2:* **Search area:** In Step 2, the sensor computes a search area for its location. Let $X_{min}, Y_{min}, X_{max}, Y_{max}$ denote the minimum and the maximum locator coordinates form the set LH_s.

$$X_{min} = \min_{L_i \in LH_s} X_i, \ X_{max} = \max_{L_i \in LH_s} X_i, \ Y_{min} = \min_{L_i \in LH_s} Y_i, \ Y_{max} = \max_{L_i \in LH_s} Y(5)$$

Since every locator of set LH_s needs to be within a range R from sensor s, if s can hear locator L_i with coordinates (X_{min}, Y_i), it has to be located *left* of the vertical boundary of $(X_{min} + R)$. Similarly, s has to be located *right* of the vertical boundary of $(X_{max} - R)$, *below* the horizontal boundary of $(Y_{min} + R)$, and *above* the horizontal boundary of $(Y_{max} - R)$. The dimensions of the rectangular search area are $(2R - d_x)$x$(2R - d_y)$ where d_x, d_y are the horizontal distance $d_x = X_{max} - X_{min} \le 2R$ and the vertical distance $d_y = Y_{max} - Y_{min} \le 2R$, respectively. In Figure 3(b), we show the search area for the network setup in Figure 3(a).

- *Step 3:* **Overlapping region-Majority vote**: In Step 3, sensors determine the ROI of all sectors they hear. Since it would be computationally expensive for each sensor to analytically determine the ROI based on the line intersections, we employ a grid scoring system that defines the ROI based on majority vote.

Grid score table: The sensor places a grid of equally spaced points within the rectangular search area as shown in Figure 4(a). For each grid point, the sensor holds a score in a grid score table, with initial values equal to zero. For each grid point, the sensor executes the *grid-sector test* detailed in the following, to decide if the grid point is included in a sector heard by a locator of set LH_s. If the grid score test is positive the sensor increments the corresponding grid score table value by one, otherwise the value remains unchanged. This process is repeated for all locators

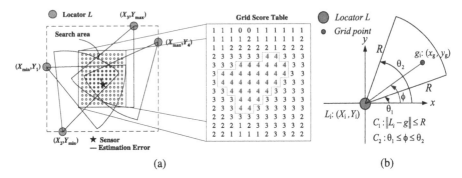

Fig. 4. (a) Steps 3,4: Placement of a grid of equally spaced points in the search area, and the corresponding grid score table. The sensor estimates its position as the centroid of all grid points with the highest score, (b) Step 3: Grid-sector test for a point g of the search area.

heard LH_s, and all the grid points. The ROI is defined by the grid points that have the highest score in the grid score table. In Figure 4(a), we show the grid score table and the corresponding ROI.

Note that due to the finite grid resolution, error is induced in the calculation. The resolution of the grid can be increased to reduce the error at the expense of energy consumption due to the increased processing time.

Grid-sector test: A point $g : (x_g, y_g)$ is included in a sector of angles $[\theta_1, \theta_2]$ originating from locator L_i if it satisfies two conditions:

$$C_1 : \ \|g - L_i\| \leq R, \qquad C_2 : \ \theta_1 \leq \phi \leq \theta_2, \tag{6}$$

where ϕ is the slope of the line connecting g with L_i. Note that the sensor *does not have to* perform any angle-of-arrival (AOA) measurements. Both the coordinates of the locators and the grid points are known, and, hence the sensor can analytically calculate ϕ. In Figure 4(b), we illustrate the grid-sector test with all angles measured with reference to the x axis.

- *Step 4:* **Location estimation**: The sensor determines its location as the centroid of all the grid points that define the ROI.

$$\tilde{s} : (x_{est}, y_{est}) = \left(\frac{1}{n} \sum_{i=1}^{n} x_{g_i}, \ \frac{1}{n} \sum_{i=1}^{n} y_{g_i} \right), \tag{7}$$

where n is the number of grid points of the overlapping region, and (x_{g_i}, y_{g_i}) are the coordinates of the grid points.

4.2 HiRLoc: High-resolution Range-Independent Localization Scheme

In this section, we present the High-resolution Range-independent Localization scheme (*HiRLoc*) that allows sensors to determine their location with higher accuracy compared to SeRLoc at the expense of more complex hardware at the locator side.

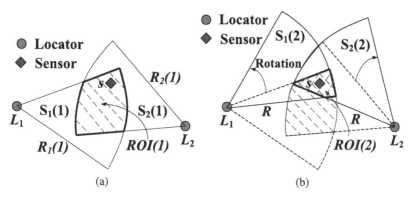

Fig. 5. (a) The sensor is located within the intersection of the sectors $S_1(j)$, $S_2(j)$, which defines the region of intersection ROI. (b) The ROI is reduced by the rotation of the antenna sectors by some angle α.

4.3 Location Determination in HiRLoc

In HiRLoc, localization accuracy in improved by having locators either rotate their antenna system, or change their communication range in order to define new sectors where transmission takes place. Superimposing the sectors indicated by the beacons not only in space but also in time provides the extra location resolution. Based on the beacon information the sensors define the sector area $S_i(j)$ as the confined area covered by the jth transmission of a locator L_i.

By collecting beacons from the locators $L_i \in LH_s$, the sensor can compute its location as the ROI of all the sectors $S_i(j)$. Note that a sensor can hear beacons from multiple locators, and multiple beacons generated by the same locator. Hence, the ROI after the mth round of beacon transmissions can be expressed as the intersection of all the sectors corresponding to the beacons available at each sensor:

$$ROI(m) \overset{(i)}{=} \bigcap_{i=1}^{|LH_s|} \bigcap_{j=0}^{m} S_i(j) \overset{(ii)}{=} \bigcap_{j=0}^{m} \left(\bigcap_{i=1}^{|LH_s|} S_i(j) \right), \qquad (8)$$

Since the ROI indicates the confined region where the sensor is located, reducing the size of the ROI leads to an increase in the localization accuracy. Based on equation (8), we can reduce the size of the ROI by, (a) reducing the size of the sector areas $S_i(j)$ and, (b) increase the number of intersecting sectors $S_i(j)$.

In HiRLoc, reduction of the ROI is achieved by exploiting the temporal dimension. The locators provide different localization information at consecutive beacon transmissions by, (a) varying the direction of their antennas and, (b) varying the communication range of the transmission via power control. We now explore how both these methods lead to the reduction of the ROI.

1. Varying the antenna orientation: The locators are capable of transmitting at all directions (omnidirectional coverage) using multiple directional antennas. Every antenna has a specific orientation and hence corresponds to a fixed sector area

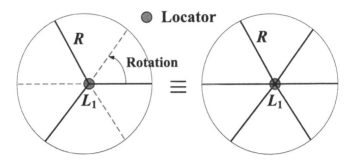

Fig. 6. Locator L_1 is equipped with three directional antennas of beamwidth $\frac{2\pi}{3}$ each. The transmission of beacons at each sector, followed by antenna rotation by $\frac{\pi}{3}$, followed by a transmission of update beacons, is equivalent to equipping L_1 with six directional antennas of beamwidth $\frac{\pi}{3}$.

$S_i(j)$. The antenna orientation is expressed by the angle information contained in the beacon $\theta_i(j) = \{\theta_{i,1}(j), \theta_{i,2}(j)\}$, where $\theta_{i,1}(j), \theta_{i,2}(j)$ denote the lower and upper bounds of the sector $S_i(j)$.

Instead of reducing the size of the intersecting sectors by narrowing the antenna beamwidth, locators can change the orientation of their antennas and re-transmit beacons with the new sector boundaries. A change in the antenna orientation can occur either by changing the orientation of the locators, or by rotation of their antenna system. A sensor collects multiple sector information from each locator over a sequence of transmissions: $S_i(j) = S_i(\theta_i(j), j), j = 1 \ldots Q$. As expressed by equation (8), the intersection of a larger number of *distinct* sectors leads to a reduction in the size of the ROI. As an example, consider Figure 5 where a sensor s hears locators L_1, L_2. In Figure 5(a), we show the first round of beacon transmissions by the locators L_1, L_2, and the corresponding $ROI(1)$. In Figure 5(b), the locators L_1, L_2 rotate their antennas by an angle α and transmit the second round of beacons with the new sector boundaries. The ROI in the two rounds of beacon transmissions, can be expressed as:

$$ROI(1) = S_1(1) \cap S_2(1) \qquad ROI(2) = ROI(1) \cap S_1(2) \cap S_2(2). \qquad (9)$$

The antenna rotation over time can be interpreted as an increase on the number of antenna sectors of each locator via superposition over time. For example, consider Figure 6, where a locator is equipped with three directional antennas of beamwidth $\frac{2\pi}{3}$. Transmission of one round of beacons, followed by antenna rotation by $\frac{\pi}{3}$ and re-transmission of the updated beacons is equivalent to transmitting one round of beacons when locators are equipped with six directional antennas of beamwidth $\frac{\pi}{3}$.

2. Varying the Communication range: A second approach to reduce the area of the ROI, is to reduce the size of the intersecting sectors. This can be achieved by allowing locators to decrease their transmission power and re-broadcast beacons with the new communication range information. In such a case, the sector area $S_i(j)$ is dependent upon the communication range $R_i(j)$ at the jth transmission, i.e. $S_i(j) =$

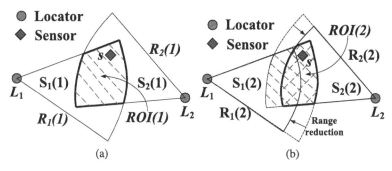

(a)   (b)

Fig. 7. (a) The sensor is located within the intersection of the sectors $S_1(j), S_2(j)$, which defines the ROI, (b) the locators reduce their communication range and transmit updated beacons. While s is outside the communication range of L_1, it can still hear the transmission of L_2. The new beacon information leads to the reduction of the ROI.

$S_i(R(j), j)$. To illustrate the ROI reduction, consider Figure 7(a), where locators L_1, L_2 transmit with their maximum power; sensor s computes: $ROI(1) = S_1(1) \cap S_2(1)$. In Figure 7(b), locators L_1, L_2 reduce their communication range by lowering their transmission power and re-transmit the updated beacons. While locator L_1 is out of range from sensor s and, hence, does not further refine the sensor's location, s can still hear locator L_2 and therefore, reduce the size of the ROI.

3. Hybrid approach: The combination of the variation of the antenna orientation and communication range leads to a dual dependency of the sector area $S_i(\theta_i(j), R(j), j)$. Such a dependency can also be interpreted as a limited mobility model for the locators. For a locator L_i moving in a confined area, the antenna orientation and communication range with respect to a static sensor varies, thus providing the sensor with multiple sector areas $S_i(j)$. The mobility model is characterized as limited, since the locator has to be within the range of the sensor for at least a fraction of its transmissions in order to provide the necessary localization information. We now present the algorithmic details of HiRLoc.

4.4 Securing the Beacon Transmissions

We now describe the mechanisms used to secure the beacons transmitted by the locators.

Encryption: All beacons transmitted from locators are encrypted with a globally shared symmetric key K_0. Although K_0 can easily be compromised with the compromise of a single sensor, this solution is adopted for resource efficiency reasons. Using K_0, Locators are able to broadcast the localization information, instead of unicasting the information to each sensor. Stronger broadcast authentication algorithms known for ad hoc networks, require the existence of a central authority and time synchronization among all nodes of the network [29]. In Section 5, we show

SeRLoc: Secure Range-Independent Localization Scheme

L : broadcast L_i : $\{ (X_i, Y_i) \| (\theta_1, \theta_2) \| (H^{n-j}(PW_i)) \| j \| IDL_i \}_{K_0}$
$LH_s = \{L_i : \|s - L_i\| \le R\} \cap \{H(H^{n-j}(PW_i)) = H^{n-j+1}(PW_i)\}$
s : define $A_s = [X_{max} - R, \ X_{min} + R, \ Y_{max} - R, \ Y_{min} + R]$
for k=1:res
  for w=1:res
    $g(k, w) = (x_{g_i}, y_{g_i}) = \left(X_{max} - R + k\frac{X_{max} - X_{min}}{res}, \ Y_{max} - R + w\frac{Y_{max} - Y_{min}}{res}\right)$
    for $z = 1 : |LH_s|$
      if $\{\|g(k, w) - L_z\| \le R\} \cap \{\theta_1 \le \angle g(k, w) \le \theta_2\}$
      $GST(k, w) = GST(k, w) + 1$
$MG_s = \{g(k, w) : \{k, w\} = \arg\max GST\}$

$$\tilde{s} : (x_{est}, y_{est}) = \left(\frac{1}{|MG_s|} \sum_{i=1}^{|MG_s|} x_{g_i}, \ \frac{1}{|MG_s|} \sum_{i=1}^{|MG_s|} y_{g_i}\right)$$

Fig. 8. The pseudocode of SeRLoc.

that sensors are able to detect attacks even if K_0 has been compromised, using consistency checks.

In addition to K_0, every sensor s shares a symmetric pairwise key K_{s,L_i} with every locator L_i, also preloaded. Since the number of locators deployed is relatively small, the storage requirement at the sensor side is within the storage constraints (a total of $|L|$ keys). For example, mica motes [24] have 128Kbytes of programmable flash memory. Using 64-bit RC5 [32] symmetric keys and for a network with 400 locators, a total of 3.2Kbytes of memory is required to store all the keys of the sensor with every locator. In order to save storage space at the locator (locators would have to store $|S|$ keys), pairwise keys K_{s,L_i} are derived by a master key K_{L_i}, using a pseudorandom function h [37], and the unique sensor ID_s: $K_{s,L_i} = h_{K_{L_i}}(ID_s)$.

Locator ID Authentication: We use the following scheme based on *efficient one-way hash chains* [15], to provide locator ID authentication. Each locator L_i has a unique password PW_i, blinded with the use of a collision-resistant hash function such as SHA1 [37]. Due to the collision resistance property, it is computationally infeasible for an attacker to find a PW_j, such that $H(PW_i) = H(PW_j)$, $PW_i \ne PW_j$. The hash sequence is generated using the following equation:

$$H^0 = PW_i, \ \ H^i = H(H^{i-1}), \ \ i = 1, \cdots, n,$$

with n being a large number and H^0 never revealed to any sensor. Each sensor is preloaded with a table containing the ID of each locator and the corresponding hash value $H^n(PW_i)$. For a network with 400 locators, we need 9 bits to represent locator IDs. In addition, collision-resistant hash functions such as SHA1 [37] have a 160-bit output. Hence, the storage requirement of the hash table at any sen-

HiRLoc: High-resolution Robust Localization Scheme

L : **broadcast** L_i : $\{ (X_i, Y_i) \parallel (\theta_{i,1}(1), \theta_{i,2}(1)) \parallel R_i(1) \}$
s : **define** $LH_s = \{ L_i : \|s - L_i\| \le R_i(1) \}$
s : **define** $A_s = [X_{max} - R_i(1), X_{min} + R_i(1), Y_{max} - R_i(1), Y_{min} + R_i(1)]$
s : **store** $S \leftarrow S_i(1) : \{ (X_i, Y_i) \parallel (\theta_{i,1}(1), \theta_{i,2}(1)) \parallel R_i(1) \}, \ \forall L_i \in LH_s$
$j = 1$
for $k = 1 : Q - 1$
  for $w = 1 : N - 1$
    $j + +$
    L **reduce** $R(j) = R(j-1) - \frac{R(1)}{N}$
    L : **broadcast** L_i : $\{ (X_i, Y_i) \parallel (\theta_{i,1}(j), \theta_{i,2}(j)) \parallel R_i(j) \}$
    s : **replace** $S \leftarrow S_i(j) : \{ (X_i, Y_i) \parallel (\theta_{i,1}(j), \theta_{i,2}(j)) \parallel R_i(j) \},$
      $\forall L_i : \|s - L_i\| \le R_i(j) \bigcap L_i \in LH_s$
  endfor
  $j + +$
  $R_i(j) = R_i(1), \ \forall L_i \in LH_s$
  L **rotate** $\theta_i(j) = \{ \theta_{i,1}(j-1) + \frac{2\pi}{MQ}, \theta_{i,2}(j-1) + \frac{2\pi}{MQ} \}$
  L : **broadcast** L_i : $\{ (X_i, Y_i) \parallel (\theta_{i,1}(j), \theta_{i,2}(j)) \parallel R_i(j) \}$
  s : **store** $S \leftarrow S_i(j) : \{ (X_i, Y_i) \parallel (\theta_1(j), \theta_2(j)) \parallel R_i(j) \}, \ \forall L_i : \|s - L_i\| \le$
$R(j) \bigcap L_i \in LH_s$
endfor
s : **compute** $ROI = \bigcap_{i=1}^{|S|} S_i$

Fig. 9. The pseudocode of HiRLoc.

sor is 8.45Kbytes^2. To reduce the storage needed at the locators, we employ an efficient storage/computation method for hash chains of time/storage complexity $\mathcal{O}(\log^2(n))$ [8].

The jth broadcasted beacon from locator L_i includes the hash value $H^{n-j}(PW_i)$, along with the index j. Every sensor that hears the beacon accepts the message only if $H(H^{n-j+1}(PW_i)) = H^{n-j}(PW_i)$. After verification, the sensor replaces $H^{n-j+1}(PW_i)$ with $H^{n-j}(PW_i)$ in its memory and increases the hash counter by one so as to perform only one hash operation in the reception of the next beacon from the same locator L_i. The index j is included in the beacons so that sensors can resynchronize with the current published hash value in case of loss of some intermediate hash values. The beacon of locator L_i has the following format:

$$L_i : \{ (X_i, Y_i) \parallel (\theta_1, \theta_2) \parallel (H^{n-j}(PW_i)) \parallel j \parallel ID_{L_i} \}_{K_0},$$

where $\parallel$ denotes the concatenation operation and $\{m\}_K$ denotes the encryption of message m with key K. Note that our method does not provide end-to-end locator authentication, but only guarantees authenticity for the messages received from locators directly heard to a sensor. This condition is sufficient to secure our localization

---

[2] The required storage at each sensor in order to store 400 64-bit RC5 keys, 400 160-bit SHA1 hash values for secure communication with 400 locators is now 11.65Kbytes.

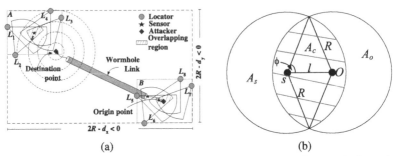

(a)   (b)

Fig. 10. (a) Wormhole attack: an attacker records beacons in area B, tunnels them via the wormhole link in area A and rebroadcasts them. (b) Computation of the common area A_c, where locators are heard to both s, O.

scheme against possible attacks. The pseudocode for SeRLoc is presented in Figure 8. The pseudocode for HiRLoc is presented in Figure 9.

5 Threat Analysis

In this section, we show how SeRLoc and HiRLoc are resilient to the attacks described in Section 2. Note that our goal to allow sensors to determine their location, even in the presence of attacks and not to prevent attacks that may be harmful in other network protocols.

5.1 The Wormhole Attack

Threat Model

In the case of our location estimation process an attacker launching a wormhole attack records the beacons transmitted from locators at the origin point of the attack and replays them at the destination point, thus providing false localization information to the sensors attacked. In Figure 10(a), the attacker records beacons at region B, tunnels them via the wormhole link in region A, and replays them, thus leading sensor s to believe that it can hear locators $\{L_1 \sim L_8\}$.

Detecting Wormholes

In the case of a wormhole attack, the cryptography used to secure the beacon transmissions, and to authenticate the source of the information is not violated. Wormholes violate the geometry of the network by enabling the propagation of messages at a distance longer than the communication range [30]. Hence, in the case of the wormhole attack, additional non-cryptographic mechanisms are needed to detect the geometry violation. We now show how a sensor can detect a wormhole attack using two consistency check properties: the *single message/sector per locator* property and the *communication range constraint* property.

Single Message/Sector per Locator Property: The origin point O of the wormhole attack defines the set of locators LH_s^r replayed to the sensor s under attack. The location of the sensor defines the set of locators LH_s^d directly heard to the sensor s, with $LH_s = LH_s^r \cup LH_s^d$. Based on the single message/sector per locator property we show that the wormhole attack is detected when $LH_s^r \cap LH_s^d \neq \emptyset$.

Lemma 1. *Single message per locator/sector property: reception of multiple messages authenticated with the same hash value is due to replay, multipath effects, or imperfect sectorization.*

Proof. In the absence of any attack, a sensor can hear multiple sectors due to multipath effects. In addition, a sensor located at the boundary of two sectors can also hear multiple sectors even if there is no multipath or attack. We assume that the same but fresh hash value is used to authenticate them per beacon transmission. Hence, sensors will only accept the first message arriving from any sector of the same locator, per transmission. Due to the use of an identical but fresh hash in all sectors per transmission, if an adversary replays a message from any sector of a locator directly heard by the sensor under attack, the sensor will have already received the hash via the direct path and, hence, detect the attack and reject the message.

If we consider reception of multiple messages containing the same hash value due to multipath effects or imperfect sectorization to be a replay attack, a sensor will always assume it is under attack when it receives messages with the same hash value. Hence, an adversary launching a wormhole attack will always be detected if it replays a message from locator $L_i \in LH_s^d$, that is, if $LH_s^r \cap LH_s^d \neq \emptyset$. In Figure 11(a), A_s denotes the area where, $L_i \in LH_s^d$ (circle of radius R centered at s), A_o denotes the area where $L_i \in LH_s^r$ (circle of radius R centered at O), and the shaded area A_c denotes the common area $A_c = A_s \cap A_o$.

Claim. The detection probability $P(SG)$ due to the single message/sector per locator property is equal to the probability that at least one locator lies within an area of size A_c, and is given by:

$$P(SG) = 1 - e^{-\rho_L A_c}, \quad \text{with } A_c = 2R^2\phi - Rl\sin\phi, \quad \phi = \cos^{-1}\frac{l}{2R}. \quad (10)$$

with l as the distance between the origin point and the sensor under attack.

Proof. If a locator L_i lies inside A_c, it is less than R units away from a sensor s and, therefore $L_i \in LH_s^d$. Locator L_i is also less than R units away from the origin point of the attack O, and therefore, $L_i \in LH_s^r$. Hence, if a locator lies inside A_c, $LH_s^r \cap LH_s^d \neq \emptyset$, and the attack is detected due to the single message/sector per locator property. The detection probability $P(SG)$ is equal to the probability that at least one locator lies within A_c. If LH_{A_c} denotes the set of locators located within area A_c then:

200   Loukas Lazos and Radha Poovendran

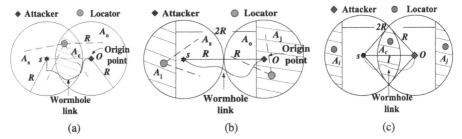

Fig. 11. (a) Single message/sector per locator property: a sensor s cannot hear two messages authenticated with the same hash value. (b) Communication range violation property: a sensor s cannot hear two locators more than $2R$ apart. (c) Combination of the two properties for wormhole detection.

$$P(SG) = P(|LH_{A_c}| \geq 1) = 1 - P(|LH_{A_c}| = 0) = 1 - e^{-\rho_L A_c}, \quad (11)$$

where A_c can be computed from Figure 10(b) to be:

$$A_c = 2R^2\phi - Rl\sin\phi, \quad \phi = \cos^{-1}\frac{l}{2R}, \quad (12)$$

with $l = \|s - O\|$.

Figure 12(a) presents the detection probability $P(SG)$ vs. the locator density ρ_L and the distance $\|s - O\|$ between the origin point and the sensor under attack, normalized over R. We observe that if $\|s - O\| \geq 2R$, then $A_c = 0$, and the use of the single message/sector per locator property is not sufficient to detect a wormhole attack. For distances $\|s - O\| \geq 2R$, a wormhole attack can be detected using the following communication range constraint property.

Communication Range Violation Property: Given the coordinates of node s, all locators LH_s heard by s should lie within a circle of radius R, centered at s. Since node s is not aware of its location it relies on its knowledge of the locator-to-sensor communication range R to verify that the set LH_s satisfies Lemma 2.

Lemma 2. *Communication range constraint property: A sensor s cannot hear two locators $L_i, L_j \in LH_s$, more than $2R$ apart, that is, $\|L_i - L_j\| \leq 2R, \forall L_i, L_j \in LH_s$.*

Proof. Any locator $L_i \in LH_s$ has to lie within a circle of radius R, centered at the sensor s (area A_s in Figure 11(b)), $\|L_i - s\| \leq R, \forall L_i \in LH_s$. Hence,

$$\|L_i - L_j\| = \|L_i - s + s - L_j\| \leq \|L_i - s\| + \|s - L_j\| \leq R + R = 2R. \quad (13)$$

Using the coordinates of LH_s, a sensor can detect a wormhole attack if the communication range constraint property is violated. We now compute the detection probability $P(CR)$ due to the communication range constraint property.

Claim. A wormhole attack is detected due to the communication range constraint property, with a probability:

$$P(CR) \geq \left(1 - e^{-\rho_L A_i^*}\right)^2, \quad A_i^* = x\sqrt{R^2 - x^2} - R^2 \tan^{-1}\left(\frac{x\sqrt{R^2 - x^2}}{x^2 - R^2}\right)(14)$$

where $x = \frac{\|s - O\|}{2}$.

Proof. Consider Figure 11(b), where $\|s - O\| = 2R$. If any two locators within A_s, A_o have a distance larger that $2R$, a wormhole attack is detected. Though $P(CR)$ is not easily computed analytically, we can obtain a lower bound on $P(CR)$ by considering the following event. In Figure 11(b), the vertical lines defining shaded areas A_i, A_j, are perpendicular to the line connecting s, O, and have a separation of $2R$. If there is at least one locator L_i in the shaded area A_i and at least one locator L_j in the shaded area A_j, then $\|L_i - L_j\| \rangle 2R$ and the attack is detected. Note that this event does not include all possible locations of locators for which $\|L_i - L_j\| \rangle 2R$, and hence it yields a lower bound. If $\mathcal{LH}_{A_i,A_j}$ denotes the event $\left(|LH_{A_i}|\rangle 0 \cap |LH_{A_j}|\rangle 0\right)$ then,

$$P(CR) = P(\|L_i - L_j\|\rangle 2R, L_i, L_j \in LH_s)$$
$$\geq P(CR \bigcap \mathcal{LH}_{A_i,A_j}) \tag{15}$$
$$= P\left(CR \mid \mathcal{LH}_{A_i,A_j}\right) P(\mathcal{LH}_{A_i,A_j}) \tag{16}$$
$$= P(\mathcal{LH}_{A_i,A_j}) \tag{17}$$
$$= (1 - e^{-\rho_L A_i})(1 - e^{-\rho_L A_j}), \tag{18}$$

where (15) follows from the fact that the probability of the intersection of two events is always less or equal to the probability of one of the events, (16) follows from the definition of the conditional probability, (17) follows from the fact that when $\mathcal{LH}_{A_i,A_j}$ is true, we always have a communication range constraint violation ($P(CR \mid \mathcal{LH}_{A_i,A_j}) = 1$), and (18) follows from the fact that A_i, A_j are disjoint areas and that locators are randomly deployed.

We can maximize the lower bound of $P(CR)$, by finding the optimal values A_i^*, A_j^*. In fact it can be shown that the lower bound in (18) attains its maximum value when $A_i^* = \max_i\{A_i\}$ subject to the constraint $A_i = A_j$ (A_i, A_j are symmetric) [17]. and is given by:

$$A_i^* = A_j^* = x\sqrt{R^2 - x^2} - R^2 \tan^{-1}\left(\frac{x\sqrt{R^2 - x^2}}{x^2 - R^2}\right), \quad \text{and } x = \frac{\|s - O\|}{2}. \tag{19}$$

Inserting (19) into (18) yields the required result: $P(CR) \geq \left(1 - e^{-\rho_L A_i^*}\right)^2$.

In Figure 12(b), we show the maximum lower bound on $P(CR)$ vs. the locator density ρ_L, and the distance $\|s - O\|$ normalized over R. The lower bound on

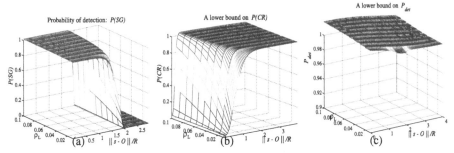

Fig. 12. Wormhole detection probability based on, (a) the single message/sector per locator property: $P(SG)$. (b) A lower bound on the wormhole detection based on the communication range violation property: $P(CR)$. (c) A lower bound on the wormhole detection probability for SeRLoc.

$P(CR)$ increases with the increase of $\|s - O\|$ and attains its maximum value for $\|s-O\| = 4R$ when $A_i^* = A_j^* = \pi R^2$. For distances $\|s-O\|\rangle 4R$ a wormhole attack is always detected based on the communication range constraint property, since any locator within A_o will be more than $2R$ apart from any locator within A_s.

Detection Probability P_{det} of the Wormhole Attack: We now combine the two detection mechanisms, namely the single message/sector per locator property and the communication range constraint property for computing the detection probability of a wormhole attack.

Claim. The detection probability of a wormhole attack is lower bounded by $P_{det} \geq (1 - e^{-\rho_L A_c}) + (1 - e^{-\rho_L A_i^*})^2 e^{-\rho_L A_c}$.

Proof. In the computation of the communication range constraint property, by setting $A_i = A_j$ and maximizing A_i regardless of the distance $\|s - O\|$, the areas A_i, A_j, and A_c do not overlap as shown in Figure 11(c). Hence, the corresponding events of finding a locator at any of these areas are independent and we can derive a lower bound on the detection probability P_{det} by combining the two properties.

$$P_{det} = P(SG \cup CR) = P(SG) + P(CR) - P(SG)P(CR)$$
$$= P(SG) + P(CR)\,(1 - P(SG))$$
$$\geq (1 - e^{-\rho_L A_c}) + (1 - e^{-\rho_L A_i^*})^2 e^{-\rho_L A_c}. \qquad (20)$$

The left side of (20) is a lower bound on P_{det} since $P(CR)$ was also lower bounded.

In Figure 12(c), we show the lower bound on P_{det} vs. the locator density ρ_L and the distance $\|s - O\|$ normalized over R. For values of $\|s - O\|\rangle 4R$, $P_{CR} = 1$, since any $L_i \in LH_s^d$ will be more than $2R$ away from any $L_j \in LH_s^r$ and hence, the wormhole attack is always detected. From Figure 12(c), we observe that a wormhole attack is detected with a probability very close to unity, independent of the origin and destination point of the attack.

Attach to Closer Locator Algorithm (ACLA)

s : **broadcast** $\{ \eta_s \parallel ID_s \}$
if L_i *hears* $\{ \eta_s \parallel ID_s \}$ **reply**
   L_i : $\{ \eta_s \parallel (X_i, Y_i) \parallel (\theta_1, \theta_2) \parallel (H^{n-j}(PW_i)) \parallel j \parallel ID_{L_i} \}_{K_{s,L_i}}$
L'_i : *first authentic reply from a locator.*
$LH^d_s = \{ L_i \in LH_s : sector\{L_i\} \; intersects \; sector\{L'_i\}\}$
s : **execute** *SeRLoc with* $LH_s = LH^d_s$

Fig. 13. The pseudocode of ACLA.

Location Resolution Algorithm: Although a wormhole can be detected using one of the two detection mechanisms, a sensor s under attack cannot distinguish the set of locators directly heard LH^d_s from the set of locators replayed LH^r_s and hence, estimate its location. To resolve the location ambiguity sensor s executes the *Attach to Closer Locator Algorithm* (ACLA).

Assume that a sensor authenticates a set of locators $LH_s = LH^d_s \cup LH^r_s$, but detects that it is under attack.

- *Step 1:* Sensor s broadcasts a randomly generated nonce η_s and its ID_s.
- *Step 2:* Every locator hearing the broadcast of sensor s replies with a beacon that includes localization information and the nonce η_s, encrypted with the pairwise key K_{s,L_i} instead of the broadcast key K_0. The sensor identifies the locator L'_i that replies first with an authentic message that includes η_s.
- *Step 3:* Sensor s identifies the set LH^d_s as all the locators whose sectors overlap with the sector of L'_i, and executes SeRLoc with $LH_s = LH^d_s$.

The pseudocode of ACLA is presented in Figure 13. Note that the closest locator to sensor s will always reply first if it directly hears the broadcast from s, and not through a replay from an adversary. In order for an adversary to force sensor s to accept set LH^r_s as the valid locator set, it can only replay the nonce η_s to a locator $L_i \in LH^r_s$, record the reply, tunnel via the wormhole and replay it in the vicinity of s. However, a reply from a locator in LH^r_s will arrive later than any reply from a locator in LH^d_s, since locators in LH^r_s are further away from s than locators in LH^d_s.

To execute ACLA, a sensor must be able to communicate bidirectionally with at least one locator. The probability $P_{s \to L}$ of a sensor having a bidirectional link with at least one locator and the probability P_{bd} that *all* sensors can bidirectionally communicate with at least one locator can be computed as:

$$P_{s \to L} = 1 - e^{-\rho_L \pi r^2 G^{\frac{2}{\gamma}}}, \quad P_{bd} = (1 - e^{-\rho_L \pi r^2 G^{\frac{2}{\gamma}}})^{|S|}. \tag{21}$$

Hence, we can select the system parameters ρ_L, G so every sensor has a bidirectional link with at least one locator with any desired probability.

5.2 Impersonation (Sybil) Attack

An adversary can launch an impersonation attack against SeRLoc or HiRLoc if it successfully impersonates locators. Since sensors are pre-loaded with valid locator IDs along with the hash values corresponding to the head of the reversed hash chain for each locator, an adversary can only impersonate locators by compromising the globally shared key K_0.

Once K_0 has been compromised, the adversary has access to both locators IDs, the hash chain values published by the locators, as well as the coordinates of the locators. Since sensors always have the latest published hash values from the locators that they directly hear, an adversary can only impersonate locators that are not directly heard to the sensors under attack. The adversary can generate bogus beacons, attach an already published hash value from a locator not heard by the sensor under attack, and encrypt it with the compromised K_0.

Depending on the type of locators used, static or mobile, an adversary can impersonate locators in different ways. If the locators are static and their location is known before deployment, the coordinates of all locators can be preloaded to every sensor. Hence, the adversary cannot advertise a location that is different from the actual coordinates of an impersonated locator. In such a case, the Sybil attack is equivalent to a replay attack since the adversary cannot alter the content of the beacons[3]. If the locators are mobile, or their coordinates cannot be preloaded to the sensors before deployment, the adversary can place the impersonated locators to arbitrary positions. Hence, by impersonating a higher number of locators than the ones directly heard by the sensor under attack, the adversary can compromise the majority vote scheme of SeRLoc and displace the sensor.

Defense against the Sybil Attack: Though we do not provide a mechanism to prevent an adversary from impersonating locators except for the ones directly heard by a sensor, we can still determine the position of sensors in the presence of Sybil attack. In the case where sensors know a priori the coordinates of the locators, the sensor can detect the Sybil attack with the same mechanisms used for the wormhole attack, since the Sybil attack becomes a beacon replay. In the case where the coordinates of the locators are not preloaded to the sensors, an adversary can manipulate the coordinates of the impersonated locators, so that neither of the wormhole defense mechanisms detect an anomaly. The adversary needs to impersonate more than LH_s^d locators in order to displace the sensor s. To avoid sensor displacement we rely on the invariability of the locator deployment statistics to detect locator impersonation.

Since the locator density ρ_L is known before deployment, we can select a threshold value L_{max} as the maximum allowable number of locators heard by each sensor. If a sensor hears more than L_{max} locators, it assumes that it is under attack and executes ACLA to determine its position. The probability that a sensor s hears more than L_{max} locators is given by:

---

[3] The adversary can alter the angle information contained in the beacon. However, this is equivalent to replaying the beacon of another sector.

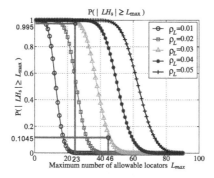

Fig. 14. $P(|LH_s| \geq L_{max})$, vs. L_{max} for varying locator densities ρ_L. When $\rho_L = 0.03$, a choice of $L_{max} = 46$ allows a sensor to localize itself when under Sybil attack with a probability $P(|LH_s| \geq 23) = 0.995$, while the false positive alarm probability is $P(|LH_s|\rangle 46) = 0.1045$.

$$P(|LH_s| \geq L_{max}) = 1 - P(|LH_s|\langle L_{max}\rangle)$$

$$= 1 - \sum_{i=0}^{L_{max}-1} \frac{(\rho_L \pi R^2)^i}{i!} \, e^{-\rho_L \pi R^2}. \qquad (22)$$

Using (22), we can select the value of L_{max} so that there is a very small probability for a sensor to hear more than L_{max} locators, while there is a very high probability for a sensor to hear more than $\frac{L_{max}}{2}$ locators. If a sensor hears more than L_{max} locators without being under attack, the detection mechanism will result in a false positive alarm and force the sensor to execute ACLA to successfully locate itself. However, if a sensor hears less than $\frac{L_{max}}{2}$, the sensor is vulnerable to a Sybil attack. Therefore, we must select a threshold L_{max} so that any sensor hears less than $\frac{L_{max}}{2}$ locators with a probability very close to zero.

In Figure 14, we show $P(|LH_s| \geq L_{max})$ vs. L_{max}, for varying locator densities ρ_L. Based on Figure 14, we can select the appropriate L_{max} for each value of ρ_L. For example, when $\rho_L = 0.03$, a choice of $L_{max} = 46$ allows a sensor to localize itself when under Sybil attack with a probability $P(|LH_s| \geq 23) = 0.995$, while the false positive alarm probability is $P(|LH_s|\rangle 46) = 0.1045$.

5.3 Compromised Network Entities

In this section, we examine the robustness of SeRLoc and HiRLoc to compromised network entities. We consider a sensor node or a locator node to be compromised if an attacker assumes full control over the behavior of the node and knows all the keys stored at the compromised node.

Compromised Sensors: Though sensors are assumed to be easier to compromise, an attacker has no incentive to compromise sensors, since they do not actively participate in the localization procedure. The only benefit in compromising a sensor is to gain access to the globally shared key K_0.

Compromised Locators: An adversary that compromises a locator L_i gains access to the globally shared key K_0, the pairwise keys K_{s,L_i} shared between the locator and every sensor, as well as all the hash values of the locator's hash chain. By compromising a single locator, the adversary can displace any sensor, by impersonating the compromised locator from a position closer to the sensor under attack compared to the closest legitimate locator. The adversary impersonates multiple locators in order to force location ambiguity to the sensor under attack. Once the attack is detected, sensor s executes ACLA to resolve its location ambiguity. Since the adversary is closer to the sensor s than the closest legitimate locator, its reply will arrive to s first. Hence, s will assume that the impersonated set of locators is the valid one and will be displaced.

To avoid sensor displacement by a single locator compromise, we can intensify the resilience to locator compromise by involving more than one locators in the location resolution algorithm at the expense of higher communication overhead. A sensor s under attack, can execute the *Enhanced Location Resolution Algorithm* (ELRA) that follows.

- *Step 1:* Sensor s broadcasts a randomly generated nonce η_s, the set of locators heard LH_s and its ID_s.

$$s : \{ \eta_s \parallel LH_s \parallel ID_s \}. \tag{23}$$

- *Step 2:* Every locator L_i receiving the broadcast from s appends its coordinates, the next hash value of its hash chain and its ID_{L_i}, encrypts the message with K_0 and re-broadcasts the message to all sectors.

$$L_i : \{\eta_s \parallel LH_s \parallel ID_s \parallel (X_i, Y_i) \parallel H^{n-k}(PW_i) \parallel \parallel j \parallel ID_{L_i} \}_{K_0}. \tag{24}$$

- *Step 3:* Every locator receiving the rebroadcast, verifies the authenticity of the message, and that the transmitting locator is within its range. If the verification is correct and the receiving locator belongs to LH_s, the locator broadcasts a new beacon with location information and the nonce η_s encrypted with the pairwise key K_{s,L_i} with sensor s.

$$L_i : \{ \eta_s \parallel (X_i, Y_i) \parallel (\theta_1, \theta_2) \parallel H^{n-k}(PW_i) \parallel j \parallel ID_{L_i} \}_{K_{s,L_i}}. \tag{25}$$

- *Step 4:* The sensor collects the first L_{max} authentic replies from locators and executes SeRLoc with $LH_s = L_{max}$.

The pseudocode for the enhanced location resolution algorithm is presented in Figure 15. Note that for a locator to hear the sensor's broadcast, it has to be within a range $r_{sL} = rG^{\frac{1}{\gamma}}$ from the sensor. Furthermore, in order for a the sensor to make the correct location estimate, all locators within a range R from s need to provide new beacon information.

Claim. Every locator positioned within R from a sensor s is within the range of any locator positioned at a distance r_{sL} from the sensor s.

Enhanced Location Resolution Algorithm (ELRA)

---

s : **broadcast** $\{\ \eta_s\ \|\ LH_s\ \|\ ID_s\ \}$

$RL_s = \{L_i : \|s - L_i\| \leq r_{sL}\}$

RL_s : **broadcast** $\{\ \eta_s\ \|\ LH_s\ \|\ ID_s\ \|\ (X_i, Y_i)\ \|\ H^{n-k}(PW_i)\ \|\ j\ \|\ ID_{L_i}\ \}_{K_0}\ BL_s = \{L_i : \|RL_s - L_i\| \leq r_{LL}\} \bigcap LH_s$

BL_s : **broadcast** $\{\ \eta_s\ \|\ (X_i, Y_i)\ \|\ (\theta_1, \theta_2)\ \|\ H^{n-k}(PW_i)\ \|\ j\ \|\ ID_{L_i}\ \}_{K_{s,L_i}}$

s : **collect** *first L_{max} authentic beacons from BL_s*

s : **execute** *SeRLoc with collected beacons*

Fig. 15. The pseudocode for the enhanced location resolution algorithm (ELRA).

Proof. For any locator positioned at a distance r_{sL} from the sensor s to reach any locator positioned at a distance R from sensor s, the following condition has to hold: $r_{LL} \geq R + r_{sL}$.

$$RG^{\frac{2}{\gamma}} \geq R + rG^{\frac{1}{\gamma}} \Rightarrow \frac{R}{rG^{\frac{1}{\gamma}}}(G^{\frac{2}{\gamma}} - 1) \geq 1. \qquad (26)$$

Since $R \geq rG^{\frac{2}{\gamma}}$ by assumption, and $G^{\frac{2}{\gamma}} \geq 1$, the left side of (26) is always greater than one.

Each beacon broadcast from a locator has to include the nonce η_s initially broadcasted by the sensor and be encrypted with the pairwise key between the sensor and the locator. Hence, given that the sensor has at least $\frac{L_{max}}{2}$ locators within range R with very high probability (see Figure 14), the adversary has to compromise at least $\left(\frac{L_{max}}{2} + 1\right)$ locators, in order to compromise the majority vote scheme of SeR-Loc. In addition, the attacker has to possess the hardware capabilities to process and transmit $\left(\frac{L_{max}}{2} + 1\right)$ replies before $\frac{L_{max}}{2}$ replies from valid locators reach the sensor under attack. Our enhanced location resolution algorithm significantly increases the resilience of SeRLoc to locator compromise at the expense of higher communication overhead at the locators.

6 Performance Evaluation

In this section, we evaluated the performance of SeRLoc and HiRLoc with respect to their localization accuracy. To emulate the conditions of a real deployment, we also evaluated SeRLoc under error in the locators' coordinates and false estimation of the antenna sector that includes the sensors and empirically showed that SeRLoc is robust against both sources of error.

6.1 Simulation Setup

We randomly distributed 5,000 sensors within a $100 \times 100m^2$ rectangular area. We also randomly placed locators within the same area and computed the average localization error as:

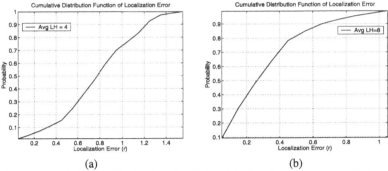

Fig. 16. The cumulative distribution function (cdf) of the localization error of SeRLoc when $M = 3$ and, (a) $\overline{LE} = 4$, (b) $\overline{LH} = 8$.

$$\overline{LE} = \frac{1}{|S|} \sum_i \frac{\|\tilde{s}_i - s_i\|}{r}, \tag{27}$$

where S is the set of sensors, $\tilde{s}_i$ is the sensor estimated position, s_i is the real position and r is the sensor-to-sensor communication range.

6.2 Localization Error vs. Locators Heard

In our first experiment, we investigated the impact of the average number of locators heard $\overline{LH}$ on the localization error. In Figures 16(a) and (b), we show the cumulative distribution function (cdf) of the localization error for SeRLoc when 3-sector antennas are used at the locators and the average number of locators heard are $\overline{LH} = 6$ and $\overline{LH} = 8$, respectively. We observe that for $\overline{LH} = 4$, the error is more evenly distributed among its possible values with 90% of the sensors having an error of less than 1.2r, while for $\overline{LH} = 8$, more than 90% of the sensors have an error smaller than 0.7r.

The highest localization error occurs when a sensor hears only one locator L_i and is R units away from L_i. The probability for such an event to occur can be set to an arbitrary small value by deploying a sufficient number of locators. For example, when $\overline{LH} = 8$, the probability for a sensor to hear just one locator is $P(|LH| = 1) = 2.7 \times 10^{-3}$.

In Figure 17(a) we show the ROI vs. the number of antenna rotations, and for varying $\overline{LH}$, when 3-sector antennas are used at each locator. Note that the ROI is normalized over the size of the ROI given by SeRLoc denoted by ROI(1) (no antenna rotation). From Figure 17(a), we observe that even a single antenna rotation, reduces the size of the ROI by more than 50%, while three antenna rotations reduce the size to $ROI(4) = 0.12 ROI(1)$, when $\overline{LH} = 5$. A reduction of 50% in the size of the ROI by a single antenna rotation means that one can deploy half the locators compared to SeRLoc and achieve the same localization accuracy by just rotating the locators' antennas once. The savings in locators are significant considering that the reduction in hardware requirements comes at no additional communication cost.

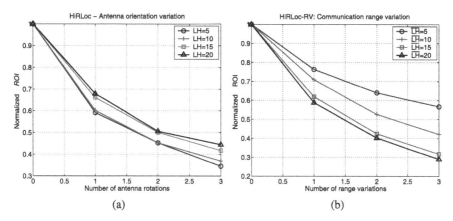

Fig. 17. Normalized ROI vs. number of antenna rotations for varying $\overline{LH}$. The ROI is normalized with respect to the ROI acquired with no variation of the antenna orientation (SeRLoc). (b) ROI vs. number of range reductions for varying $\overline{LH}$.

We also observe that as $\overline{LH}$ increases, HiRLoc provides diminishing returns. This is due to the fact that when the number of locators heard at each sensor is high, SeRLoc already provides a good location estimate (small ROI) and, hence, the margin for reduction of the ROI size is limited. In Figure 17(b) we show the normalized ROI vs. the number of communication range reductions, and for different $\overline{LH}$ values, when locators are equipped with 3-sector antennas.

From Figure 17(b), we observe that the communication range variation, though significantly improves the system performance, does not achieve the same ROI reduction as the antenna orientation variation[4]. This behavior is explained by the fact that the gradual reduction of the communication range reduces the number of beacons heard at each sensor, in contrast with the antenna orientation variation case where the same number of locators is heard at the sensors at each antenna rotation. In addition, we observe that greater ROI reduction occurs when the $\overline{LH}$ at each locator is high. This is justified by considering that a higher $\overline{LH}$ allows for more sectors with lower communication range to intersect and hence, smaller ROI.

6.3 Localization Error vs. Sector Error

Sensors may be located close to the boundary of two sectors of a locator, or be deployed in a region with high multipath effects. In such a case, a sensor may falsely assume that it is located in another sector, than the actual sector that includes it. We refer to this error as sector error (SE) defined as:

$$SE = \frac{\#\ \text{of sectors falsely estimated}}{LH}. \tag{28}$$

---

[4] The comparison is valid for the same number of $\overline{LH}$, the same number of antenna sectors and the same number of variations in the antenna rotation and communication range, respectively.

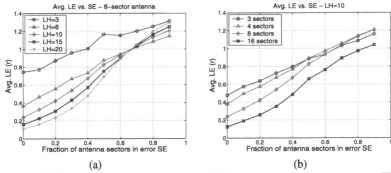

(a)   (b)

Fig. 18. (a) $\overline{LE}$ vs. sector error SE for varying $\overline{LH}$. (b) Average localization error $\overline{LE}$ vs. sector error SE for varying number of antenna sectors for a network of $|S| = 5,000$ and $\frac{R}{r} = 10$.

A sector error of 0.5 indicates that *every* sensor falsely estimated the sectors of half the locators heard. In Figure 18(a), we show the $\overline{LE}$ vs. the SE for varying $\overline{LH}$, and 8-sector antennas. We observe that the $\overline{LE}$ does not grow significantly large (larger than the sensor communication range r), until a fraction of 0.7 of the sectors are falsely estimated.

SeRLoc is resilient to sector error due to the majority vote scheme employed in the determination of the overlapping region. Even if a significant fraction of sectors are falsely estimated, these sectors do not overlap in the same network area and hence a score low in the grid-sector table.

Note that for a $SE \rangle 0.7$, $\overline{LE}$ increases with $\overline{LH}$. When the SE grows beyond a threshold, the falsely estimated sectors dominate in the location determination. As $\overline{LH}$ grows, the falsely estimated overlapping region, shrinks due to the higher number of overlapping sectors. Therefore, the CoG that defines the sensor's location gets further apart than the actual sensor location.

In Figure 18(b), we show the $\overline{LE}$ vs. SE for $\overline{LH} = 10$ and varying number of antenna sectors. We observe that the narrower the antenna sector the smaller the $\overline{LE}$, even in the presence of SE. For a small SE the overlapping region is dominated by the correctly estimated sectors and shrinks with increasing antenna sectors. For large SE the overlapping region is dominated by the false sectors and an increase in $\overline{LH}$ does not reduce the $\overline{LE}$.

7 Related Work

7.1 Related Work

An extensive literature exists for location estimation schemes for WSN in a benign environment [4, 10, 12, 25, 27, 31, 34–36]. Recently, a number of articles have appeared addressing the problem of sensor location estimation and verification in an adversarial setting [3, 5, 7, 14, 17–22, 33].

Sastry et al. [33] proposed the $ECHO$ protocol for verifying the location claim of a node, using a challenge response scheme and a combination of RF and Ultrasound signals. $ECHO$ is based on a distance bounding protocol proposed by Brands and Chaum [3]. Čapkun and Hubaux proposed Verifiable Multilateration (VM) for securing range-based localization schemes [5]. In VM, a node must verify its distance to at least three reference points in order to securely estimate its position. Čapkun et al. also proposed a location verification method based on hidden reference points that can verify the validity of the location claims of nodes [7].

Liu et al. [23] proposed an attack-resistant location estimation technique that can filter bogus beacon information provided that the majority of significant majority of beacons is benign. Li et al. [21] discuss a variety of attacks specific to the localization process and propose robust statistical methods that provide attack resistant localization. Finally, Kuhn [14] has proposed an asymmetric security mechanism for securing GPS-like navigation signals.

7.2 Open Problems

While the schemes that have been proposed for secure location estimation in WSNs [5, 7, 17–22, 33] are a significant step forward in providing a transparent and secure localization service, several problems remain open. The dependency of the location estimation schemes to physical characteristics such as received signal strength [1], time of arrival or time difference of arrival [27, 34], allows side-channel attacks not related to the strength of the cryptographic primitives used to secure the communication [19, 21, 22].

To combat side-channel attacks a series of consistency checks have been proposed [17–19, 22]. It remains an open problem which of the modalities of a sensor network used to detect attacks against the localization process are invariant to side-channel attacks. The ability of an adversary to alter the physical properties used for localization and distort the environment can significantly impact the localization accuracy.

Furthermore, current secure location estimation techniques do not provide any guarantee on the localization accuracy. The analytical evaluation of the localization error in the presence of adversaries is a problem requiring further investigation. Finally, most secure localization schemes studied localization for static sensor networks. Securing the location estimation process when the reference points, the sensors or both are mobile remains an open problem.

8 Conclusion

In this chapter, we have studied the problem of location estimation for WSN in an adversarial environment. We have demonstrated a series of attacks relevant to range-independent localization methods, such as the relay attack, the impersonation attack and compromise of reference points. We showed that securing the location estimation process requires not only securing the communication link between the reference points and the sensors, but also additional non-cryptographic consistency checks based on invariant properties such as the communication range or the network deployment statistics.

We proposed a range-independent, decentralized localization scheme called SeR-Loc that allows sensors to determine their location in an untrusted environment. We also proposed HiRLoc, a secure location estimation algorithm that relies on the superposition of location information over time to improve the location estimation accuracy. We analytically evaluated the probability of sensor displacement due to security threats in WSNs such as the wormhole attack, the Sybil attack, and compromise of network entities and showed that SeRLoc and HiRLoc provide accurate location estimation even in the presence of these threats. In doing so, we used the geometric and radio range information to detect the attacks on the localization.

Our performance evaluation studies showed that our algorithm are resilient to sources of error such as location error of reference points as well as error in the sector determination. We identified the integration of new modalities for consistency checks, the analytical evaluation of the location estimation error in the presence of adversaries and the secure location estimation for mobile sensor networks as areas of future research.

Acknowledgements

This work was supported in part by the following grants: CAREER grant from NSF ANI-0093187; ARO grant W911NF-05-1-0491; Collaborative Technology Alliance (CTA) from ARL DAAD19-01-2-0011[5].

References

1. P. Bahl and V. Padmanabhan, RADAR: An In-Building RF-Based User Location and Tracking System, In *Proc. of the IEEE INFOCOM 2000,* Tel-Aviv, Israel, March 2000.
2. S. Basagni, I. Chlamtac, V. Syrotiuk, and B. Woodward, A Distance Routing Effect Algorithm for Mobility (DREAM), In *Proc. of MOBICOM 1998,* Dallas, TX, USA, October 1998.
3. S. Brands and D. Chaum, Distance-bounding protocols, in Workshop on the Theory and Application of Cryptographic Techniques on Advances in Cryptology, Springer-Verlag New York, Inc., 1994, pp. 344-359.

---

[5] This document was prepared through the collaborative participation in the Communication and Networks Consortium sponsored by the U.S. Army Research Laboratory under the Collaborative Technology Alliance Program DAAD19-01-2-0011. The U.S. Government is authorized to reproduce and distribute reprints for Government purposes notwithstanding any copyright notation thereon. The views and conclusions contained in this documents are those of the authors and should not be interpreted as representing the official policies, either expressed or implied, of the Army Research Laboratory or the U.S. Government.

4. N. Bulusu, J. Heidemann and D. Estrin, GPS-less Low Cost Outdoor Localization for Very Small Devices, In *IEEE Personal Communications Magazine*, 7(5):28–34, October 2000.

5. S. Čapkun and J.-P. Hubaux, Secure Positioning in Wireless Networks, IEEE Journal on Selected Areas in Communications, vol. 24, no. 2, February 2006.

6. S. Čapkun, L. Buttyan, J. Hubaux, SECTOR: Secure Tracking of Node Encounters in Multi-hop Wireless Networks, in *Proc. of SASN 2003*, Fairfax, Virginia, October 2003.

7. S. Čapkun, M. Cagalj, and M. Srivastava, Secure Localization with Hidden and Mobile Base Stations, in Proceedings of the IEEE Conference on Computer Communications (InfoCom), 2006.

8. D. Coppersmith and M. Jakobsson, Almost optimal hash sequence traversal, In *Proc. of the FC 2002*, Lecture Notes in Computer Science, IFCA, Springer-Verlag, Berlin Germany, 2002.

9. N. Cressie, *Statistics for Spatial Data*, John Wiley & Sons, 1993.

10. L. Doherty, L. Ghaoui and K. Pister, Convex Position Estimation in Wireless Sensor Networks, In *Proc. of the IEEE INFOCOM 2001*, Anchorage, April 2001.

11. J. Douceur, The Sybil Attack, In *Proc of IPTPS 2002*, Cambridge, MA, USA, March 2002.

12. T. He, C. Huang, B. Blum, J. Stankovic and T. Abdelzaher, Range-Free Localization Schemes in Large Scale Sensor Network, In *Proc. of MOBICOM 2003*, San Diego, CA, USA, September 2003.

13. Y. Hu, A. Perrig, and D. Johnson, Packet Leashes: A Defense Against Wormhole Attacks in Wireless Ad Hoc Networks, In *Proc. of INFOCOM 2003*, San Francisco, CA, USA, April 2003.

14. M. G. Kuhn, An Asymmetric Security Mechanism for Navigation, in Proceedings of the Information Hiding Workshop, 2004.

15. L. Lamport, Password Authentication with Insecure Communication, In Communications of the ACM, 24(11):770-772, November 1981.

16. L. Lazos and R. Poovendran, Energy-Aware Secure Multicast Communication in Ad-hoc Networks Using Geographic Location Information, In *Proc. of IEEE ICASSP 2003*, Hong Kong, China, April 2003.

17. L. Lazos and R. Poovendran, SeRLoc: Robust Localization for Wireless Sensor Networks, ACM Transactions on Sensor Networks (TOSN), August 2005, vol. 1, pp. 73–100.

18. L. Lazos and R. Poovendran, HiRLoc: High Resolution Localization for Wireless Sensor Networks, IEEE Journal on Selected Areas in Communications (JSAC), Special Issue on Network Security, February 2006, Vol. 24 (2), pp. 233–246.

19. L. Lazos and R. Poovendran, SeRLoc: Secure Range-Independent Localization for Wireless Sensor Networks, ACM Workshop on Wireless Security, October 2004, (WiSe '04), pp. 21–30.

20. L. Lazos, S. Čapkun and R. Poovendran, ROPE: Robust Position Estimation in Wireless Sensor Networks, 4th International Symposium on Sensor Networks, April 2005, (IPSN '05), pp. 324–331.

21. Z. Li, W. Trappe, Y. Zhang, and B. Nath, Robust Statistical Methods for Securing Wireless Localization in Sensor Networks, in Proceedings of the International Conference on Information Processing in Sensor Networks (IPSN), 2005.

22. D. Liu and P. Ning, Location-based pairwise key establishments for static sensor networks, In *Proc. of SASN 2003*, Fairfax, VA, October 2003.

23. D. Liu, P. Ning, and W. Du, Attack-Resistant Location Estimation in Sensor Networks, in Proceedings of the International Conference on Information Processing in Sensor Networks (IPSN), 2005.

24. MICA Wireless Measurement System, available at: http://www.xbow.com /Products/Product_pdf_files /Wireless_pdf/MICA.pdf.
25. R. Nagpal, H. Shrobe and J. Bachrach, Organizing a Global Coordinate System from Local Information on an Ad Hoc Sensor Network, In *Proc. of IPSN 2003*, Palo Alto, USA, April, 2003.
26. J. Newsome, E. Shi, D. Song and A. Perrig, The Sybil Attack in Sensor Networks: Analysis and Defenses, In *Proc, of IPSN 2004*, Berkeley, CA, April 2004.
27. D. Nicolescu and B. Nath, Ad-Hoc Positioning Systems (APS), In *Proc. of IEEE GLOBECOM 2001*, San Antonio, TX, USA, November 2001.
28. P. Papadimitratos and Z. J. Haas, Secure Routing for Mobile Ad Hoc Networks, in *Proc. of CNDS 2002*, January 2002.
29. A. Perrig, R. Szewczyk, V. Wen, D. Culler and J. Tygar, SPINS: Security Protocols for Sensor Networks, In *Proc of MOBICOM 2001*, Rome; Italy, July 2001.
30. R. Poovendran and L. Lazos, A Graph Theoretic Framework for Preventing the Wormhole Attack in Wireless Ad Hoc Networks, to appear in ACM/Springer Journal on Wireless Networks (WINET).
31. N. Priyantha, A. Chakraborthy and H. Balakrishnan, The Cricket Location-Support System, In *Proc. of MOBICOM 2000*, Boston, MA, USA, August 2000.
32. R. L. Rivest, The RC5 encryption algorithm, In *Proc. of the first Workshop on Fast Software Encryption*, pp. 86-96, 1995.
33. N. Sastry, U. Shankar and D. Wagner, Secure Verification of Location Claims, In *Proc. of WISE 2003*, San Diego, CA, USA, September 2003.
34. A. Savvides, C. Han and M. Srivastava, Dynamic Fine-Grained Localization in Ad-Hoc Networks of Sensors, In *Proc. of MOBICOM 2001*, Rome, July 2001.
35. Y. Shang, W. Ruml, Y. Zhang and M. Fromherz, Localization from Mere Connectivity, In *Proc. of MOBIHOC 2003*, Annapolis, MD, USA, June 2003.
36. B. Hofmann-Wellenhof, H. Lichtenegger and J. Collins, Global Positioning System: Theory and Practice, Fourth Edition, Springer-Verlag, 1997.
37. D. Stinson, *Cryptograhpy: Theory and Practice,* 2nd edition, CRC Press, 2002.
38. R. Pickholtz, D. Schilling, and L. Milstein. Theory of Spread Spectrum Communications - A Tutorial, In the *IEEE Transactions on Communications,* 30(5):855–884, May 1982.
39. S. B. Wicker and M.D. Bartz, Type-II Hybrid-ARQ Protocols Using Punctured MDS Codes, In *Proc. of IEEE Transactions on Communications*, April 1994.

TRaVarSeL–Transmission Range Variation based Secure Localization

Santosh Pandey[1], Farooq Anjum[2], and Prathima Agrawal[1]

[1] Auburn University, AL, USA. {pandesg, agrawpr}@auburn.edu
[2] Telcordia Technologies, NJ, USA. fanjum@telcordia.com

1 Abstract

Summary. Location based services (LBS) enable the wireless network operators to provide specific services to wireless users, based on their current locations. In a wireless network, users can connect from various physical locations, but their current location information is required to enable LBSs. Several localization schemes have been proposed earlier to estimate the location of a wireless user. However these schemes assume that none of the users in the system is malicious which might not always be true. This chapter presents a Secure Localization Algorithm (SLA) that can be implemented on current 802.11 infrastructure based wireless networks without the need for any additional specialized hardware. The scheme is based on the transmission of unique messages at different power levels from the access points. The user device receives a certain set of messages which it will have to transmit back to the access point. The location of the user device can be estimated securely based on this set of messages. We have investigated the properties of SLA using simulations.

2 Introduction

Location based services (LBS) are expected to be the next "killer" application. The range of these personalized services vary from traditional services such as determining the nearest place of interest to emergency services such as wireless 911. In addition, information about location is also expected to enable newer services. One such service that has not received much attention is a location based authorization service [1]. In this case, an entity will not only have to prove its identity but also provide evidence of being in the right location in order to get access to the network resources. For example, a user might have to be present in his office in order to access top-secret documents over WLAN or to participate in a conference call. Similarly, a cafe might want to provide WLAN service only to customers sitting inside and not the others. As another example, the network provider can provide a service whereby a user's cell phone is not allowed to be turned on in premises such as health clubs, company campuses etc to prevent misuse of the camera phone capabilities[3].

---

[3] There are several details involved in implementing such a service which we skip here

In addition, location information can also be used to validate some of the mobile e-commerce transactions. Thus, information about the exact location from where the transaction was initiated could be used along with other pieces of information for corroboration. Location based access control can also facilitate applications such as the interactive dance club [2] where it is necessary to determine that a user is in "valid" locations before allowing her to communicate (interact) with the system. Mesh network based applications, where it would be necessary to verify the location of the next hop before connecting, would also benefit from this. Information about location has also been used to enhance the operation of 802.11 MAC leading to an increase in the system throughput by about 22 percent while also ensuring better fairness [3].

Such services in which location determination is a major component would attract attention of adversaries whose goal would be to try to deceive the localization system. An adversary could seek to achieve this while making use of special hardware, power variation etc. Given such an adversary (also referred to as an "intruder" or "attacker"), it would be necessary to design schemes that are secure and hence provide correct information about the location of the end user in spite of the attempts by the intruder to cheat the localization system. We refer to localization techniques that achieve this objective as secure localization techniques and the applications that depend on these techniques as secure location based applications.

While some of these secure location based applications can be enabled by GPS or similar technology the others would require a different solution. This is because not only is GPS costly, since it adds a couple of hundred dollars to the solution, but also is not secure[4] [4]. Thus, it is very easy to spoof GPS signals thereby causing the GPS receivers to report wrongly on the location. In addition, with GPS the network will have to trust the end device to report the correct location. This would require the use of tamper proof GPS receivers. Tamper proof devices while adding to the costs are also not foolproof [5]. Hence, it is necessary to investigate alternative technologies for determining the location of an end-user securely. The proposed solution should ensure that the end user will not be able to spoof his location easily.

A simple solution to this problem is to use the time of flight technique. Here each of the user devices whether cell phones or laptops[5] needs to have equipment that can reflect back transmitted signals without any delay. Then three or more base stations can determine the location of this device by transmitting signals to the user device and triangulating. Techniques like verifiable multilateration [6] can then be combined with this basic idea to ensure that the end user will not be able to falsify his location. A problem though with these approaches is the additional costs associated with deployment of a localization infrastructure. The high degree of reliability in the estimated location provided by such systems comes at a high cost. In many cases, especially those geared towards non-critical applications, this tradeoff might not be justifiable thereby making such systems impractical. What might be desired

---

[4] More precisely, GPS for non-military usage is not secure

[5] We consider cellphones to represent the cellular interface and laptops to represent the WLAN interface though in reality these devices can also have more than one interface

is a lower degree of reliability in the estimated location at a far lower cost. Attention has not been paid to this problem at all.

We focus our attention on this problem of low cost secure location determination in this chapter. A solution which depends on using the same radio hardware and capabilities for both communication and localization without the need for any special hardware would result in tremendous cost savings and is hence very much desirable. We propose such an approach to secure localization in this chapter. This problem is important both in cellular as well as WLAN networks. In this chapter though our focus is on WLAN networks. The normal range of a 802.11b Access Point (AP) spans from about 80 to 1750 ft in an office environment [7]. In many situations as explained earlier the localization resolution requirements are much smaller than this area.

The widely used approach for localization in WLANs based on signal strength measurement satisfies the requirements of low cost. But we will see in Section 3 that such an approach can be easily compromised by an adversary. In this chapter we investigate a low cost technique to achieve secure localization based on current WLAN devices and capabilities. The idea behind the proposed scheme is as follows. We exploit the property of current access points (AP) which enables an AP to transmit at different power levels. Use of a different power level will result in a different transmission range for the AP. The proposed scheme assumes that each location in the system under consideration is within the maximum transmission range of multiple APs. An AP in the system at a given time associates a message with each power level and securely transmits each message at that power level to the user whose location is to be determined. As a result, a user device at any location will receive a unique set of messages from multiple APs at any given point in time. This set depends on the power levels at which each AP has transmitted the messages and the location of AP with respect to that of the user device. The user device is expected to securely transmit back the messages received. The location of the user device can be determined based on the set of messages transmitted back.

The objective of this work is to provide secure location based access control for 802.11 networks. There are two approaches for achieving this. In the first approach, called location verification, the user device determines its location itself and the infrastructure will have to verify the location claim of the user device. In the second approach, called location determination, the infrastructure itself determines the location of the user device. A user device that is not in the proper location or whose location claim cannot be verified (for the first approach) or which cannot be successfully located (for the second approach) is not granted access to the system. In both these approaches it is necessary to ensure that both accidental and malicious attempts to fool the system do not succeed. In this work we take the second approach.

The remainder of the chapter is organized as follows. Section 3 describes the previous work done on secure localization in wireless networks. Section 4 contains details of the system being considered as well a description of the proposed localization scheme. Section 4.4 studies the security properties of the proposed scheme. Section 6 presents results from simulating the proposed secure localization scheme.

Several open issues are discussed in Section 7. Finally, we present our conclusions in Section 8.

3 Previous Work

There have been many localization schemes proposed for wireless networks. These schemes are typically based on the features of the underlying physical layer. For example various schemes based on ultrasound [8], infrared [9], Bluetooth [10], 802.11 RF networks [10–13] have been proposed. These schemes infer the location of users by measuring various parameters such as received signal strength indicator (RSSI), time of flight [6], and angle of arrival [14]. Some of these schemes are client based schemes where the user can determine his location and the network would have no knowledge of the user's location [8, 11] while the others are network based schemes where the network infrastructure is used to determine the location of the user [12, 15]. Note that the former approach might be preferred when user privacy concerns abound, however the latter is preferred in this case due to security considerations.

Localization schemes proposed for WLAN (802.11) systems are normally based on measuring the signal strength (SS) parameter [10–13, 16, 17]. The idea is to initially determine the SS map, representing the SS at various locations. The system then tries to determine the location of the user based on the best match between the observed signal strength and the SS map. The match can be done based on deterministic or probabilistic techniques so as to improve the location accuracy and resolution. In a vast majority of these cases the emphasis is on localization while neglecting presence of any malicious user.

There has been some work recently on secure localization. This problem has been mainly looked at in the context of sensor networks [6, 14, 18, 19]. The authors in [14] focused on secure positioning in a network of sensors and proposed techniques based on the use of directional antennas. Localization in the presence of an attacker in a wireless network has been discussed in [6], where explicit RF distance bounding was used in order to obtain a verifiable localization scheme. [19] uses a combination of directional antennae and explicit RF distance bounding. In Casper [18] the use of covert base stations is recommended. In all these cases the schemes depend on the availability of special hardware such as directional antennae or hardware with very tight time constraints required for RF distance bounding. [20] proposes statistical methods for secure localization in wireless sensor networks. Typically, the RSSI value from a device are used to compare with the expected RSSI values and the location of the device is that value that minimizes the mean squared error. This though is shown to give wrong results when faced with a malicious user. Hence the authors in [20] propose to determine the location based on minimizing the median squared error. In [21], the authors propose techniques for the detection of malicious attacks against beacons in sensor networks. The detection is based on the use of time of flight techniques to verify the consistency of information in the received beacons.

An approach to location based access control was provided in [22]. Here the focus is on location verification using ultrasound and time of flight techniques.

Schemes for secure localization in 802.11 networks based on the signal strength measurement approach are provided in [23, 24]. In [23] the authors propose a secure localization scheme using the SS values. The SS lookup table is built efficiently but this scheme assumes an enterprise like environment with cooperating users. The paper indicated that using a simple trilateration based on averaged signal strength lookup table, an accuracy of 85 % with a location error range of about 10 ft was obtained. Using the data from [23], we studied the impact of a simple attack that can be easily launched by a single attacker. We emulated the scenario of an intruder transmitting at higher power level by increasing the received SS values by 25 % and using the regular matching techniques based on least mean squares error. We observed that the accuracy of the localization scheme dropped to 19 %. A way to address this is to use the difference approach as in [24] or to use more robust mapping techniques. The 'difference method' developed in [24] detects the location of an intruder who can transmit at different power levels. However, the accuracy of the estimated locations is poor with 70 % probability of correct location estimate and a resolution of about 10 ft.

4 Secure Localization

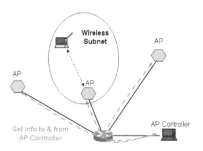

Fig. 1. WLAN system

We consider a WLAN system with several access points. All these access points are managed by a single AP controller as shown in Figure 1. This fits the current trend towards controlling the cheaper and lightweight access points via standard protocols from a central location [25]. We consider that each AP has a capability to vary the power level. For example, five different transmission power levels are available on the DLINK DWL-2100AP [26]. The different power levels supported are 15dBm, 12dBm, 9dBm, 6dBm, and 3dBm. By using attenuators and amplifiers it is further possible to vary the number of power levels. Each power level is assumed to correspond to a different transmission range. For this preliminary work to explain the proposed scheme and to investigate its characteristics, we assume an ideal environment, whereby the transmission range of the AP is assumed to correspond to a circle of radius 'r' where 'r' is different for the different power levels. We assume that all the APs deployed in the system are valid (not rogue APs) and have identical capabilities in terms of the number and value of each of the available power levels. We do not assume the presence of any specialized hardware on either the access points or on the client devices.

Communication between the access points and the users is protected by the 802.11i group of protocols. Thus, authentication, key management and encryption mechanisms are all present in the system. Given such a scenario, the system will try to locate users anytime after a user device is associated with an access point. Note that data traffic starts flowing between a user device and an access point in the WLAN network only after the user device is associated with an access point. Association follows successful authentication and key exchange. The need for localization could be in response to detection of malicious traffic or at instances as decided upon by the AP controller.

In this work we consider a threat model that consists of an intruder (malicious user) acting alone, thereby ruling out collaborative attacks. The intention of the intruder is to convince the location determination system that the intruder is present in a region different from his actual physical location. We assume that the intruder does not have specialized hardware such as high gain antennae or directional antennae. The system components such as the various APs and the AP controller are all trusted. The various APs are expected to communicate with each other via the AP controller over wireline links; hence such communication is considered secure from the intruder. Note that given the dependency on 802.11i, an intruder has to be an insider since otherwise the intruder will not be able to access the network.

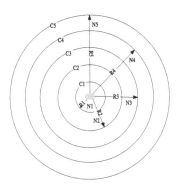

Fig. 2. Access point with 5 transmission power levels

4.1 Localization Algorithm

We next describe the localization algorithm[6]. When the AP controller wants to locate a user device, it determines the AP with which the user device is associated and requests this AP and its neighboring APs to send a location query securely to the device at different power levels. Secure communication would be the result of using the keys derived from the usage of 802.11i[7]. The location query is an encrypted message consisting of a nonce (random number) that is unique for each power level at each AP, AP identifier, power level information and the ID of the user device. The localization process is then based on the client devices listening for such messages from different APs and responding back to the access point that the user device is associated with. The response contains all the messages that the user device hears from the various APs. This set is then sent to the AP controller which is able to determine the location of the user device from this set of messages.

An illustration of the transmitted range and the messages from a single AP is shown in Figure 2. The figure indicates that at different power levels the transmission range of the AP will be limited by the radii R1 to R5. The regions are indicated in the figure as circles C1 to C5 respectively. The AP would be broadcasting the message N1 to N5 at the respective power level. Thus N1 will be heard only within the circle C1, while N5 will be heard at all points within C5. Thus a user device close to the AP would be able to hear all the transmitted messages. The location of a user device may hence be discriminated based on the message heard by the user device.

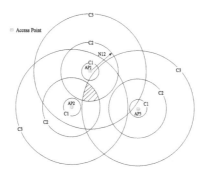

Fig. 3. Three access points with 3 transmission power levels

When a single AP is used, a user device can only be located to within an annulus. To locate the user device more precisely, we would need more than a single AP.

---

[6] Our description assumes two dimensional space and extends trivially to the case of three dimensional space

[7] While this addresses secure communication with the AP the user device is associated with, questions arise as to how do the neighboring APs communicate securely with the user device. This could be done using a key provided by the AP controller to the neighboring APs and the user device. These issues are orthogonal to our objective and hence we do not address them.

Consider three APs as shown in Figure 3. Let each AP have three transmission power levels. The user device's location would hence be based on the messages from all the three APs received by it. The message at each AP would be chosen at random. Let Nij represent the jth message from the ith AP. For example, N12 represents the second message corresponding to circle C2 of the first AP. A user device present in the shaded region of the Figure 3 will receive the set {N12, N13, N22, N23, N33} of message from different APs. The user device would then respond by retransmitting the set of messages (received by it from all the three APs) to the AP the user device is associated with. It can be easily verified from the figure that the set of message heard at the shaded region is unique to the region. The AP controller would then be able to determine the location of the user device based on this set of messages received. A formal statement of the algorithm is given next.

Secure Location Algorithm (SLA)

1. When the AP controller wants to determine the location of a user device, it determines the AP, X, the user device is associated with. It then requests X as well as the neighboring APs of X to transmit messages securely at different power levels. If the APs are properly placed then the user device will be within range of at least three of these APs.
2. Each user device thus receives a set U of messages from at least three APs. The node will retransmit all the received messages back securely to X which forwards U to the AP controller.
3. The AP controller then determines the location of the user device from the set U received.

In step 1 above, we propose that each AP along with all its neighboring APs transmit messages towards the user device which is to be located. A more efficient scheme would require that the AP only coordinate with the neighboring APs which cover the region where the user device node is. This can be done by using the SS values from the node and using the SS values to select the set of APs to transmit the message. We do not consider this aspect in this chapter. In addition, this step also requires placement of APs such that every location is covered by at least three APs, the minimal number needed to locate a user device. We will address deployment of APs later in this section. Note that the set U corresponds to the union of maximal number of messages received from each AP by the user device. We will investigate the security properties of SLA in Section 4.4.

It should be noted that SLA can be used in conjunction with the other conventional distance estimation schemes. For example, the SS of the user device transmission may be recorded along with the message sent by the user device. A SS based localization scheme may then be used in addition to the proposed scheme in order to enhance the distance estimation. But this is not our focus in this work.

At this point several questions arise with respect to SLA. Some of these are 1) AP deployment: How should the APs be deployed in the entire network, 2) Localization resolution: What is the average resolution of localization. We address these questions next.

4.2 AP Deployment

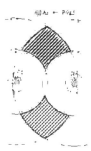

Fig. 4. Two access points with 3 transmission power levels

In this section we address the first question related to deployment of APs. It is obvious that every location in the given deployment area needs to be covered by at least three APs to ensure accurate localization. We see this from Figure 4 which represents the regions formed by two APs. The shaded area in the figure represents the region which would hear the same set of messages {N13, N23}. Due to the separation of the regions, the localization scheme can only determine that the user device is in one of the regions. Thus, there would be ambiguity in locating a user device even for honest users. A location covered by more than three APs would increase the location resolution but at the cost of interference if the network is not properly engineered. Note that for WLAN networks, the three APs could be on the three non-overlapping channels thereby preventing interference. Of course, this would require that the user device would have to switch channels in order to hear the messages transmitted by the other two APs. An alternative solution would be for the three APs to be on the same channel but for only one of these APs to carry data (and hence participate in association process) while the purpose of the other APs would be to help in localization. This would enable the user device to more easily receive the transmitted messages. This would not be a major impact on the system costs given the costs (in the range 10-50USD) associated with WLAN APs. We are looking into all these alternatives as part of the prototype implementation of SLA which is outside the scope of this chapter. Therefore, in the sequel we consider a deployment scheme where each user device can hear exactly three APs.

An efficient way to deploy the minimum number of APs while ensuring that every location is within the maximum range of at least three APs is as shown in Figure 5. Here the APs are placed equidistant from each other and at the edge of the maximum transmitting range. Since the APs are assumed to be identical, this scenario results in deploying APs at the vertices of an equilateral triangle whose length equals the maximum transmitting range of the AP. It should be noted that the localization of users is possible only within the region bounded by the maximum transmitting range from each of the APs.

Fig. 5. Efficient deployment of access point

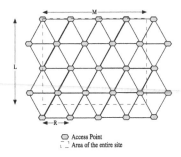

○ Access Point
⌐ ⌐ Area of the entire site

Fig. 6. Deployment of access point for the entire site

Figure 6 illustrates the possible deployment of APs throughout the site by replicating the configuration of Figure 5. Given such a deployment of APs, a user device would be located in one of the triangles with the vertices of the triangle corresponding to APs[8]. The localization of user devices would then be based on the messages transmitted by the APs corresponding to the vertices of the triangle in which the user device is present. Note that in this case the number of APs, n, needed to cover any area L x M (assuming $M \rangle L$) is given by

$$n = \lfloor \frac{\left(\lceil \frac{2M}{R} \rceil + 3 \right) \left(\lceil \frac{2L}{\sqrt{3}R} \rceil + 1 \right)}{2} \rfloor \qquad (1)$$

where R corresponds to the maximum transmission range of an AP. It can be observed from Figure 5 that the actual region defined by the three APs (bounded by maximum transmission range arcs from each APs) is slightly larger than the equilateral triangle. Thus there will be small overlaps amongst the adjacent triangular regions in Figure 6. This however is not significant and is hence ignored.

4.3 Localization Resolution

We next address the second question related to localization resolution of the proposed scheme. We refer to the region bounded by the maximum transmission range of each of the three APs as shown in Figure 5 as the 'area of interest' (AOI) and

---

[8] We ignore boundary cases here

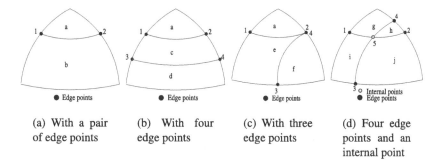

(a) With a pair of edge points

(b) With four edge points

(c) With three edge points

(d) Four edge points and an internal point

Fig. 7. Partitioning of sub-regions in the AOI with different number of edge points and internal points

the several areas formed inside the area of interest by the intersecting circles as sub-regions. Thus, a sub-region is the smallest bounded area in the area of interest which corresponds to a unique subset of messages. Greater the size of the sub-region, more coarse would the location resolution be. Thus, our focus is on obtaining the expression for the average size of a sub-region in the area of interest. We achieve this by dividing the area of the AOI by the number of sub-regions within the AOI.

We first seek to obtain the number of sub-regions that would be formed within the AOI for the scenario of Figure 5. We first define a few terms. The intersection point of a curve with the boundary of the AOI is called an 'edge points'. The intersection points of curves within the AOI are called 'internal points'. Further, a non-intersecting curve is a curve that does not intersect any other curve within the AOI. Given this, we have the following theorem.

Theorem 1. *Each non-intersecting curve would intersect the boundary of the AOI in exactly two edge points and result in exactly one additional sub-region.*

Consider a simple case as shown in Figure 7, where the AOI is divided by the curve 12. The curve 12 intersects the boundary of the AOI in exactly two edge points (1 and 2) and divides the AOI into two sub-regions indicated as 'a' and 'b' in the figure.

Now add a second non-intersecting curve to the AOI in Figure 7(a). The non-intersecting curve cannot intersect curve 12 already present in the AOI and thus would have to lie entirely in either the sub-region 'a' or the sub-region 'b'. We arbitrarily choose sub-region 'b', which is bounded by boundary of the AOI and curve 12. A new curve 34 is to be drawn in the sub-region 'b'. This is exactly the same as the previous case wherein a curve 12 was added to the AOI. Thus the curve 34 would intersect the boundary of region 'b' at exactly two points (3 and 4) and further divide 'b' into two more sub-regions. Since the curve 34 is a non-intersecting curve, points 3 and 4 cannot lie on curve 12 and hence have to lie on the boundary of the AOI. Figure 7(b) and Figure 7(c) show the two ways in which curve 34 can be created.

Although the points 2 and 4 coincide in Figure 7(c), they are treated as two different points.

This process can be carried out recursively for each added non-intersecting curve. Hence the proof by induction.

Corollary: Each pair of the edge points lying on the boundary of the AOI would form one additional region.

This follows from the previous theorem. It should be noted that the edge points may be non-distinct as in Figure 7(c). Using the above theorem, the total number of sub-regions formed may be calculated based on the number of edge points and internal points.

Theorem 2. *The number of sub-regions formed within the AOI is given by $\frac{N_e}{2} + N_i + 1$, where N_e is the number of edge points and N_i is the number of internal points.*

Proof: The proof is by induction. In the case of non-intersecting curves ($N_i = 0$), the corollary above indicates that each pair of edge points will result in an additional sub-region. Thus in addition to the original single sub-region, a total of $\frac{N_e}{2}$ additional sub-regions would be added.

Consider an addition of curve 34 to Figure 7(a) intersecting curve 12 at point 5 as shown in Figure 7(d). The effect of the curve 34 can be considered in parts, namely, the effect of curve 35 on sub-region 'b' and effect of curve 54 on sub-region 'a'. For each of these parts, corollary of Theorem 1 implies that curve 35, with edge points 3 and 5 for sub-region 'b', would result in one additional sub-region in 'b'. Similarly curve 54, with edge points 5 and 4 for sub-region 'a', would result in one additional sub-region in 'a'. Thus the total number of additional sub-regions formed (2 in this case) is equal to half the total number of edge nodes for sub-regions 'a' and 'b', i.e. $\frac{4}{2}$. It should be noted that the internal point 5 is considered twice as edge node for adjacent sub-regions 'a' and 'b'. Thus, the theorem is true for one intersecting and one non-intersecting curve.

Now assume that the AOI is divided into $\frac{N_{e(m)}}{2} + N_{i(m)} + 1$ sub-regions due to some m curves. Let us add a new curve to the AOI. Considering the case described for Figure 7(d), the number of additional sub-regions formed would be equal to half the total number of edge points when the additional curve is considered in parts (each part formed due an intersecting curve). Each of the internal points would be counted twice as edge points. Thus the additional sub-regions would be given as

$$\frac{N_e' + 2N_i'}{2} = \frac{N_e'}{2} + N_i' \tag{2}$$

where N_e' denotes the number of new edge points and N_i' denotes the number of new internal points formed due to the additional curve. It should be noted that the new edge and intersection points may coincide with the earlier points, but are treated as distinct points. Thus the total number of sub-regions due to the m+1 curves is given by

$$\frac{N_{e(m)}}{2} + N_{i(m)} + 1 + \frac{N_e'}{2} + N_i' = \frac{N_{e(m+1)}}{2} + N_{i(m+1)} + 1 \tag{3}$$

Setting $N_{e(m+1)} = N_{e(m)} + N'_e$ and $N_{i(m+1)} = N_{i(m)} + N'_i$ we obtain the result.

It is interesting to note that this equation is equivalent to the Eulers formula relating the number of faces to the number of edges and vertices in a graph [27].

We next need to determine the area of AOI. We assume that this is given by the area of an equilateral triangle with side R^9. Therefore, considering the area of the AOI and Theorem 2, we have the following result.

Theorem 3. *The average resolution using the proposed scheme is given by*

$$\frac{\sqrt{(3)}R^2/4}{N_e/2 + N_i + 1}$$

Remarks: We can see from the above that as the number of power levels increases while keeping R the same, so does N_e and N_i. As a result the location resolution becomes finer. This indicates that larger number of power levels for the same maximum transmission range is desirable. Note though that in practice, too many power levels might also be undesirable since two different power levels might not lead to different transmission ranges. There will hence be an optimum number of power levels for an AP. We do not investigate this here as it is part of a separate study related to implementation of the proposed Secure Location Algorithm.

5 Security Analysis

Several attacks on previous localization schemes have been discussed in [6]. In this section, our focus will be on attacks on the proposed localization scheme. Our threat model specified earlier consists of an attacker working singly (no collusion) and a network protected by 802.11i. We start by proving that such an attacker will be unable to spoof his location under SLA.

Theorem 4. *Consider an AP with large number of power levels for a given transmission range R. In this case, SLA is secure in that any attacker acting alone cannot falsify its location.*

Proof: Assume that the APs are spread as the vertices of equilateral triangles as given in Figure 5. Then a user device will be located in an AOI formed by 3 basestations[10].

First consider a single AP of the three that make up the AOI. A user device is expected to transmit a set of messages to the AP. Now the user device will hear one or more messages from this AP depending on the distance. Let the AP have k transmission circles (each circle of radius jr, $j = 1, 2, ...k$; note $kr = R$) corresponding to k power levels. Then a user device present at distance ir from the AP will hear (k-i+1) messages (i.e. the ir, $(i + 1)r$, $...(k − 1)r$ and kr messages).

---

[9] This is slightly imprecise as we remarked earlier and is only used for the area calculation

[10] As earlier, we ignore the boundary cases

Now if this user device wants to spoof its distance, it can suppress some messages. The user device cannot just select any of the messages to transmit back to the AP. For e.g. if the user device only transmits back to the AP controller messages ir and kr, then the AP controller can immediately conclude maliciousness as any user device that can obtain ir, must be able to obtain all other messages between ir and kr[11]. Therefore, all that a malicious user device can do is to pretend it did not hear the ith message and the $(i+1)^{th}$ message and so on. Therefore, by spoofing the user device can only increase its distance from the AP.

Now consider 3 APs A, B and C. Let a user device receive at most na, nb and nc messages from each of the APs respectively. If the user device wants to spoof its location then the only thing it can do is increase its distance from each of the three APs. Therefore, it can pretend to have heard $nka\langle na$ and $nkb\langle nb$ messages from A and B but this means that it will have to be closer to node C. This implies that it will have to transmit back $nkc\rangle nc$ messages from the access point C[12] but this is a contradiction since the user device cannot hear more than nc messages from the access point C. Therefore, a user device cannot spoof its location under the given SLA.

Remarks: When the number of power levels is small, then a malicious user device can appear to be in a neighboring region. For example, consider Figure 8. Here a user device present in location 1 hears the set of messages {N13,N14,N15,N23,N24, N25,N33,N34,N35}. Such a user device can drop message $N23$ from the set of messages it hears and thereby appear to be in location 2 which corresponds to the message set {N13,N14,N15,N24,N25,N33,N34,N35}. If the user device is present in location 2 though, it cannot pretend to be in location 1 although it can pretend to be in location 3 by skipping message N24. On the other hand a user device present in location 4 can hear a set of messages {N13,N14,N15,N24,N25,N34,N35} and such a user device cannot spoof its location at all. Similarly, a user device in location 3 also cannot spoof its location. Statements of Theorem 4 apply precisely to such locations which we call the non-spoofable locations. When the number of power levels increases, such non-spoofable locations dominate.

One way to make it difficult for the intruder to spoof an area by using the above procedure (of ignoring some messages) is to make it difficult for the intruder to relate the message transmitted to the power level. This could be done by transmitting the messages randomly and not in the sequence of increasing or decreasing power levels. Note that if the intruder studies and tries to infer the power levels of all the transmissions at different regions, then the probability for the intruder being able to spoof his location increases. Later in Section 6, we will investigate via simulations the relationship between the area that can be spoofed and the number of power levels. We would also like to remark here that another way of preventing any spoofing could

---

[11] An implicit assumption here is that the messages are transmitted reliably. There are many ways to ensure this and is hence justified

[12] This is not strictly true if the number of power levels is small and hence the need for large number of power levels

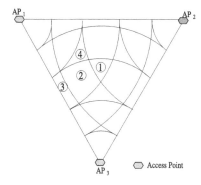

Fig. 8. Possible spoofing for the proposed scheme

be to deploy the APs such that each location is covered by more than three APs. This will be considered as part of future work.

A malicious user device can attempt the replay attack. Here it can store the set of messages recorded at multiple locations and then select the messages to transmit to the AP based on the location the user device wants to appear in. But such a scheme will not succeed due to the fact that the messages would change at each location for each location query. Further, given our assumption of a network protected by 802.11i mechanisms, a malicious user device cannot overhear the set of messages sent to a different user device by walking around but has to use the message set based on location queries directed to itself. Attacks based on delaying the response, signal strength manipulation etc would also not work with SLA as SLA does not depend on these properties.

6 Results

In this section we investigate the performance of SLA using simulations. The simulations are performed using MATLAB. We assume that the APs are deployed such that the distance between them is equal to the maximum transmission range. A user device present in the AOI is localized using SLA. For such a scenario we assume two cases. In the first case, the case of uniform APs, each AP is assumed to have transmission ranges that vary in uniform increments. This implies that the adjacent concentric circles symbolizing the range of each AP for different power levels are equidistant. We relax this assumption to get the second case, the case of non-uniform APs. Here each AP has transmission ranges that vary in non-uniform increments. Since the APs deployed in the system have the same capabilities, the values of non-uniform increments are the same for all APs. This implies that the adjacent concentric circles symbolizing the range of each AP for different power levels are not equidistant. We look at each of these cases next.

6.1 Uniform APs

As explained earlier, in this case the transmitting range increments corresponding to increasing power levels are equal. Thus for five different power levels, the radii corresponding to the five transmission regions will be $r, 2r, 3r, 4r$ and $5r$. The AOI is the region formed by APs placed at the vertex of an equilateral triangle in a 100 x 100 square unit grid. Each side of the triangle is of length 57 units corresponding to a maximum range (R) of 57 units for each AP.

We first consider the number of sub-regions formed within the AOI as given by Theorem 2. In Figure 9 we show the variation of the number of sub-regions within the AOI as the number of power levels is increased. Note that the maximum transmission range (R) is the same in all cases. We see from this figure that the number of sub-regions increases almost exponentially as a function of the number of power levels. The number of sub-regions indicate the resolution of localization in the proposed scheme. As expected, the resolution of localization increases with the increase in the number of power levels of an AP. We will see this next.

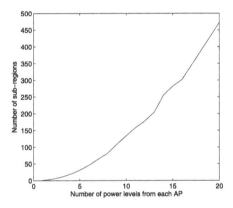

Fig. 9. Number of sub-regions formed in the AOI corresponding to three access points

In Figure 10 we show the maximum, minimum and average area of the sub-regions as obtained via simulations. We also show the average area of the sub-regions given by Theorem 3 as the number of power levels varies. We plot the number of power levels of each AP on the x-axis. On the y-axis we plot the percentage area of the sub-region which is the ratio of the area of the sub-region to the area of the AOI. Thus, if we use a single power level, then the sub-region corresponds to the entire area. In such a case, the percentage area of the sub-region will be 100% for the maximum, minimum and average values. We see from this figure that the area of the sub-regions decreases exponentially with the increase in the number of power levels. Further, we also see from this figure that as the number of power levels increase, the maximum, minimum and average area of sub-regions converge, indicating that the sub-regions of approximately equal areas are formed which is very much desirable.

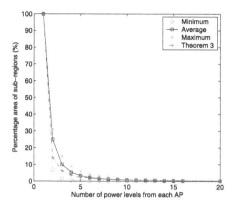

Fig. 10. Area of sub-regions as a function of the number of power levels

We next consider the number of messages heard at different sub-regions within the AOI. As is clear from our earlier discussion, the number of messages heard in each sub-region is different. Hence, in Figure 11 we plot the maximum and minimum number of messages that are heard within the AOI. A client present in the sub-region where the minimum number of messages are heard will just have to transmit back these messages while a client present in the sub-region corresponding to the maximum number of messages will have to transmit all these back to the AP. Thus, the larger the variation between the minimum and maximum number of messages heard in the AOI, the harder it would be for the attacker to guess the correct set of messages that correspond to a location and hence more secure would the scheme be. Figure 11 shows that these values increase linearly with the number of power levels. More importantly, the spread between the maximum and the minimum number of messages also increases with power levels. This indicates that with the increase in the number of power levels, SLA becomes more robust to the presence of intruders.

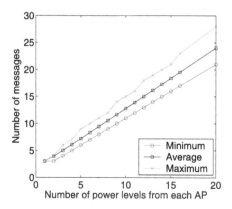

Fig. 11. Number of messages heard in the AOI

We next consider the possibility of an attacker spoofing his location as the number of power levels of an AP increase. We have seen earlier that one possible way for a lone attacker to spoof her location is to ignore one of the messages heard in a sub-region. With increase in the number of power levels, the minimum number of message heard at a sub-region increases and hence the possibility of such a forged region is expected to decrease. This is also indicated in Figure 12 where we show the maximum, minimum and average area of region that can be spoofed by the intruder. In the x-axis we plot the number of power levels while on the y-axis we plot the percentage area of the sub-region that can be spoofed. This is the ratio of the area of the sub-region which can be spoofed to the area of the AOI. For example, in Figure 8, an attacker present in location 1 can spoof to be either in areas 2 or in area 3. So the corresponding value of the percentage area of the sub-region that can be spoofed is the value of the areas such as 2, 3 etc divided by the area of the AOI. As seen from the figure, in general as the number of power levels increase, the area that can be forged decreases. Thus based on Figures 11 and 12, it can be concluded that with the increase in the number of power levels the scheme would be more robust against attacks.

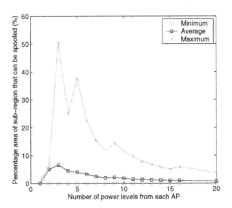

Fig. 12. Area of regions that may be forged in terms of percentage of common transmission area

Figure 12 also indicates that the deployment with three power levels results in a high value for the percentage of area that can be forged. This is mainly due to the use of equidistant circles, which leads to the formation of a huge spoofable sub-region at the center of the AOI. This implies that equidistant transmission ranges are not desirable which leads to the non-uniform APs case that we investigate in the next sub-section. But if equidistant transmission ranges have to be used, then this figure may be used to avoid using the number of power levels that result in a large area that can be forged. We would also like to remark here that almost every choice for the number of power levels has some regions where location spoofing is not possible (such as location 3 and 4 in Figure 8). This corresponds to a value of zero for the spoofed area. Thus an attacker present in such regions will not be able to spoof his location at all. This indicates that the network deployment can be done in such a manner so as to ensure that "physically" less secure locations such as a lobby would correspond to one of these non-spoofable sub-regions.

6.2 Non-uniform APs

We have so far considered that the adjacent concentric circles signifying the transmission ranges of an AP are equidistant. We next consider the case whereby each AP has transmission ranges that vary in non-uniform increments. Different values for the ratios of different radii (for e.g. 1:2:3:4:5[13] corresponding to circles of radii r,3r,6r,10r and 15r=R or 1:1:3:4:5 corresponding to circles of radii r',2r',5r',9r' and 14r'=R) would result in different number of sub-regions as well as in different number of messages heard in each sub-region. The area of each sub-region and the region that may be forged will also depend on the chosen radius ratio.

Consider the deployment with an AP which can transmit at five different power levels. The ratio of these five power levels may be varied in order to obtain optimum performance. Figure 13 shows the percentage area that may be forged for different radius ratios of the concentric regions. In this figure, we consider 5 transmission ranges and about 7500 unique ratios for these ranges. This is obtained by having 6 possible values for each the value of each transmission range which will give us $6^5 = 7776$ different ratios. Some of these will be identical. We remove these identical ratios and consider all the rest of the ratios which approximates to around 7500 different ratios. Each ratio corresponds to a number which is plotted on the x-axis of this figure. On the y-axis we plot the percentage area of the sub-region that can be spoofed. As earlier, this is the ratio of the area of the sub-region which can be spoofed to the area of the AOI. We plot the maximum, average and minimum area that can be spoofed for a given ratio. Thus, given a point on the x-axis, we can determine the maximum, average and minimum areas that can be spoofed for the ratio that this point represents. As seen from the figure depending on the radius ratio, the maximum forged area may be anywhere between 3.8 % to 99.65 %. This corresponds to ratio increments of (4:1:1:1:4) and (6:1:1:1:1) respectively. By choosing equidistant regions the maximum forged area obtained is about 21.94 %. Hence, from security perspective, choosing an appropriate non-uniform radius ratio (non-uniform AP case) results in better performance as compared to uniform AP case. Similar results were observed for the other parameters, such as number of messages received and number of sub-regions formed. Thus, we see that it is more beneficial to have non-uniform increments for the transmission ranges of an AP. Note that a combination of attenuators and amplifiers can be used along with the inbuilt capabilities of an AP to achieve non-uniform radius ratios (non-uniform AP case).

---

[13] Note that this corresponds to the ratio for the increments as opposed to the ratio of the radii which is obtained from this

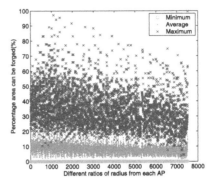

Fig. 13. Percentage area of regions that may be forged for different radius ratios of the concentric regions

7 Discussion

There are two main abstractions that will have to be addressed in order to implement this scheme. The first abstraction is the assumption that every power level of an AP corresponds to a coverage given by a circle of some radius. This will not be true in practice. The area covered for a given power level is not a strict circle [11]. Further, the transmission range corresponding to a given power level from an AP could also be time-varying. We plan to address this by making use of sniffers (acting as reference points) placed at the boundaries corresponding to a particular power level transmission from an AP. These sniffers would be connected to the AP controller and any variation with time in the transmission range for the same power level from access point A could then be reported to the AP controller. The AP controller could in turn request A to modify the power levels accordingly. Note that such a time varying range would make it difficult even for an attacker with specialized hardware to infer the relationship between power level and the location which thereby increases the security of SLA. We are currently working on these aspects and expect to report on these in the future.

We are also currently considering a system based on the proposed scheme that is secure against colluding attackers. We are currently investigating the combination of SLA with the other conventional schemes such as schemes based on SS measurements (which should be done automatically without the need for laborious bootstrapping) in order to provide a secure scheme that can tolerate multiple colluding intruders. Finally although the localization scheme in this chapter is developed in the context of WiFi networks, it can be extended to other wireless networks such as, wireless sensor networks.

8 Conclusion

In this chapter we focused on the location based access control problem in WLAN networks. We see that a solution to this problem is dependent on the design of secure localization schemes. Hence, in this chapter we propose a secure localization scheme that can work with current WLAN hardware. The proposed scheme is based on the capability of a WLAN AP to vary its power level. We propose that multiple APs cover each location from which a user device can access the wireless network. Further, each AP is expected to securely transmit a message at each different power level. A user device is expected to reflect back all the messages heard at its location back to the AP. Based on the set of messages heard back from the user device, the AP can determine the location of the user device. Assuming a user acting alone, we have shown the robustness of this scheme to several user misbehaviors. We have also shown that if the power level of an AP can be continuously varied from the minimum to maximum level, the proposed scheme would result in highly precise and secure localization of users.

9 Acknowledgment

The authors would like to thank Dr. Wlodzimierz Kuperberg of the Department of Mathematics and Statistics, Auburn University, for his valuable comments.

References

1. D. E. Denning and P. F. MacDoran, "Location-based authentication: grounding cyberspace for better security," *Internet besieged: countering cyberspace scofflaws*, pp. 167–174, 1998.
2. D. Hromin, M. Chladil, N. Vanatta, D. Naumann, W. Wetzel, F. Anjum, and R. Jain, "CodeBLUE: A bluetooth interactive dance club system," *IEEE Global Telecommunications Conference*, vol. 5, pp. 2814 – 2818, December 2003.
3. T. Nadeem, L. Ji, A. Agrawala, and J. Agre, "Location enhancement to IEEE 802.11 DCF," *Proceedings of IEEE INFOCOM*, vol. 1, pp. 651–663, March 2005.
4. M. G. Kuhn, "An asymmetric security mechanism for navigation signals," *Proceedings of the Information Hiding Workshop*, 2004.
5. R. Anderson and M. Kuhn, "Tamper resistancea cautionary note," *Second USENIX Workshop on Electronic Commerce Proceedings*, pp. 1–11, November 1996.
6. S.Capkun and J. Hubaux, "Secure positioning of wireless devices with application to sensor networks," *Proceedings of IEEE INFOCOM*, vol. 3, pp. 1917–1928, March 2005.
7. Proxim Corporation, *ORiNOCO AP-2000 User Guide*, http://www.proxim.com/products/wifi/ap/ap2000/.
8. N. B. Priyantha, A. Chakraborty, and H. Balakrishnan, "The cricket location-support system," *Proceedings of the Annual International Conference on Mobile Computing and Networking, MOBICOM*, pp. 32–43, August 2000.
9. R. Want, A. Hopper, V. Falco, and J. Gibbons, "The active badge location system," *ACM Transactions on Information Systems*, vol. 10, no. 1, pp. 91–102, January 1992.
10. Y. Gwon, R. Jain, and T. Kawahara, "Robust indoor location estimation of stationary and mobile users," *Proceedings of IEEE INFOCOM*, vol. 2, pp. 1032–1043, March 2004.
11. S. Ganu, A. S. Krishnakumar, and P. Krishnan, "Infrastructure-based location estimation in wlan networks," *IEEE Wireless Communications and Networking Conference (WCNC 2004)*, vol. 1, pp. 465–470, March 2004.

12. P. Bahl and V. N. Padmanabhan, "RADAR: An in-building rf-based user location and tracking system," *Proceedings of IEEE INFOCOM*, vol. 2, pp. 775–784, March 2000.
13. A. M. Ladd and et. al., "On the feasibility of using wireless ethernet for indoor localization," *IEEE Transactions on Robotics and Automation*, vol. 20, no. 3, pp. 555–559, June 2004.
14. L. Lazos and R. Poovendran, "SeRLoc: Secure range-independent localization for wireless sensor networks," *Proceedings of the 2004 ACM Workshop on Wireless Security, WiSe*, pp. 21–30, October 2004.
15. P. Castro, P. Chiu, T. Kremenek, and R. Muntz, "A probabilistic room location service for wireless networked environments," *Proceedings of the 3rd international conference on Ubiquitous Computing*, pp. 18 – 34, September 2001.
16. D. Madigan, E. Elnahrawy, R. P. Martin, W.-H. Ju, P. Krishnan, and A. S. Krishnakumar, "Bayesian indoor positioning systems," *Proceedings of IEEE INFOCOM*, vol. 2, pp. 1217–1227, March 2005.
17. M. Youssef and A. Agrawala, "The Horus WLAN location determination system," *Proceedings of the 3rd international conference on Mobile systems, applications, and services*, pp. 205–218, June 2005.
18. S.Capkun, M. Srivastava, M. Cagalj, and J. Hubaux, "Securing positioning with covert base stations," NESL-UCLA Technical Report, Tech. Rep., March 2005.
19. L. Lazos, S. Capkun, and R. Poovendran, "ROPE: Robust position estimation in wireless sensor network," *Proceedings of IPSN*, April 2005.
20. Z. Li, W. Trappe, Y. Zhang, and B. Nath, "Robust statistical methods for securing wireless localization in sensor networks," *Proceedings of the International Conference on Information Processing in Sensor Networks (IPSN)*, April 2005.
21. D. Liu, P. Ning, and W. Du, "Attack-resistant location estimation in sensor networks," *In Proceedings of the International Conference on Information Processing in Sensor Networks (IPSN)*, pp. 99 – 106, April 2005.
22. N. Sastry, U. Shankar, and D. Wagner, "Secure verification of location claims," *Proceedings of the Workshop on Wireless Security*, pp. 1–10, 2003.
23. S. Pandey, B.Kim, F. Anjum, and P. Agrawal, "Client assisted location data acquisition scheme for secure enterprise wireless network," *IEEE Wireless Communications and Networking Conference, WCNC*, vol. 2, pp. 1174–1179, March 2005.
24. P. Tao, A. Rudys, A. M. Ladd, and D. S. Wallach, "Wireless LAN location-sensing for security applications," *Proceedings of the Workshop on Wireless Security*, pp. 11–20, September 2003.
25. P. Calhoun, B. OHara, S. Kelly, R. Suri, D. Funato, and M. Vakulenko, "Light weight access point protocol (LWAPP)," *IETF Internet Draft, draft-calhoun-seamoby-lwapp-03*, June 2003.
26. D-Link, *DWL-2100AP Data Sheet*, ftp://ftp10.dlink.com/pdfs/products/DWL-2100AP/DWL2100AP_ds.pdf.
27. C. Dodson, P. Parker, and P. E. Parker, *A User's Guide to Algebraic Topology (Mathematics and Its Applications)*. Springer, 1997.

Secure Sequence-based Localization for Wireless Networks

Bhaskar Krishnamachari and Kiran Yedavalli

Department of Electrical Engineering-Systems
Viterbi School of Engineering
University of Southern California
Los Angeles, CA 90089
{bkrishna, kyedaval}@usc.edu

Summary. We present a unique sequence-based approach to localization that allows for automatic error correction. In dense deployments, this technique can provide accurate localization even in the face of malicious falsification of reference signals.

1 Introduction

Providing mobile devices with accurate location information is a fundamental service in many current and envisioned deployments of wireless networks, including wireless local area networks, ad hoc networks and wireless sensor networks. Besides being useful in itself in providing end users with location awareness, localization can be a key building-block for other wireless network protocols such as those for routing, sleep-scheduling, call-admission, etc. As we enter a world where wireless devices are deployed pervasively in industrial and military settings as well as public and private buildings, there is increasing interest in providing robust localization techniques that are not only functional (in terms of providing the desired level of accuracy and speed), but also secure with respect to possible malicious attacks.

In the abstract, of course, there are a large number of different security threats and concerns in a wireless network. In some settings, of greatest concern is protecting the privacy and confidentiality of localization, ensuring that the location of unknown nodes in the network is not revealed to outsiders. In others, a significant concern is preventing denial of service attacks that involve disabling location reference nodes, exploiting vulnerabilities associated with the MAC protocol, or physical-layer jamming. Another key concern, the one that we focus on primarily in this work, is that of preventing degradation of the accuracy of the location service itself. This involves detecting, correcting, and mitigating the deliberate insertion of false signals (spoofing) by attackers.

We propose a novel methodology for secure localization that is inspired by error control coding techniques that have been used for many decades to provide robustness to noisy channels in traditional digital communications [8]. In error control coding techniques, the message to be sent is first encoded into a higher-dimensional codeword. When the message passes through a communication channel, it may be distorted due to noise and interference effects. At the receiver end, errors introduced by the channel can be detected (to an extent that depends upon the code-rate, the ratio of message bits to the bits in the codeword) by exploiting the fact that there are a relatively small number of correct/feasible codewords in the codeword space. Erroneous messages can be corrected, again to an extent depending upon the code-rate, by mapping the distorted encoded message to the the nearest feasible codeword and then decoding back to the original message.

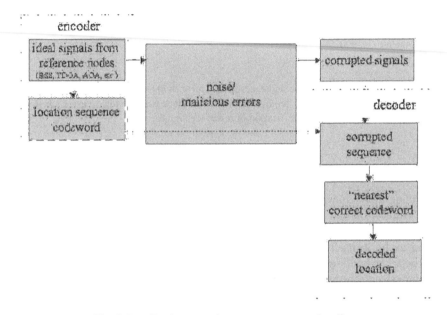

Fig. 1. Localization as analogous to error control coding

We present a sequence-based self-error-correcting localization mechanism, that is depicted as being analogous to error control coding in Figure 1. In this mechanism, the codeword that is used for localization is formed as a sequence by ranking the distances from the mobile node (with unknown location) to the various nearby reference nodes. For a given deployment of reference nodes, different location regions in the deployment area have distinct and unique codewords that can be tabulated in advance. Given such a codeword, the location of the node is then determined by a reverse table-lookup.

A key to the security of this technique is that the the density of feasible codewords (i.e. the codewords that correspond to an actual location region in the deployment area) decreases steeply with the number of location reference nodes. In particular, if N represents the number of reference nodes in a two-dimensional area, only $O(N^4)$

of the $O(N!)$ possible sequences are feasible codewords. This low density of feasible location codewords provides the ability to detect, correct and mitigate errors in the encoded location sequence that may be deliberately introduced by malicious attackers by the spoofing of ranging signals.

We present a brief numerical evaluation of the robustness of the proposed technique to signal spoofing by a powerful attacker. While the results from this technique are indeed encouraging, it is possible that the particular mechanism we propose may be just a tip of the iceberg. Extensions of this technique may reveal a large space of related secure location coding techniques that provide even greater immunity to malicious attacks.

2 Related Work

Localization of nodes in the context of densely deployed wireless networks in benign settings has been the object of intense study in the past decade. Several articles have provided a thorough survey of the subject (e.g., see [2], [3, Ch. 3]). However, it has been only even more recently that researchers have developed localization techniques suitable for settings involving malicious attackers. We briefly survey some of this work on secure localization:

In [7], the authors propose a secure localization technique that offers protection against spoofing of reference node locations by attackers. In this technique, localization is performed using a set of public base stations as reference nodes, whose location claims are verified by another set of covert base stations. The probability of the attacker's success is shown to grow linearly with ranging error, inversely to the square-root of the area of the localization space, and inversely to the number of covert base stations. In [1], another work that studies secure localization against spoofing of locations of reference nodes, two techniques based on majority-vote are proposed. In the first technique, the spoofed location is assumed to appear as a "outlier" in the set of locations of all reference nodes. The outlier is detected and discarded by verifying the unknown node location estimate with different subsets of reference nodes. In the second technique, reference nodes vote on the unknown node location likelihood at each grid point in the localization space and the grid point with the highest number of votes is chosen as the location estimate.

A statistical method of secure localization is proposed in [9]. In this method, the authors suggest that using the median of localization data is inherently more secure than using its average. Based on this, they propose that triangulation based localization techniques such as least sum of squares estimator should instead use least median of squares. They also argue that RF finger-printing based localization techniques should use minimum median distance instead of minimum average distance as the metric to determine the nearest neighbor.

In a work that addresses other attacker models, the authors in [4] propose a secure localization technique that is robust to *Wormhole* and *Sybil* attacks. In this technique, secure localization is provided using a combination of encryption keys, power control, and directional antennas that can transmit in different directions at the same time. In [5], the authors propose a combination of distance bounding, exchange of challenge-response type of messages, and authentication keys to ensure security of location determination. The authors of [6] address the problem of location verification in which the location of a wireless node is verified for its authenticity. This is done by using the inherent constraints in round trip propagation times of RF and ultra sound signals.

The technique that we describe here is quite different from prior work on secure localization in its use of ordered sequences as a location code. It is most similar to the Ecolocation technique that we previously proposed for localization using pure RF signals [10]. A key difference from that work is that we focus here on errors introduced by malicious attackers, rather than RF channel errors due to multi-path fading and other environmental factors.

3 Secure Sequence-Based Localization

Consider a two-dimensional location area in which N reference nodes with known locations are deployed. The goal is to enable a mobile node in the environment with an unknown location to locate itself. The mechanism can be implemented in either a node-centric or an infrastructure-centric mode. In the former implementation, the node obtains distance estimates to each reference node and computes its location according to the mechanism described below. In the latter, the reference nodes each obtain a distance estimate to the unknown node and transmit this information to a centralized node where the computation is performed.

The exact mechanism for obtaining the distance estimate can vary in either case. For instance, it could be based on time difference of arrival between a radio and acoustic signal, or based on an estimate using pure radio signal strength, or possibly even via phase interferometry). For a detailed discussion of how sequence-based localization provides robustness to errors in the distance estimation in the context of pure-RF localization, we refer the reader to the evaluation of the Ecolocation technique presented in [10]. In the discussion below, we shall assume that the ranging estimates are sufficiently accurate that they affect the accuracy of localization minimally. Thus our focus with regard to errors will be solely on those introduced by a malicious attacker.

3.1 Sequence Encoding of Locations

We now give a description of how the ranging signals are interpreted for localization in the deployment region. First, the distance estimates between the unknown node and the reference nodes is converted to a sequence that represents the relative distance ranking of each reference node (from nearest to farthest).

A correct sequence represents a valid location region in the deployment area where the distances to the different reference nodes are ranked accordingly. The two-dimensional deployment area can be partitioned into all valid location regions by drawing all lines that are perpendicular bisectors of lines joining all pairs of reference nodes. The two sides of each such line represent the distance inequalities with respect to a particular pair of reference nodes. The faces formed by the intersection of these lines represent all feasible regions, each forming unique combinations of the inequalities that result in the corresponding sequence.

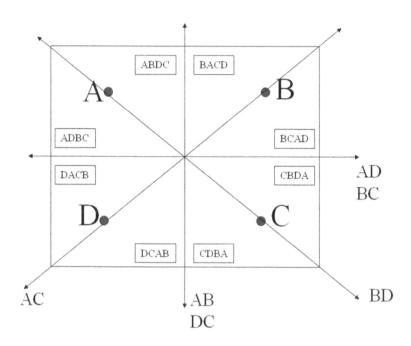

Fig. 2. Illustration of location regions and sequences for four nodes in a square region

For example, consider figure 2. It shows a square region in which there are four reference nodes labeled A, B, C, and D. Because the perpendicular bisectors for the node pairs AB, DC and that if node pairs AD, BC overlap respectively, there are four bisecting lines in all, dividing the region into eight location regions. The unique sequences corresponding to each of the eight regions area also shown. For instance the region on the bottom of the top-left quadrant corresponds to the sequence ADBC because all points in this region are closest to A, closer to D than B and C, and finally closer to B than C.

This mapping between location regions and location sequences can be encoded in a table *a priori*, during or before network initialization. In the absence of any errors, the unknown node can be located by performing a reverse look-up on this table to obtain the location given the sequence. Localization using a table look-up is quite feasible for intermediate-sized networks (we show in the following section that the size of the table is polynomial); the speed can be further improved using multi-resolution and gradient-based grid search techniques.

3.2 Density of Feasible Sequences

We now investigate a bound on the number of feasible sequences for a given deployment. Given that there are $N!$ possible sequences in all, how many of these actually correspond to feasible locations? The following proposition addresses this question, which is of core relevance not only to the time and space complexity of the algorithm, but also its ability to detect and correct errors.

Proposition 1. *In a two-dimensional area with N reference nodes, the number of feasible location sequences (i.e. sequence codewords that correspond to location regions in the 2D area) is $O(N^4)$.*

Proof: Each feasible location sequence represents a combination of pairwise inequality relationships with respect to the distance to two reference nodes. Each corresponding location region results from the intersection of these inequalities, each of which is represented by a line in the 2D plane that is a perpendicular bisector of the line joining each pair of reference nodes. It suffices to show that the number of faces formed by these intersecting lines is $O(N^4)$. We can do this in two steps.

- First, note that the number of bisecting lines is $O(N^2)$, since there is at most one line for each pair of reference nodes and there are $C_2^N = \frac{n(n-1)}{2}$ pairs of nodes.
- Next, we can prove that the arrangement of k lines in the plane can result in at most $O(k^2)$ faces. This can be proved by induction by showing that each line added to an existing arrangement of $i - 1$ lines adds at most i additional faces, which bounds the total number of faces by $k(k + 1)/2 + 1$.

Combining these together, we get that the total number of faces with N reference nodes must be $O(N^4)$.

$\diamond$

Note that we have assumed here a strict ordering of references, ignoring any lines or intersection points in the region where two or more references are equidistant, otherwise there could be potentially N^N total sequences (not $N!$), though the number of feasible sequences remains $O(N^4)$ even in that case.

3.3 Decoding Falsified Sequences

When there are no errors in the observed sequence, the localization algorithm is simple as it requires only a table look-up; it also provides excellent performance in terms of accuracy. Since there are $O(N^4)$ regions in a given deployment area A, the average area of location regions decreases roughly as $O(\frac{A}{N^4})$, providing a rapid gain in location accuracy as the number of reference nodes increases.

However, in the face of an attacker or even errors due to the environment, the original sequences may not be received correctly. The error detection and correction capabilities of the sequence-based localization mechanism are called for. These capabilities rely on the proposition we described in the last section: the ratio of feasible sequences (that correspond to valid locations in the deployment area) and the ratio of all possible sequences is very small for even moderate values of N, the number of reference nodes.

Most changes to the reference nodes' signals will not result in a sequence with a feasible location — this provides the basis for error detection. If the ranging mechanism used is inherently highly accurate (such as TDoA techniques in some settings), then a sequence that does not correspond to any of the feasible location region provides an indicator that a malicious attack may be in progress. Another way to view this is that the location encoding provides a consistency check because they represent fundamental geographic constraints; when a received sequence fails this consistency check, one can infer that errors have been introduced into the system.

Given that there is an error introduced into the sequence, a location can still be obtained from the sequence. This is done by mapping the erroneous sequence to the "nearest" feasible codeword sequence corresponding to a valid location region. The notion of "nearest" can be defined using different metrics suitable for sequences. A simple choice is to count the number of pairwise inequality constraints (pertaining to reference node distances) that are violated in the given erroneous sequence with respect to each feasible region, and pick the region that minimizes the number of unsatisfied constraints. Consider again the example deployment in figure 2. If the sequence ABCD is received, it can be detected immediately that an error has been introduced into the sequence since this sequence does not correspond directly to any of the location regions. The nearest feasible codeword to this sequence is ABDC, which corresponds to the region on the top of the top-left quadrant. This is because there is a difference of only one constraint violation between these two sequences (the inversion of the 'D is closer than C' constraint). If this mapping is assumed to be correct based on limitations on the attacker's capability, this may suggest that the attacker is either manipulating the signals of C to make it appear closer to the mobile node, or the signals of reference node D to make it appear farther than it really is to the mobile node.

While there is no guarantee that this approach will necessarily result in a correct solution in all cases, it can mitigate significantly the impact of the errors in the location sequence. The likelihood of correction and the obtained accuracy both increase with the number of reference nodes.

4 Evaluation

We briefly evaluate the proposed sequence-based localization mechanism though a set of simulations involving malicious attackers.

Evaluating the performance of a localization mechanism with respect to a malicious attacker requires some care. In particular, we need to be explicit about the capabilities of the attacker and strike a balance between providing the attacker with sufficient power and imposing some costs for the attack. Our approach is to impose such a cost by limiting the number of reference nodes that the attacker can manipulate to k (we provide numerical simulations for $k = 1$ and $k = 2$ here, but these can be extended to other values of k). At the same time, we provide an advantage to the attacker by allowing the attacker to know the 'true' location of the mobile node and switch accordingly at each time to the set of k reference nodes that hurts the localization technique the most in terms of accuracy. This allows the attacker to impose a worst-case penalty on the performance of our localization mechanism at all locations in the deployment area. In this sense, the provided results represent an upper bound on the damage that could be suffered at the hands of an intelligent attacker. In particular, the location accuracy in real attack scenarios may be substantially better than what is shown in these figures. This is because realistically, in some settings, the attacker may not able to observe the unknown node, and therefore may not be able to change which reference node is being spoofed or manipulated on the fly as the mobile node moves through the network.

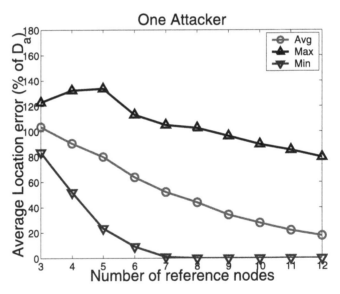

Fig. 3. Impact of one powerful attacker on secure sequence-based localization in a square area

We measure location error as the absolute distance between the true and estimated location as a percentage of D_a, the average distance between pairs of reference nodes; this provides normalization with respect to the density of reference node deployment. In case of a single strong attacker, whose impact is evaluated in Figure 3, the attacker inflicts the most damage on the localization scheme by introducing $N - 1$ constraint violations; this is done by moving the first node on the sequence to the end of the sequence (by spoofing the nearest reference node to the

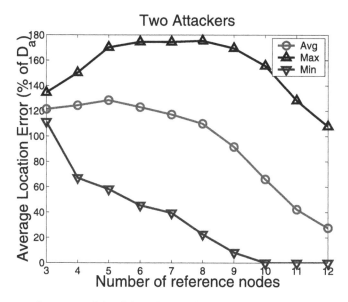

Fig. 4. Impact of two powerful collaborating attacker on secure sequence-based localization in a square area

mobile nodes' true location to make it seem the farthest). The three curves represent the normalized location error as a function of the number of reference nodes. The max, min and average shown are taken over all locations chosen uniformly from a square deployment area. The overall decreasing trends in all three curves is highly encouraging (although the max curve shows a small increase for less than five references). The steep falloff in the curve for min and the slower falloff for max suggests that there are some locations of the unknown node where a single attacker is quite powerless to affect the localization accuracy, and others where it has greater impact. On average, the proposed technique provides significant improvements (error less than 20% of inter-node spacing) when more than 12 reference nodes are present.

Figure 4 shows the corresponding results when two intelligent attackers collaborate to spoof the readings from two different reference nodes. In this case the worst damage to the location technique is inflicted by moving the first node in the sequence to the end, and bringing the last node in the sequence to the front (by adjusting the distance estimates for these two reference nodes accordingly). We observe again that there are decreasing trends in max, min and average location error once there are sufficiently many reference nodes (though there is an initial increase for a small number of references). We also see that there exist locations where two attackers have negligible impact on localization error when there are at least 10 references.

When relatively small-to-medium numbers of reference nodes are considered, the location regions obtained tend to exhibit some skewed boundary effects (the regions near the boundary are larger than those in the center). Eliminating these edge effects gives an idea of the performance in the context of a seamless deployment

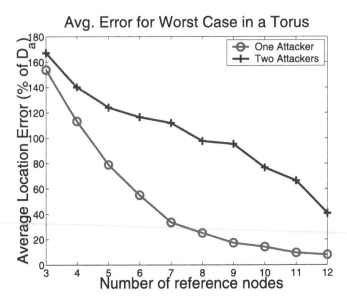

Fig. 5. Impact of one and two attackers on secure sequence-based localization in a torus (no edge effect)

of a large number of reference nodes (only a subset of which are within range at any location of the unknown node). We therefore examine the performance of the normalized location error (averaged over all locations) on a torus, for both single and two-reference-node attacks, in Figure 5. We observe that for the single attacker case, while the initial localization error for a torus is higher compared to the grid, the falloff is faster so that by the time there are 12 reference nodes, the localization error is lower (less than 10% of the baseline inter-reference spacing). For two attackers, the average localization error also shows a steady (almost linear) decline for the torus, though it appears slightly worse than in the case of the grid.

These simple experiments suggest on the whole that localization using sequences provides robust protection against malicious attacks in dense deployments.

5 Conclusions

We have presented a novel sequence-based secure localization mechanism. The key to its performance (which improves rapidly with increasing density of reference nodes) is that the number of feasible sequences is considerably smaller than the set of all sequences that can be generated, allowing for robust detection and correction of errors in the sequence. In future work it would be of interest to evaluate this technique for attacks involving an even greater number of compromised nodes, and compare the security provided by this technique with other state-of-the-art approaches. Still, this technique is perhaps just the tip of an iceberg; much remains to be learned about such self-error-correcting localization techniques, particularly with respect to malicious attacks.

6 Acknowledgements

The research in this work has been supported in part by National Science Foundation through grants NeTS-NOSS CNS-0435505, CAREER CNS-0347621, ITR CNS-0325875, and CCF-0430061, and through a gift from Bosch Research.

References

1. Donggang Liu, Peng Ning, and Wenliang Du. Attack-resistant location estimation in sensor networks. In *Proceedings of the Fourth International Symposium on Information Processing in Sensor Networks, IPSN*, pages 99–106, Los Angeles, CA, USA, 2005.
2. Jeffrey Hightower and Gaetano Borriello. Location systems for ubiquitous computing. *Computer*, 34(8):57–66, August 2001.
3. Bhaskar Krishnamachari *Networking Wireless Sensors* Cambridge University Press, 2005.
4. Loukas Lazos and Radha Poovendran. SeRLoc: secure range-independent localization for wireless sensor networks. In *WiSe '04: Proceedings of the 2004 ACM workshop on Wireless security*, pages 21–30, New York, NY, USA, 2004. ACM Press.
5. Loukas Lazos and Radha Poovendran and Srdjan Capkun. ROPE: Robust Position Estimation in Wireless Sensor Networks. In *Proceedings of the Fourth International Symposium on Information Processing in Sensor Networks, IPSN*, Los Angeles, CA, USA, 2005.
6. Naveen Sastry, Umesh Shankar and David Wagner. Secure verification of location claims. In *WiSe '03: Proceedings of the 2003 ACM workshop on Wireless security*, pages 1–10, New York, NY, USA, 2003. ACM Press.
7. Srdjan Capkun, Mario Cagalj and Mani Srivastava. Secure Localization With Hidden and Mobile Base Stations. In *IEEE INFOCOM*, Barcelona, Spain, 2006.
8. Stephen Wicker, *Error Control Systems for Digital Communication and Storage*, Prentice-Hall, 1995.
9. Zang Li, Wade Trappe, Yanyong Zhang and Badri Nath. Robust Statistical Methods for Securing Wireless Localization in Sensor Networks. In *Proceedings of the Fourth International Symposium on Information Processing in Sensor Networks, IPSN*, Los Angeles, CA, USA, 2005.
10. Kiran Yedavalli, Bhaskar Krishnamachari, Sharmila Ravula and Bhaskar Srinivasan. Ecolocation: A Sequence Based Technique for RF-only Localization in Wireless Sensor Networks In *Proceedings of The Fourth International Conference on Information Processing in Sensor Networks, IPSN*, Los Angeles, CA, April 2005.

Securing Localization in Wireless Networks *(using Verifiable Multilateration and Covert Base Stations)*

Srdjan Čapkun

Informatics and Mathematical Modelling Department, Technical University of Denmark (DTU), DK-2800 Lyngby, Denmark; sca@imm.dtu.dk

1 Introduction

In the last decade, researchers have proposed a number of localization and ranging techniques for wireless networks [2, 5, 12, 26, 39, 40]. The use of these techniques is broad and ranges from enabling networking functions (i.e., location-based routing) to enabling location-related applications (e.g., access control, data harvesting).

The proposed techniques were mainly studied in non-adversarial settings. Ranging and localization techniques are, however, highly vulnerable to attacks from internal attackers (e.g., dishonest nodes) and external attackers; internal attackers can report false location and distance information in order to cheat on their locations; external attackers can spoof measured locations of network nodes. Localization and ranging techniques in wireless networks mainly rely on measurements of the times of flight of radio (RF ToF) or ultrasound signals (US ToF), and on the measurements of received strengths of radio signals of devices (RF RSS). An attacker can generally influence all these measurements by jamming and delaying signals, and by modifying their signal strengths. Localization systems based on ultrasound time of flight (US ToF) and those based on measurements of signal strength of radio signals (RF RSS) are particulary vulnerable to location spoofing attacks; systems based on radio time of flight measurements are less vulnerable to attacks because of the high speed of signal propagation,

In this work, we first review attacks on ranging and localization techniques in wireless networks. We then describe two approaches for securing localization: Verifiable Multilateration (VM) and Secure Localization with Covert Base Stations.

Verifiable Multilateration (VM) is based on the measurements of the time of radio (or sound) signal propagation (i.e., time-of-flight (ToF)); it consists of conventional multilateration with distance bounding or with authenticated ranging. This protocol enables verification of node locations by a set of (at least three) trusted base stations, which do not need to be tightly synchronized.

---

[0] The author acknowledges contributions from Mario Čagalj, Jean-Pierre Hubaux and Mani Srivastava to original publications [36–38] from which this chapter is compiled.

Secure Localization with Covert Stations is based on the unpredictability of locations of base stations performing localization. Notably, locations of covert base stations represent a secret input *(a key)* to the system. Covert base stations can be realized by hiding or disguising static base station or by the random motion of mobile base stations. Typically, covert base stations are passive.

The organization of this chapter is as follows. In Section 2, we review localization techniques and analyze attacks against them. In Section 3, we describe secure localization with Verification Multilateration (VM). In Section 4, we present a scheme for secure localization based on hidden base stations. In Section 5, we present an review current proposals and techniques for secure localization in wireless networks. We conclude the paper in Section 6.

2 Attacks against location and distance estimation techniques

We now review localization and distance estimation techniques and analyze their vulnerabilities. First, we briefly present our attacker model.

2.1 Attacker model

We call an attacker *external* if it cannot authenticate itself as an honest network node to other network nodes or to a central authority. We call an attacker *internal* if the node is *compromised* or if the user controlling the node is *malicious*. We assume that malicious and compromised nodes can authenticate themselves to the authority and to other network nodes. When a node is compromised, its secret keys and other secrets that it shares with other nodes are known to the attacker. Furthermore, users have full access to their devices, meaning also to their authentication material.

Similarly, we observe two types of attacks: internal and external. Internal attacks are those in which an internal attacker reports a false location or convinces the localization infrastructure that it is at a false location. External attacks are those in which an (external) attacker convinces an honest node and the localization infrastructure that the node is at a different location from its true location (i.e. the attacker *spoofs* node's location).

We distinguish two types of localization systems: infrastructure-centric and node-centric. By a node-centric localization system we mean that a node computes its location by observing signals received from public base stations with known locations. If the localization system is *node-centric*, internal attacks are generally straightforward: the attacker simply lies about the location that it computed. *Infrastructure-centric* localization systems are those in which the infrastructure computes locations of nodes based on their mutual communication.

2.2 Attacks on Global Positioning System (GPS)

The Global Positioning System is today the most widespread outdoor localization system for mobile devices. The system is based on a set of satellites that provide a three dimensional localization with an accuracy of around 3 m. GPS also provides devices with an accurate time reference. GPS, however, has several limitations: it cannot be used for indoor localization nor for localization in dense urban regions: in those cases, because of the interferences and obstacles, satellite signals cannot reach the GPS devices. Furthermore, civilian GPS was never designed for secure localization. Civilian GPS devices can be "spoofed" by GPS satellite simulators, that produce fake satellite radio signals that are stronger than the real signals coming from satellites. Most current GPS receivers can be totally fooled, accepting these stronger signals while ignoring the weaker, authentic signals. GPS satellite simulators are legitimately used to test new GPS products and can be bought for $10k-$50k or rented for just $1k per month. Some simple software changes to most GPS receivers would permit them to detect relatively unsophisticated spoofing attacks [41]. Nevertheless, more sophisticated spoofing attacks would still be hard to detect. Military GPS are protected from location spoofing by codes that cannot be reproduced by the attackers.

Even if a mobile node is able to obtain its correct location from the GPS satellites, the authority or another mobile node have no way to verify the correctness of node's location, unless the mobile node is equipped with a trusted software or hardware module [1], providing the correct location.

2.3 Attacks on Ultrasound (US)-based localization

Ultrasound-based systems operate by measuring ToF of the sound signal measured between two nodes. An interesting feature of these systems is that, if used with RF signals, they do not require any time synchronization between the sender and the receiver. The limitations of the US-based systems are that, due to outdoor interferences, they can be mainly used indoors.

US-based systems are vulnerable to distance reduction and distance enlargement attacks by external and internal attacks. To reduce the measured distance between two honest nodes, two attackers can use a radio link, as it transmits the signal several orders of magnitude faster than the US. Furthermore, by jamming and replaying the signals at a later time, attackers can enlarge the measured distances between honest nodes. With US-based techniques, an internal attacker can also reduce or enlarge the measured distance by laying about the signal sending/reception times or by simply delaying its response to honest nodes. Recently, Sastry, Shankar and Wagner [31] have proposed a US-based distance bounding technique which resists to distance reduction attacks from internal attacks; it does not, however, resist to attacks from external nodes.

2.4 Attacks on Radio (RF)-based localization

In techniques based on the Received Signal Strength (RSS), the distance is computed based on the transmitted and received signal strengths. To cheat on the measured distance, an internal attacker therefore only needs to report a false power level to an honest node. Malicious attackers can also modify the measured distance between two honest nodes by jamming the nodes' mutual communication and by replaying the messages with higher or lower power strengths.

RF time-of-flight-based systems exhibit the best security properties. In these systems, nodes measure their mutual distance based on the time of propagation of the signal between them. Because RF signals travel at the speed of light, an attacker can, by jamming and replaying the signals, only increase, but not decrease the measured time-of-flight between the nodes. An internal attacker can further cheat on the distance by laying about the signal transmission and reception times.

An RF distance bounding technique proposed by Brands and Chaum [3] exhibits better security properties than conventional RF ToF distance estimation; it allows the nodes to upper bound their distances to other nodes, meaning that it prevents an internal attacker from reducing the measured distance. As we will show in Section 3.1 in more detail, with RF ToF distance-bounding protocols, attackers can only increase, but not decrease the measured distances to honest nodes.

	Internal attackers	External attackers
RSS (Received Signal Strength)	Distance enlargement and reduction	Distance enlargement and reduction
US time-of-flight (ToF)	Distance enlargement and reduction	Distance enlargement and reduction
RF time-of-flight (ToF)	Distance enlargement and reduction	Distance enlargement only
US distance bounding	Distance enlargement only	Distance enlargement and reduction
RF distance bounding	Distance enlargement only	Distance enlargement only
Civilian GPS	False location reports	Location spoofing

Table 1. Vulnerabilities of localization and distance estimation techniques to distance and location spoofing attacks.

2.5 Conclusion

Our review of vulnerabilities of localization systems is summarized Table 1. This table illustrates that the RF ToF-based localization solutions are best suited for secure localization. The RF ToF distance estimation and distance bounding techniques are the most effective techniques to counter attacks. The reason is that with RF it is generally possible to perform precise non-line-of-sight distance estimations; the precision of the system can be very high (15 cm error with Ultra Wide Band systems at a distance of 2 km [8]). A potential drawback of these systems is that, because they operate with the speed of light, the devices require fast-processing hardware.

u : **Generate** random nonce N_u
    : commitment $(c, d) = \text{commit}(N_u)$
$u \rightarrow v : c$
    v : **Generate** random nonce N_v
$v \rightarrow u : N_v$ *(bits sent from MSB to LSB)*
$u \rightarrow v : N_u \oplus N_v$ *(bits sent from LSB to MSB)*
    v: **Measure** time t_{vu} between sending N_v
       and receiving $N_u \oplus N_v$
$u \rightarrow v : N_u, N_v, d, MAC_{K_{uv}}(u, N_u, N_v, d)$
    v: **Verify** MAC and verify if
       $N_u = \text{open}(c, d)$

Fig. 1. Distance bounding protocol.

3 Secure Localization with Verifiable Multilateration

In this section, we describe Verifiable Multilateration (VM) algorithm. Before describing the algorithm, we first introduce two constructs used in VM: distance bounding and authenticated ranging.

3.1 Distance bounding and Authenticated ranging

Distance bounding techniques are used to upper-bound the distance of one device to another (compromised) device. As we indicated in Table 1, RF-based distance bounding protocols are vulnerable to distance enlargement attacks but not to distance reduction attacks. Distance bounding protocols are used by a verifier v to verify that a claimant node u being at a distance d_{uv} from a verifier node v, cannot claim to be at a distance $d'_{uv} \langle d_{uv}$. These protocols were first introduced by Brands and Chaum [3] to prevent Mafia Fraud attacks.

The pseudocode of the distance bounding protocol is shown in Figure 1. In the first step of the protocol, the claimant u commits to a random value N_u. The verifier replies with a challenge nonce N_v, sends it to u in a reverse bit order and starts its timer as soon as the last bit of the challenge has been sent. The claimant u responds immediately with $N_v \oplus N_u$, upon receiving the challenge from v. Once the verifier has received the response from u it stops the timer and converts the challenge-response time t_{vu} to a distance d_{vu}. In the last step of the protocol, u authenticates itself to v and reveals the decommit value $\hat{d}$. The authentication and the authenticity of d is ensured with a message authentication code (MAC), using a secret key K_{vu} that u and v share. Finally, v verifies if the value N_u received in the time-measuring phase corresponds to the received (commit, decommit) pair $(c, \hat{d})$.

The commitment scheme needs to satisfy two properties: (i) a user who commits to a certain value cannot change this value afterwards (we say that the scheme is *binding*), (ii) the commitment is hidden from its receiver until the sender "opens" it (we say that the scheme is *hiding*). A commitment scheme transforms a value m into a commitment/opening pair (c, d), where c reveals no information about m, but (c, d) together reveal m, and it is infeasible to find $\hat{d}$ such that $(c, \hat{d})$ reveals $\hat{m} \neq m$. Simple

$$u : \textbf{Generate} \text{ random nonce } N_u$$
$$: \text{commitment } (c, d) = \text{commit}(N_u)$$
$$u \rightarrow v : c$$
$$v : \textbf{Generate} \text{ random nonce } N_v$$
$$v(t_s^v) \rightarrow (t_r^u)u : N_v$$
$$u(t_s^u) \rightarrow (t_r^v)v : N_u \oplus N_v$$
$$u \rightarrow v : t_s^u, t_r^u, d, MAC_{K_{uv}}(u, N_u, N_v, t_s^u, t_r^u, d)$$
$$v : \textbf{Verify} \text{ MAC and verify if}$$
$$N_u = \text{open}(c, d)$$
$$v : \textbf{Compute } d = (t_r^v - t_s^v - t_s^u + t_r^u)s$$

Fig. 2. Authenticated ranging protocol.

commitment schemes can be realized with hash functions, which do not impose high computational requirements on sensor nodes.

The described protocol is suitable for devices that can perform rapid message exchanges, execute XOR operations rapidly, and perform encryption. In the case of RF-based distance bounding, the most important assumptions are that the claimant needs to be able to bound its processing (XOR) to a few nanoseconds, and that the verifier v needs to be able to measure time with nanosecond precision ($1ns$ corresponds to the time that it takes an electromagnetic wave to propagate over 30 cm). This requirement allows the node to perform distance bounding with radio signals with an uncertainty of 30 cm. We are aware that a nanosecond processing and time measurements are achievable only with dedicated hardware. Recent developments in location system show that RF time of flight systems based on Ultra Wide Band (UWB) can achieve nanosecond precision of measured times of signal flight (and consequently of the distances). The tests with Multispectral solution's UWB Precision Asset Location system [9] consisting of active tags and tracking devices show that this system can provide two- and three-dimensional location of objects to within a few centimeters. The range of the system is 100 m indoor and 2km outdoor. The used UWB tags are active and roughly the size of a wristwatch, weighing approximately 40 grams each.

In the case of a US-based distance bounding, node processing speed and clock accuracy can be of the order of milliseconds. Thus, US distance bounding can be easily implemented with off-the-shelf components such as microphones and 802.11 wireless cards [31].

Authenticated ranging protocols enable two honest and trusted parties to measure their mutual distance in an authenticated manner. Figure 2 shows one possible realization of the authenticated distance ranging protocol, inspired by Brand's and Chaum's distance bounding protocol. Here, t_s^u, t_r^u and t_s^v, t_r^v are the message sending and reception times at nodes u and v, respectively; s is the speed of light.

In this protocol, unlike in the distance bounding protocol, it is not required that the claimant replies within a nanosecond time, but only that it is able to measure time with that precision. Given that the claimant and the verifier are mutually trusted, the claimant (u) reports its processing time to the verifier (v) which then computes the range based on the reported times using speed of light.

Like in the distance bounding protocol, in this protocol, the processing at the nodes is minimized during the ranging phase, and most processing (MAC and commitment verification) is performed a posteriori to the ranging.

The advantages of the ranging protocol over distance bounding are in that the nodes do not need to have high-speed hardware to perform XOR and that the channel does not need to be reserved during the ranging phase (as processing and channel access times are measured and reported). One disadvantage is that ranging is not resistant to distance reduction by internal attackers.

A very important observation here is that, essentially, authenticated ranging and distance bounding have the same resistance to external attackers: the only attack that the external attackers can successfully perform is distance enlargement. In case of internal attackers, distance bounding prevents distance reduction, whereas the authenticated ranging is vulnerable to this attack.

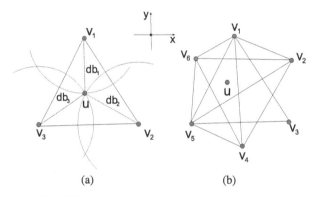

(a)   (b)

Fig. 3. Examples of Verifiable Multilateration. a) with three verifiers. b) with six verifiers.

3.2 Verifiable Multilateration

In Section 2, we described security problems related to various localization and distance estimation techniques and in Section 3.1 we showed how the devices can upper-bound their mutual distances. We now propose a technique for location verification that we call *Verifiable Multilateration* (VM). This technique enables a secure computation and verification of the locations of mobile devices in the presence of attackers. Here, by *secure location computation* we mean that base stations compute the correct location of a node in the presence of attacker, or that a node can compute its own location in the presence of an attacker; by *secure location verification* we mean that the base stations can verify the location reported by the node.

Multilateration is a technique for determining the location of a (mobile) device from a set of reference points whose locations are known, based on the ranges measured between the reference points and the device. The location of the device in two (three) dimensions can be computed if the device measured its distance to three (four) reference points. As we already detailed in Section 2, distance estimation techniques are vulnerable to attacks from internal and external attacks, which can maliciously modify the measured distances. Multilateration is equally vulnerable to the same set of attacks because it relies on distance estimations.

Algorithm

Verifiable Multilateration relies on distance bounding (or on authenticated ranging). It consists of distance bound measurements from at least three reference points (verifiers) to the mobile device (the claimant) and of subsequent computations performed by an authority. In this description, we will assume that the verification is performed with distance bounding. For simplicity, we show the algorithm for two dimensional localization; at the end of the section, we briefly comment on how a similar algorithm can be applied to the three dimensional case.

The intuition behind the verifiable multilateration algorithm is the following. Because of the distance bounding property, the claimant can only pretend that it is more distant from the verifier than it really is. If it increases the measured distance to one of the verifiers, in order to keep the location consistent, the claimant needs to prove that at least one of the measured distances to other verifiers is shorter than it actually is, which it cannot because of the distance bounding. This property holds only if the location of the claimant is determined within the triangle formed by the verifiers. This can be explained with a simple example: if an object is located within the triangle, and it moves to a different location within the triangle, it will certainly reduce its distance to at least one of the triangle vertices. The same properties hold if an external attacker enlarges distances between verifiers and an honest claimant. This basic intuition is illustrated in Figure 3a.

More precisely, the verifiable multilateration algorithm is executed by the verifiers as shown on Figure 4.

In step 1 of the algorithm, the verifiers $v_1, ..., v_n$ which are in the power range of the claimant u perform distance bounding to the claimant u and obtain distance bounds $db_1, ..., db_n$. These distance bounds as well as the locations of the verifiers (which are known) are then reported to the central authority. In step 2, the authority computes an estimate (x', y') of the claimant's location; this location is computed by using distance bounds from all verifiers in u's neighborhood, typically by the Minimum Mean Square Estimate (MMSE):

$$\text{Let } f_i(x'_u, y'_u) = db_i - \sqrt{(x_i - x'_u)^2 + (y_i - y'_u)^2}$$

The location of u is obtained by minimizing
$$F(x'_u, y'_u) = \sum_{v_i \in T} f_i^2(x'_u, y'_u)$$
over all estimates of u

In step 3 of the algorithm, the authority runs the following two tests: (i) δ-test: for all v_i, does the distance between (x'_u, y'_u) and v_i differ from the measured distance

$\mathcal{T} = \emptyset$; set of verification triangles enclosing u

$\mathcal{V} = \{v_1, ..., v_n\}$; set of verifiers in the power range of u

1 *For all* $v_i \in \mathcal{V}$, perform distance bounding
from v_i to u and obtain db_i

2 With all $v_i \in \mathcal{V}$, compute the estimate (x'_u, y'_u) of the location
by MMSE

3 *If for all* $v_i \in \mathcal{V}$, $|db_i - \sqrt{(x_i - x'_u)^2 + (y_i - y'_u)^2}| \leq \delta$ *then*
for all $(v_i, v_j, v_k) \in \mathcal{V}^3$, if $(x'_u, y'_u) \in \triangle(v_i, v_j, v_k)$
then $\mathcal{T} = \mathcal{T} \cup (v_i, v_j, v_k)$
if $|\mathcal{T}|\rangle 0$ *then* location is accepted and $x_u = x'_u$, $y_u = y'_u$
else the location is rejected

else the location is rejected

Fig. 4. Verifiable multilateration

bound db_i by less than the expected distance measurement error δ and (ii) *point in the triangle test:* does (x'_u, y'_u) fall within at least one physical triangle formed by a triplet of verifiers. Note also that we call the triangle formed by the verifiers the *verification triangle*. If both the δ and the point in the triangle tests are positive, the authority accepts the estimated location (x'_u, y'_u) of the claimant as correct; else, the location is rejected.

The expected error δ is a system parameter that depends on the number of verifiers and on the distance estimation techniques used. This error becomes smaller as more verifiers are used to compute (x'_u, y'_u).

If both the δ and the point in the triangle tests are positive, this means that the claimant falls in at least one verification triangle v_i, v_j, v_k, and that distance bounds (db_i, db_j, db_k) are consistent with the estimated location and with each other (Figure 3a). This means that none of the distance bounds (db_i, db_j, db_k) were enlarged.

If any of the distance-bounds db_i differs from the estimated location (x'_u, y'_u) by more than δ, this indicates that there is a possible distance enlargement attack on one or more of the distance bounds that caused such an unexpectedly high error to occur. If a larger number of verification triangles can be formed around u, the authority can try to detect which of the distances are enlarged. Those distances can then be filtered-out and the location can be computed with the remaining set of distances. This detection is performed such that the location of u is computed independently in each triangle. If in a given triangle the computation is successful, then all the distance bounds from the verifiers forming that triangle are considered correct; otherwise, all three distance bounds are considered suspicious (see Figure 5).

In this algorithm, the number of verification triangles and the number of enlarged distances will determine if the algorithm can detect which distance(s) is(are) enlarged. Nevertheless, in all cases, even if the number of verifiers is strictly equal to three, the Verifiable Multilateration algorithm will detect any distance enlargement attack (even if only one distance is enlarged), but it will not always be able to detect which distance it is.

$\mathcal{C} = \emptyset$; set of verifiers with correctly measured bounds
$\mathcal{NC} = \emptyset$; set of verifiers whose bounds are suspicious
1 *For all* $v_i \in \mathcal{T}$
  if in at least one of the verification triangles
     with v_i the location of u is computed correctly
  then db_i is correct, $\mathcal{C} = \mathcal{C} \cup \{v_i\}$
  else $\mathcal{NC} = \mathcal{NC} \cup \{v_i\}$
2 *For all* $v_i \in \mathcal{NC}$
  if v_i can create a verification triangle
     with any pair $(v_j, v_k) \in \mathcal{C}^2$
  then db_i *is subject to an enlargement attack*
3 With all $v_i \in \mathcal{C}$, compute the estimate (x'_u, y'_u) of the location
  by MMSE
4 *For all* $v_i \in \mathcal{NC}$, if $|db_i - \sqrt{(x_i - x'_u)^2 + (y_i - y'_u)^2}| \leq \delta$
  then db_i *is subject to an enlargement attack*

Fig. 5. Detection of enlarged distances.

Verifiable Multilateration can be also applied to three dimensional localization. For this, the system requires a minimum of four verifiers, that form a triangular pyramid, within which the secure determination of the claimant's location is possible. The algorithm is then executed in a way similar to the two-dimensional case.

3.3 Security Analysis

In this section, we analyze the security properties of verifiable multilateration in various scenarios. We observe verifiable multilateration with distance bounding or with authenticated ranging, assuming trusted or un-trusted users, and with radio-based or ultrasound-based bounding/ranging.

Verifiable Multilateration with distance-bounding
The most important properties of the Verifiable Multilateration mechanism with distance-bounding can be summarized as follows:

1. A node located at location p within the triangle/pyramid formed by the verifiers cannot prove to be at another location $p' \neq p$ within the same triangle/pyramid.
2. A node located outside the triangle/pyramid cannot prove to be at any location p within the triangle/pyramid.
3. An external attacker performing a distance enlargement attack cannot trick the verifiers into believing that a claimant located at a location p in the triangle/pyramid is located at some other location $p' \neq p$ in the triangle/pyramid.
4. An external attacker performing a distance enlargement attack cannot trick the verifiers into believing that a claimant is located at any location p within the triangle/pyramid, if the claimant is located outside of the triangle/pyramid.

These properties hold for verifiable multilateration based on radio distance bounding (VM-RF-DB) in environments in which the signal propagates at the speed of light, and for an internal attacker that controls a single device (the claimant). VM-RF-DB therefore resists to external attacks and to internal attacks from a single un-trusted/compromised node.

However, if an attacker owns several devices and each device can authenticate to the authority as the same entity, the attacker can still successfully cheat on its location. The attacker can place three/four devices within the triangle/triangular pyramid, such that each device is close to one of the verifiers. Each of the devices can then show to its corresponding base station (by delaying the messages) that it is located at *any* distance larger than their actual distance (which is small). As to the base stations these devices appear to be a single claimant, the attacker can prove to be at any distance to the base stations, and thus at any location in the verification triangle/triangular pyramid. This attack is shown on Figure 6. Here, the attacker clones its device, or three attackers collude to appear as a single node to the verifiers. This enables the attacker (or colluding nodes) to prove that the location of the claimant is at an incorrect place within the verification triangle.

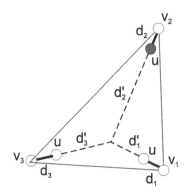

Fig. 6. Device cloning/user collusion. The attacker clones a device u and places one clone close to each verifier. The clones seem to the verifiers as a single device and can use distance enlargement to show that u is at any location within the verification triangle. Alternatively, three users collude in proving an incorrect location of node u.

A solution that prevents this attack is to make claimant devices tamper-proof such that their authentication material is not revealed to the attacker and that they cannot be cloned; however, as shown in [1], tamper-proofness has its limitations. Another possibility is that the base stations perform device fingerprinting [33] by which they identify each device as unique. In that case, the base stations can identify a claimant device by the unique "fingerprint" that characterizes its signal transmission[1].

---

[1] This process is used by cellular network operators to prevent cloning fraud; namely, a cloned phone does not have the same fingerprint as the legal phone with the same electronic identification numbers.

Verifiable multilateration with ultrasonic distance bounding (VM-US-DB) in air, exhibits only properties (1) and (2), meaning that it protects the localization system from an un-trusted claimant, but not against an external attacker nor from colluding internal attackers (claimants). However, if the devices are under water, VM-US-DB can exhibit the same properties as VM-RF-DB; this is because, underwater, the communication is limited to ultrasonic signals. VM-US-DB can be attacked if an attacker can use surface wormholes to perform distance reduction [16].

Verifiable Multilateration with authenticated ranging
Verifiable multilateration with radio authenticated ranging (VM-RF-AR) exhibits only properties 3 and 4 of the VM-RF-DB. This means that this scheme provides protection against external attacks, but not against un-trusted claimants (internal attackers). VM-RF-AR is therefore most suitable for secure localization systems in which the infrastructure (the verifiers) and the users (the claimants) are mutually trusted. In these scenarios, VM-RF-AR resists to all distance enlargement attacks by external attackers.

Verifiable multilateration with ultrasonic authenticated ranging (VM-US-AR) exhibits the same properties as VM-RF-AR, but only in underwater communications, whereas in air, it does not provide any security at all. The results of this analysis are summarized in Table 2.

	external attackers	1 internal attacker	1 internal + external attackers	colluding internal /cloning internal
VM-RF-DB + DF	yes	yes	yes	yes
VM-RF-DB	yes	yes	yes	no
VM-US-DB	no (yes UW)	yes	no	no
VM-RF-AR	yes	no	no	no
VM-US-AR	no (yes UW)	no	no	no

Table 2. Resistance of Verifiable Multilateration (VM) to attacks. RF=radio communication, US=ultrasonic communication, DB=distance-bounding, AR=authenticated ranging, DF=device fingerprinting, UW=underwater.

3.4 Maximum Attacker Impact

In this section we analyze the impact of distance measurement errors on Verifiable Multilateration. As we have already described, for the computed location to be accepted by the verifiers, each distance bound needs to be less than δ different from the distance between the computed location and the verifier measuring that distance bound. We defined δ as the expected localization error. Here, we define δ more precisely as 3σ, where σ is the expected standard deviation of the computed location. This means that we expect, with probability of 0.997 (the confidence interval corresponding to 3σ) that the real node location will lay in the circle of radius 3σ around the computed location x'_u, y'_u.

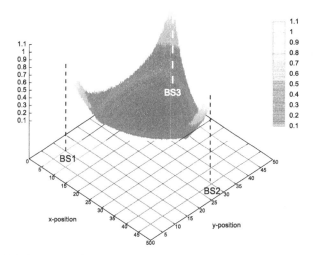

Fig. 7. Standard deviation (σ) of the location computed with MMSE of distance measurements to a UWB tag, performed by three UWB base stations. Localization is performed within a triangle formed by the three base stationss.

Within the Verifiable Multilateration this means that the maximal attacker impact on the computed location is upper-bounded by 3σ.

To estimate the expected σ of an UWB localization system, we used the results of the distance measurements between UWB base stations and UWB tags, published by Multispectral Solutions [8]. These results show that for indoor localization, the standard deviation of the measured distances increases with the distance length. The measured distances were from 0 to 50 m, and the standard deviation was from 0 to 1 m. From these distance measurements, we computed the standard deviation of the locations within a verifiable triangle formed by the three base stations. The results of this computation are shown on Figure 7. The values on the x and y axis denote the measured locations, and the z axis denotes the standard deviation of the location. We observe that the standard deviation is the highest at locations close to the base stations. This is an expected result, due to geometric dilution of precision. It is also important to observe that the value of σ is lower than 0.5 m in the center part of the triangle, and that it increases to 1 m for locations closer to the base stations.

Given that the verifiers do not know a priory if the location that they computed is correct or not, VM cannot operate with δ, as it depends on the computed location. This is notably because we do not want to give any advantage to the attacker by allowing him to modify the distances (the location) in order to influence the choice of δ. VM therefore needs to operate in a "worst case" scenario with a fixed value for δ. This also means that δ needs to be chosen such that the locations which are not spoofed are not likely to be rejected, and that the locations which are spoofed are detected.

It is important to notice that by choosing δ, the verifiers are sure that the locations at which $3\sigma \leq \delta$ will not be rejected if there was no attack on the distances. This means that by choosing different δs, the verifiers will modify the verification area; for larger δ, the verification region will be larger, but so will be the maximum attacker impact.

3.5 Location Privacy

So far, we have described the infrastructure-based verifiable multilateration (IB-VM), in which the verifiers compute the location of the claimant. IB-VM does not preserve the claimant's location privacy. There are two reasons for that. The first reason is that the verifiers compute the location of the claimant, and therefore have full knowledge of where the claimant is located. The second reason is that the described distance bounding and ranging protocols are vulnerable to attacks by fake verifiers whose goal is to detect the location of a node by initiating the execution of the distance-bounding protocol. This second problem can be eliminated If the verifiers are authenticated to the claimants prior to the distance verification. In [34], Capkun et al., proposed a protocol for mutually authenticated distance bounding (MAD) that enables two nodes to determine their mutual distance bounds at the time of encounter. This protocol can be used to prevent attacks by fake verifiers.

Still, even with mutual claimant-verifier authentication IB-VM does not fully protect the claimant's location privacy, because the infrastructure knows the location of the claimant. This problem can be solved through Node-based Verifiable Multilateration (NB-VM). In this protocol, the claimant performs distance bounding to the verifiers, and computes its location within the verification triangle in the same way as in the protocol in Figure 4. Here, the claimant trusts the verifiers about their locations, but not does allow them to find out its location. However, the verifiers could try to infer the claimant's location based on the readings of the strengths of the signals received from the claimant. This, and similar attacks on node's location privacy have been previously investigated [11, 14, 15, 27, 28, 30], but thwarting these attacks is out of the scope of this work.

4 Secure Localization with Hidden and Mobile Base Stations

In the previous section, we introduced Verifiable Multilateration, a technique that makes use of the speeds of signal propagation and geometric properties of the space to verify locations of wireless devices. If implemented with RF distance bounding or authenticated ranging, VM, however, imposes high requirement on the clock precision of the devices. In this section, we introduce a secure localization scheme that removes this assumption, and instead leverages on the unpredictability of base station locations.

4.1 Model

Our system consists of a set of covert base stations (CBS) and a set of public base stations (PBS) forming a localization infrastructure. Here, by covert base stations we mean those base stations whose locations are known only to the authority controlling the verification infrastructure. To prevent that their locations are discovered through radio signal analysis, covert base stations are silent on the wireless channel; they only listen to the on-going communication.

In our system covert and public base stations know their locations or can obtain their locations securely (e.g., through secure GPS [17]). Here, we assume that the attackers cannot tamper with these locations nor compromise the base stations.

We also assume that every legitimate node shares a secret key with the base stations, or that base stations hold an authentic public key of the node. This key is established/obtained through the authority controlling the verification infrastructure prior to location verification. Here, all communication between the authority and a node is performed through a public base station, whereas the hidden stations remain passive.

We further assume that covert base stations can measure received signal strength or have an ultrasound interfaces through which they perform ranging.

In most of this work, we assume that covert base stations are static. Thus, their mutual communication and their communication to the verification authority is performed through a channel that preserves their location privacy; this communication channel is typically wired (or infrared), such that they cannot be detected by the attackers.

Here, we use the attacker model described in Section 3.2.

4.2 Infrastructure-centric localization with hidden base stations

In this section, we describe a simple solution for securing infrastructure-centric localization systems, based on time difference of arrival (TDOA) and covert base stations.

TDOA is the process of localization a source of signal in two (respectively three) dimensions by finding the intersection of multiple hyperbolas (or hyperboloids) based on the time difference of arrival between the signal reception at multiple base stations. An hyperboloid is defined as a surface, that has a constant distance difference from two points (in our case two base stations). Using two hyperbolas (three base stations) we can obtain two dimensional device locations, and using three hyperboloids (four receivers) we can determine three dimensional locations. The operation of the TDOA technique is shown on Figure 8. Node A sends a radio signal, and the verifiers measure the difference between the times t_1, t_2, t_3, t_4 of the signal reception at each verifier and determine the location of A.

One of the main advantages of TDOA is that node localization does not require communication from the base stations to the mobile nodes: the base stations locate mobile nodes measuring signal reception times at each base station. This is why TDOA is well suited for secure localization with hidden base stations.

In our protocol, the base stations are hidden, and only listen to the beacons sent by the nodes. Upon receiving the beacons, the base stations compute node's location with TDOA, and check if this location is *well consistent* with the time differences. By well consistent we mean that the computed location is not to far from the hyperbolas constructed with measured time differences (Figure 8). TDOA with hidden base stations is designed to detect both internal and external attacks, and relies on the assumption that the attackers can guess the locations of base stations only with a very low probability. The protocol is executed as follows.

TDOA with hidden base stations

1 $PBS(t_s) \rightarrow A : N$
2 $A \rightarrow * : m = \{A, N, \mathrm{sig}_{K_A}(A, N)\}$
3 CBS_n : receive m at t_r^n
   : with $t_r^1, ..., t_r^n$, compute p with TDOA
   : if $\sum_{i)j}(|t_r^i - t_r^j| - h(p, i, j))^2 \leq \Delta$ and
      $\max_i(t_r^i - t_s) \leq T$
     then $p_A = p$; else reject p

Here, p is a location of node A computed from the measured time differences and it is the solution to the following least-square problem:

$$p = \arg \min_{p^*} \sum_{i)j}(|t_r^i - t_r^j| - h(p^*, i, j))^2$$

where $h(i, j, p^*)$ is the difference of signal reception times at CBS_i and CBS_j, if the signal is sent from location p^*. Δ is the maximal expected inconsistency between the computed location and the measured time differences. This inconsistency is caused by the errors in measurements of reception times and by pair-wise clock drifts of the base stations. T is the time within which a node needs to reply to a challenge issued by a public base station; this response time is important for the prevention of some replay attacks and to ensure message freshness. N is a fresh nonce. Note that the covert base stations know which nonce is sent by the public station.

Security analysis

Conventional TDOA schemes are vulnerable to both internal and external attacks. An internal attacker can send messages to base stations, with appropriate delays (potentially using directional antennas) and thus cheat on its location; external attackers can jam and delay node's original messages and thus spoof its location.

With covert base stations, these attacks are prevented; to successfully cheat, the attackers need to know where the base stations are located. Otherwise, the attacker needs to guess the locations of the base stations, and perform appropriate timing attacks. The attacker's cheating success depends on the system precision Δ. Essentially, Δ defines the size of attacker's guessing space. Simply, if Δ is large, a false location will be more likely accepted, as the tolerance to inconsistencies will be higher. In Section 4.4, we investigate in more detail the dependence of attacker's success on Δ.

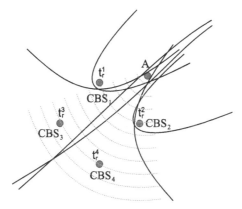

Fig. 8. An example of localization with Time Difference Of Arrival. The base stations CBS measure the differences of signal arrival times, and compute the location of node A.

In addition, we need to consider one more external attack to TDOA. This attack is performed as follows: (1) Attacker jams the original localization message (m) sent by node A; (2) Attacker replays m from a location p'_A. As a result, the base stations will be convinced that the node A is located at p'_A, whereas its true location is p_A. In order to mount this attack, an attacker needs to be able to jam all hidden base stations, which without knowing where they are located requires a lot of power and resources. Furthermore, the attacker needs to have faster processing at nodes than regular mobile nodes. Finally, in order to show that the node A is at p'_A, the attacker needs to have access to this location. Still, this attack is feasible for a resourceful attacker.

Using covert base stations, this attack is partially prevented by the challenge-response scheme. In our protocol, the node is expected to reply to a challenge nonce N within a period T, which limits the time during which the attacker can mount the attack. Here, T is estimated based on the expected signal propagation times and node processing time. We note that if our simple challenge-response scheme is replaced by a more efficient distance-bounding protocol, this and similar attacks can be completely prevented. In some implementations, this will require some specialized hardware at the side of nodes and base stations [3]. The same attacks can also be prevented through precise time synchronization.

In our protocol, node location privacy is not preserved. However, this protocol can be enhanced to include public base station authentication which prevents an attacker from challenging the node and from requesting from it to send localization signals disclosing its location. Other attacks are possible on node's location privacy [11, 14, 15, 27, 28, 30], but coping with these attacks is out of the scope of this paper.

4.3 Node-centric localization with hidden base stations

In this section, we present a protocol for secure localization in node-centric localization systems. Here, we assume that the node computed its location through a non-secure localization system. This location is then reported to the infrastructure comprised of covert base stations, which then verifies if the location is correct. In this context, internal attacks are related to nodes lying about their locations, whereas external attacks are more complex, and assume that the attacker spoofs node's location and then cheats on the location verification mechanisms.

To cope with these attacks, we propose a *location verification* protocol that relies on hidden base stations. In this protocol, node A reports a location p_F to CBS. CBS then measures its distance d_F^m to the node (passively) and verifies if the reported location p_F corresponds to the measured distance. Our protocol is executed as follows (assuming that the distance between the CBS and the node is measured using ultrasound):

Location verification with hidden base stations
1 $PBS(t_s) \rightarrow A : N$
2 $A \rightarrow$ (rf)$* : m_{rf} = p_F, \text{sig}_{K_A}(\text{rf}, p_F, N)$
(us) $: m_{us} = p_F, \text{sig}_{K_A}(\text{us}, p_F, N)$
3 $CBS :$ receive m_{rf} at t_{rf} and m_{us} at t_{us}
$: d_F^c = d(p_F, p_{CBS})$
$: d_F^m = (t_{us} - t_{rf})s$
$:$ if $|d_F^c - d_F^m| \leq \Delta$ and $(t_{rf} - t_s) \leq T$
then $p_A = p_F$; else reject p_A

Here, N is a nonce generated by the public base station, Δ is a combined localization and ranging error and T is the time within which a node needs to reply to a challenge issued by a public base station.

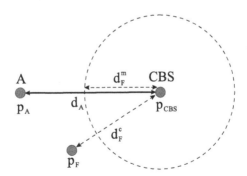

Fig. 9. False location report by node A to the covert base station. p_A is the true node location, p_F is the fake node location (reported by A to CBS), p_{CBS} is the location of CBS. $d_F^c = d(p_F, p_{CBS})$ is the (false) distance between CBS and A, computed by CBS, d_F^m is the (false) distance between A to CBS measured passively by CBS. If $|d_F^c - d_F^m| \leq \Delta$, then $p_A = p_F$.

In this protocol, the infrastructure uses a public base station to communicate with the node, and a single covert base station to verify the reported location. PBS sends a challenge to the node A, which then replies by sending a radio and an ultrasound messages, containing the alleged node location p_F. CBS then measures the time difference between the time at which it received the radio signal (t_{rf}) and the time at which it received the ultrasound signal (t_{us}), and computes the distance $d_F^c = d(p_F, p_{CBS})$ to A. If the reported (possibly fake) location corresponds to the measured (possibly fake) distance, CBS concludes that p_F is the location of A. To do this, CBS simply computes the distance $d_F^c = d(p_F, p_{CBS})$ between its own location p_{CBS} (which is unknown to the node) and the reported location p_F and compares it with the measured distance d_F^m (which A can enlarge or reduce). If two distances differ by more than the expected combined localization and ranging error Δ, then the location is rejected; else, the location is accepted as true node location. An additional verification is made by measuring the node response time T, in order to prevent replay attacks.

We note that this protocol could be similarly designed with RF RSS-based ranging techniques.

Security analysis

An internal attack in node-centric localization schemes is simply a false location report from the node to the infrastructure. Our protocol detects false location reports through checking the consistency of the reported location and of the measured distance. This detection mechanism relies on the fact that the attacker can guess the distance of p_F to the hidden base station only with a low probability. We analyze this in detail in Section 4.4.

External attacks against location verification are more complex and include location spoofing, jamming and message replays. Figure 10 shows an external attack on location verification. Node A is locationed at p_A, the attacker at location p_F. The attacker first spoofs the location of A such that A believes that it is locationed at p_F. Then, by replaying A's localization signals (radio and ultrasound) from p_F, the attacker fools the location verification mechanism. This attack enables the attacker to convince the device A that it (A) is locationed at p_F and then convinces the covert base station that A is at p_F. One limitation of this attack is that an attacker needs to have a device at the location where it wants to falsely place A and that the attacker nodes need to be fairly synchronized to perform it.

Our location verification protocol partially prevents this attack by the same technique used in the TDOA protocol with hidden base stations; the base stations request that the node replies with the RF message to the PBS challenge within a time bound T. This limits the time within which the attacker can mount the attack. With distance-bounding techniques [3], this attack can be entirely prevented, as the value of T can be reduced to nanoseconds.

Similarly to our TDOA-based protocol, the location verification protocol is also vulnerable to location privacy threats. Here, the most obvious privacy problem is that the node discloses its location to any station that issues a location verification

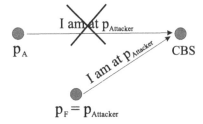

Fig. 10. Location spoofing attack. Attacker spoofs node A and CBS into believing that it (the node) is at its (attacker's) location p_F. The attacker then replays node's message from p_F to fool the location verification mechanism.

request (step 2 in the protocol). An attacker can simply listen to the node's messages and learn where the node is located. Similarly, an attacker could send a location verification request to the node to keep track of the nodes location. These attacks can be prevented by simply requiring a public base station to authenticate itself to the node, and by having a node encrypt the location information that it sends to the base stations.

4.4 Analysis

In this section, we analyze the likeliness that the attacker succeeds in cheating our secure location schemes by guessing the locations of the covert base stations. This probability will notably depend on the size of attacker's search space (which depends on base station power ranges) and on the precision of the localization system.

Here, we focus on the location verification protocol described in Section 4.3. We define the attacker's success as an event when the attacker A reports a location p_F different from its true location ($p_F \neq p_A$), and the CBS concludes that $p_A = p_F$. This event will realize only if $|d_F^c - d_F^m| \leq \Delta$. This essentially means that for a chosen location p_F an attacker needs to guess the distance to the covert base station. The probability of attackers success is therefore

$$Pr(|d_F^c - d_F^m| \leq \Delta | p_F \neq p_A) \tag{1}$$

In our analysis we assume that the localization takes place on a disk (2D), and in a ball (3D). The location of the hidden base station and the reported location of the attacker are therefore on a disk (or in the ball). We assume that the location of the base station is uniformly chosen on the disk (in the ball). Other geometries can be observed, but we have chosen the circles as they best reflect the power ranges of the devices.

Attacker's average success probability

To compute the average probability of attacker's success, we assume that the attacker chooses its fake location p_F uniformly over the disk/ball. In this case, the probability distribution function (pdf) of its distance to the uniformly chosen location of the hidden base station is given by [29]:

$$Pr_D(d_F^c = d) = \frac{4d}{\pi R^2} \cos^{-1}(\frac{d}{2R})$$
$$- \frac{2d^2}{\pi R^3}\sqrt{1 - \frac{d^2}{4R^2}} \qquad (2)$$

for a disk and by

$$Pr_S(d_F^c = d) = \frac{3d^2}{R^3} - \frac{9d^3}{4R^4} + \frac{3d^5}{16R^6} \qquad (3)$$

for a ball, where R is the radius of the disk/ball. Pr_D and Pr_S are shown on Figure 11. The maximum values of these functions are $Pr_D(d_F^c = 0.84R) = 0.809$ and

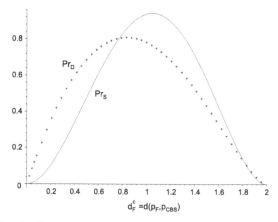

Fig. 11. Probability distribution function of the distance $d_F^c = d(p_F, p_{CBS})$ on a disk (Pr_D) and in a ball (Pr_S), when p_{CBS} and p_F are chosen uniformly over the disk and ball, respectively.

$Pr_S(d_F^c = 1.05R) = 0.942$. This means that when the attacker guesses what is the length of $d(p_F, p_{CBS})$, it will have the highest chance of success if it guesses that it is 0.84R, and hence sets $d_F^m = 0.84R$. In this case, the probability of attacker's success will be:

$$Pr_{D,uni} = \int_{0.84R-\Delta}^{0.84R+\Delta} Pr_D dd \approx 0.809 \times \frac{2\Delta}{R} \qquad (4)$$

$$Pr_{S,uni} = \int_{1.05R-\Delta}^{1.05R+\Delta} Pr_S dd \approx 0.942 \times \frac{2\Delta}{R} \qquad (5)$$

These approximations hold for $\Delta \langle\langle R$. These results are important as they show that the the probability of attacker's success grows linearly with the localization and ranging error Δ and inversely proportional to radius of the region in which the hidden base station is places. This means that the probability of attackers success is inversely proportional to the square root of the space in which localization is taking place. Simply, the more precise the localization and distance measurement is, and the larger the space is, the more secure is location verification.

The probability of attacker's success can be significantly reduced if multiple covert base stations are used for location verification. In that case, the probability of attacker's success is simply

$$Pr^n_{D,uni} \approx (0.809 \times \frac{2\Delta}{R})^n \qquad (6)$$

$$Pr^n_{S,uni} \approx (0.942 \times \frac{2\Delta}{R})^n \qquad (7)$$

The probability of attacker's success in both disk and ball can therefore be upper-bounded by $Pr^n_{uni} = (\frac{2\Delta}{R})^n$.

Attacker's maximum success probability

So far, we have assumed that the attacker chooses p_F uniformly, meaning that we have assumed that the location at which the attacker wishes to pretend to be can be anywhere within the disk/ball. Here, we observe what is location p_F, for which the attacker will have the highest probability of success. We show that the attacker has the highest probability of success ($Pmax$) if it chooses its fake location p_F at the center of the disk/ball and if it chooses $d^m_F = R$ as its fake measured distance to CBS. This probability is as follows (for disk):

$$Pr_D(d^c_F \langle d) = \frac{d^2\pi}{R^2\pi}$$

$$Pr_D(d^c_F = d) = \frac{\delta}{\delta d}Pr_D(d^c_F \langle d)$$

$$= \frac{2d\pi}{R^2\pi}$$

$$Pmax_D = Pr_D(d^c_F = R) = \frac{2}{R} \qquad (8)$$

Similarly for the ball, we obtain that $Pr_S(d^c_F = R) = \frac{3}{R}$. From this it follows that the maximum probabilities of the attacker's success $Pmax^n_D \approx (\frac{4\Delta}{R})^n$ and $Pmax^n_S \approx (\frac{6\Delta}{R})^n$. This analysis shows that in the worst-case scenario, the maximum probability of attacker's success is approx 2.5 times (disk, 2D) and 3 times (ball, 3D) the average probability of attacker's success (when $n = 1$).

Intuitive proof: It is sufficient to observe that the set with the highest number of points equidistant from a single point p in a disc/ball is the set of points on a circle (sphere) of radius R, when p is at the center of a disk/ball.

Sensitivity

In this subsection, we analyze the frequency of false positives and false negatives as a function of the expected localization and ranging error Δ. If the authority sets Δ to

0, the probability of the attacker's success will be 0, but due to the localization and ranging errors the system will reject all reported locations, even if the device is not faking its location. In this case, the frequency of false negatives will therefore be 1. Similarly, if Δ is set to 2R (maximal distance in the localization region of radius R), then the probability of the attacker's success will be 1 (if the reported location is in the center of the disk/ball). However, then, all the false locations of the attacker will be accepted and the frequency of false negatives will be 1. It is therefore important to set Δ such that it minimizes the false negatives and false positives. This means that Δ should be chosen as a minimum value that properly reflects localization and ranging errors.

As we have already noted, CBSs accept the location of the node if $|d_F^c - d_F^m| \leq \Delta$. There are two sources of error in this system. The first error is the localization error $error_P$, which is contained in the reported location p_F. The second error is the ranging error $error_R$ and it is contained in the distance measurement of d_F^m. The total error in $|d_F^c - d_F^m|$ is therefore $error = error_P + error_R$. If localization and ranging errors are already known and if we can assume that they are gaussian $error_P \sim N(0, \sigma_P^2)$ and $error_R \sim N(0, \sigma_R^2)$ the the total error of $|d_F^c - d_F^m|$ is $error \sim N(0, \sigma^2 = \sigma_P^2 + \sigma_R^2)$. If the errors are non-gaussian or even not independent, then we do assume that the joint distribution of the $error$ can be obtained experimentally.

Without any loss of generality, we can express Δ in terms of σ as follows:

$$\Delta = k\sigma \tag{9}$$

where k is a positive real number and σ is the standard deviation of $error$ ($\sigma = \sqrt{\sigma_P + \sigma_R}$) for independent gaussian errors). In the case that $error$ is gaussian, the probability that $d_F^c - d_F^m$ falls within the interval $[-k\sigma, k\sigma]$ is given by [25]:

$$Pr(-k\sigma \langle d_F^c - d_F^m \langle k\sigma) = \frac{2}{\sqrt{\pi}} \int_0^{\frac{k}{\sqrt{2}}} e^{-u^2} du$$

$$= \mathrm{erf}(\frac{k}{\sqrt{2}}) \tag{10}$$

Here, interval $[-k\sigma, k\sigma]$ is called the confidence interval. The frequency of false positives can be than computed as:

$$Pr_{FP} = 1 - Pr(-k\sigma \langle d_F^c - d_F^m \langle k\sigma) \tag{11}$$

i.e., as the probability that $d_F^c - d_F^m$ does not fall within the interval $[-k\sigma, k\sigma]$.

The frequency of false negatives is simply the probability of attacker's success given by (in 2D):

$$Pr_{FN} = \frac{4\Delta}{R} = \frac{4k\sigma}{R} \tag{12}$$

For n covert base stations, these probabilities are defined as follows. The frequency of false positives is defined as a probability that at least one of the covert

base stations rejects the reported location, even if the location is correct. This probability is given by

$$Pr_{FP}^n = 1 - (Pr(-k\sigma\langle d_F^c - d_F^m \langle k\sigma))^n \qquad (13)$$

The frequency of false negatives is defined as the probability that all the base stations
accept the reported location even if this location is false. This probability is given
simply as a probability of attacker's success for n covert base stations:

$$Pr_{FN}^n = (\frac{4k\sigma}{R})^n \qquad (14)$$

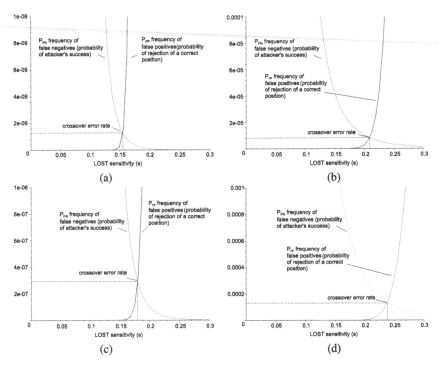

Fig. 12. The frequency of false positives and false negatives, and a crossover error rate for
$\sigma = 0.005R, n = 10$ (a), $\sigma = 0.005R, n = 5$ (b), $\sigma = 0.01R, n = 10$ (c), $\sigma = 0.01R, n = 5$ (d). $s = 1/k$ is the sensitivity. $\Delta = k\sigma$ is the tolerated localization and ranging error. σ is
the standard deviation of the localization and ranging error.

Figure 12 shows the the frequency of false positives and false negatives as a
function sensitivity s. Here, s is defined as $1/k$. Sensitivity s is thus inversely proportional to the expected error Δ and is a measure of how sensitive is the location
verification to errors; if $s = \infty$, this means that the system is very sensitive, and that
localization and ranging errors will are not tolerated, if $s = 0$, this means that the

system tolerates any error. Consequently, the frequencies of false positives and false negatives depend on s.

The same figure shows the frequencies of false positives and false negatives for 10 and 5 covert base stations, and for $\sigma = 0.005R$ (0.5% of R) and $\sigma = 0.01R$ (1% of R). The emphasis in these figures is on the crossover error rate. The crossover error rate is the error rate at which the false positive frequency equals the frequency of false negatives. From these figures we observe, as expected, that with the increase in the number of covert base stations, and with the reduction of the standard deviation of the localization and ranging error σ, the crossover error rate can significantly reduced. If the number of covert base stations is 5 and if $\sigma = 0.01$, the crossover error rate will be 0.0002. This error rate is significantly reduced to 2×10^{-9} if the σ is reduced to 0.005 and if the number of covert base stations is increased to 10.

Even if the crossover error rate is a good indicator of system performance, we emphasize that the security of the system can be significantly improved if the system can allow for a higher false positive frequency. We show on Figure 13 the frequency of false negatives (probability of attacker's success) as a function of the number of covert base stations, given that the frequency of false negatives is set to 1%. This figure shows that with the frequency of false negatives set to 1%, the probability of attacker's success is significantly lower than the crossover error rate. We therefore observe that with 5 or more covert base stations, the probability of attacker's success is lower than 10^{-5} with standard deviation of error smaller than $0.03R$.

We can also observe that with localization systems that exhibit high standard deviation of error (up to 30% or the region radius R), the probability of attacker's success can still be significantly reduced by increasing the number of covert base stations. For example, with $\sigma = 0.2R$ and 20 hidden stations, the probability of attacker's success is only 2×10^{-6}.

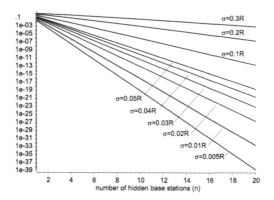

Fig. 13. The frequency of false negatives (probability of attacker's success) if the frequency of false positives is set to $0.01R$.

4.5 Integration with existing localization systems

A number of systems for localization and ranging of wireless devices have already been proposed, based on the propagation of RF, ultrasound and infrared signals. Most of these systems can be adapted to work with covert base stations. Here, we present a short overview of the precision and area sizes of existing localization and ranging systems and we discuss how they can be integrated with secure localization based on covert base.

If localization is based on GPS, the accuracy of the localization will be in 95% of cases better than $1m$. RF time of flight localization techniques aim to provide accuracy of 50-100m and 10m, in the case of UL-TOA, GSM and AGPS, CDMA, respectively. Note here that these systems are designed for area and cell sizes which can have radiuses of 500m (in highly dense urban areas) to 35km (in countryside). Indoor, localization with WiFi based on signal strength measurements with location fingerprinting can achieve localization accuracy of 2-3m, whereas ultrasound-based ranging and localization systems can be accurate up to several centimeters. Ultra wide band (UWB) time-of-flight based systems work both indoor and outdoor. Indoor they can achieve ranging precision better than 1m for ranges of up to 50m and localization accuracy of up to 15cm. Outdoor the accuracy of UWB localization and ranging systems can be also very high, approx. 1m for distances of up to 2km [9]. All the numbers presented in this paragraph are rough approximations of accuracies of these systems; each of these systems can perform better or worse, if one or more of system parameters change. Here, we use the term accuracy very loosely as the measures of accuracy vary from one system to another. For example, if GPS localization is used for providing location reference to a device, and UWB ranging is used for location verification, the standard deviation of the error can be estimated at up to 4 meters. Given that the range of UWB localization can be up to 2km than $\sigma \langle 0.005R$. Indoor, if ultrasound is used for localization and ultrasonic ranging for verification, we can assume the standard deviation of error to be of the order of 20 centimeters and ranges up to 20m, meaning that $\sigma = 0.01R$. As we have shown in Figures 12 and 13, the probability of attacker's success in these scenarios will can then be as low as 10^{-35} (in best case).

5 Related work

In the last decade, a number of indoor localization systems were proposed, based notably on infrared [39], ultrasound [26,40], received radio signal strength [2,5,12] and time-of-flight radio signal propagation techniques [8, 19]. These localization techniques were then extended and used for localization in sensor and ad hoc networks [4,6,7,22–24,32,35].

Recently, a number of secure distance and location verification have been proposed. Brands and Chaum [3] proposed a distance bounding protocol that can be used to verify the proximity of two devices connected by a wired link. Sastry,

Shankar and Wagner [31] proposed a new distance bounding protocol, based on ultrasound and radio wireless communication. In that work, the authors also propose to make use of multiple base stations to narrow down the area in which the nodes lie. However, as this proposal is based on ultrasound distance bounding, it can therefore be used only for the verification of nodes' locations, and only if external nodes have no access to the area of interest. In [13], the authors propose a mechanism called "packet leashes" that aims at preventing wormhole attacks by making use of the geographic location of the nodes (geographic leashes), or of the transmission time of the packet between the nodes (temporal leashes). Kuhn [17] proposed an asymmetric security mechanism for navigation signals. That proposal aims at securing systems like GPS [10]. Lazos et al. [18] proposed a set of techniques for secure localization of a network of sensors based on directional antennas and distance bounding. Li et al. [20] propose statistical methods for securing localization in wireless sensor networks. Liu et al. [21] propose techniques for the detection of malicious attacks against beacon-based location discovery in sensor networks, based on consistency of received beacons. Recently, a number of proposals have been made to protect the anonymity and location privacy of wireless devices [11, 14, 15, 27, 28, 30].

6 Conclusion

In this chapter, we have analyzed localization and distance estimation techniques in adversarial settings. We have shown that most proposed localization techniques are vulnerable to location spoofing attacks from internal and external attackers. We have further shown that localization and distance estimation techniques, based on radio signal propagation, exhibit the best properties for location verification.

We have proposed two novel mechanism for secure localization. The first, Verifiable Multilateration (VM) enables secure computation and verification of node locations; VM is based on distance bounding and authenticated ranging protocols. The second, Secure Localization with Hidden Base stations, makes use of unpredictability of base station locations to enable secure localization. This approach enables secure localization with a broad spectrum of localization techniques: ultrasonic or RF, based on received signal strength or on time of signal flight.

We have demonstrated that both VM and Secure Localization with Hidden Base stations can be easily integrated with several existing node-centric and infrastructure-centric localization schemes. We have further shown how security of these mechanisms depends on the precision of the localization systems, and on the number of base stations (reference points) used for localization.

References

1. R. Anderson and M. Kuhn, "Tamper resistance - a cautionary note," in *Proceedings of the Second Usenix Workshop on Electronic Commerce*, 1996.

2. P. Bahl and V. N. Padmanabhan, "RADAR: An In-Building RF-Based User Location and Tracking System," in *Proceedings of the IEEE Conference on Computer Communications (InfoCom)*, vol. 2, 2000, pp. 775–784.
3. S. Brands and D. Chaum, "Distance-bounding protocols," in *Workshop on the theory and application of cryptographic techniques on Advances in cryptology*. Springer-Verlag New York, Inc., 1994, pp. 344–359.
4. N. Bulusu, J. Heidemann, and D. Estrin, "GPS-less low cost outdoor localization for very small devices," *IEEE Personal Communications Magazine*, vol. 7, no. 5, pp. 28–34, October 2000.
5. P. Castro, P. Chiu, T. Kremenek, and R. Muntz, "A Probabilistic Room Location Service for Wireless Networked Environments," in *Proceedings of the Third International Conference Atlanta Ubiquitous Computing (Ubicomp)*, vol. 2201. Springer-Verlag Heidelberg, September 2001.
6. L. Doherty, K. Pister, and L. El Ghaoui, "Convex position estimation in wireless sensor networks," in *Proceedings of the IEEE Conference on Computer Communications (InfoCom)*, April 2001.
7. T. Eren, D. Goldenberg, W. Whiteley, Y. Yang, A. Morse, B. Anderson, and P. Belhumeur, "Rigidity, computation, and randomization in network localization," in *Proceedings of the IEEE Conference on Computer Communications (InfoCom)*, 2004.
8. R. Fontana, "Experimental Results from an Ultra Wideband Precision Geolocation System," *Ultra-Wideband, Short-Pulse Electromagnetics*, May 2000.
9. R. Fontana, E. Richley, and J. Barney, "Commercialization of an Ultra Wideband Precision Asset Location System," in *IEEE Conference on Ultra Wideband Systems and Technologies*, November 2003.
10. I. Getting, "The Global Positioning System," *IEEE Spectrum*, December 1993.
11. M. Gruteser and D. Grunwald, "Enhancing location privacy in wireless LAN through disposable interface identifiers: a quantitative analysis," in *Proceedings of WMASH*, 2003.
12. J. Hightower, G. Boriello, and R. Want, "SpotON: An indoor 3D Location Sensing Technology Based on RF Signal Strength," University of Washington, Tech. Rep. 2000-02-02, 2000.
13. Y.-C. Hu, A. Perrig, and D. B. Johnson, "Packet Leashes: A Defense against Wormhole Attacks in Wireless Networks," in *Proceedings of the IEEE Conference on Computer Communications (InfoCom)*, San Francisco, USA, April 2003.
14. L. Huang, K. Matsuura, H. Yamane, and K. Sezaki, "Enhancing Wireless Location Privacy Using Silent Period," in *Proceedings of the IEEE Wireless Communications and Networking Conference(WCNC)*, 2005.
15. J. Kong and X. Hong, "ANODR: ANonymous On Demand Routing with Untraceable Routes for Mobile Ad-hoc Networks," in *Proceedings of MobiHoc*, 2003.
16. J. Kong, Z. Ji, W. Wang, M. Gerla, R. Bagrodia, and B. Bhargava, "Low-cost Attacks against Packet Delivery, Localization and Time Synchronization Services in Under-Water Sensor Networks," in *Proceedings of the ACM Workshop on Wireless Security (WiSe)*, 2005.
17. M. G. Kuhn, "An Asymmetric Security Mechanism for Navigation Signals," in *Proceedings of the Information Hiding Workshop*, 2004.
18. L. Lazos, S. Čapkun, and R. Poovendran, "ROPE: Robust Position Estimation in Wireless Sensor Networks," in *Proceedings of IPSN*, 2005.
19. J.-Y. Lee and R. Scholtz, "Ranging in a Dense Multipath Environment Using an UWB Radio Link," *IEEE Journal on Selected Areas in Communications*, vol. 20, no. 9, December 2002.

20. Z. Li, W. Trappe, Y. Zhang, and B. Nath, "Robust Statistical Methods for Securing Wireless Localization in Sensor Networks," in *Proceedings of the International Conference on Information Processing in Sensor Networks (IPSN)*, 2005.

21. D. Liu, P. Ning, and W. Du, "Attack-Resistant Location Estimation in Sensor Networks," in *Proceedings of the International Conference on Information Processing in Sensor Networks (IPSN)*, 2005.

22. D. Moore, J. Leonard, D. Rus, and S. Teller, "Robust distributed network localization with noisy range measurements," in *Proceedings of the ACM Conference on Networked Sensor Systems (SenSys)*. ACM Press, 2004, pp. 50–61.

23. D. Niculescu and B. Nath, "Ad hoc positioning system (aps) using aoa," in *Proceedings of the IEEE Conference on Computer Communications (InfoCom)*, San Francisco, USA, April 2003.

24. ——, "DV Based Positioning in Ad hoc Networks," *Journal of Telecommunication Systems*, vol. 22, no. 4, pp. 267–280, 2003.

25. V. V. Nostrand, *Mathematics of Statistics*. Princeton, NJ, 1962.

26. N. B. Priyantha, A. Chakraborty, and H. Balakrishnan, "The Cricket location-support system," in *Proceedings of the ACM/IEEE International Conference on Mobile Computing and Networking (MobiCom)*. ACM Press, 2000, pp. 32–43.

27. A. R. Beresford and F. Stajano, "Location Privacy in Pervasive Computing," *Pervasive Computing*, January-March 2003.

28. I. W. Jackson, "Anonymous Addresses and Confidentiality of Location," in *Proceedings of International Workshop on Information Hiding*, 1996.

29. M. G. Kendall and P.A.P. Moran, *Geometrical Probability*. Hafner, New York, 1963.

30. Y.-C. Hu and H. J. Wang, "Location Privacy in Wireless Networks," in *Proceedings of the ACM SIGCOMM Asia Workshop*, 2005.

31. N. Sastry, U. Shankar, and D. Wagner, "Secure Verification of Location claims," in *Proceedings of the ACM Workshop on Wireless Security (WiSe)*. ACM Press, September 2003, pp. 1–10.

32. A. Savvides, C.-C. Han, and M. B. Strivastava, "Dynamic fine-grained localization in Ad-Hoc networks of sensors," in *Proceedings of the ACM/IEEE International Conference on Mobile Computing and Networking (MobiCom)*. ACM Press, 2001, pp. 166–179.

33. D. Shaw and W. Kinsner, "Multifractal Modeling of Radio Transmitter Transients for Clasification," in *Proceedings of the IEEE Conference on Communications, Power and Computing*, May 1997, pp. 306–312.

34. S. Čapkun, L. Buttyán, and J.-P. Hubaux, "SECTOR: Secure Tracking of Node Encounters in Multi-hop Wireless Networks," in *Proceedings of the ACM Workshop on Security of Ad Hoc and Sensor Networks (SASN)*, Washington, USA, October 2003.

35. S. Čapkun, M. Hamdi, and J.-P. Hubaux, "GPS-free Positioning in Mobile Ad-Hoc Networks," *Cluster Computing*, vol. 5, no. 2, April 2002.

36. S. Čapkun and J.-P. Hubaux, "Secure positioning of wireless devices with application to sensor networks," in *Proceedings of the IEEE Conference on Computer Communications (InfoCom)*, 2005.

37. S. Čapkun and J.-P. Hubaux, "Secure Positioning in Wireless Networks," *IEEE Journal on Selected Areas in Communications*, vol. 24, no. 2, February 2006.

38. S. Čapkun, M. Čagalj, and M. Srivastava, "Secure Localization with Hidden and Mobile Base Stations," in *Proceedings of the IEEE Conference on Computer Communications (InfoCom)*, 2006.

39. R. Want, A. Hopper, V. Falcao, and J. Gibbons, "The Active Badge Location system," *ACM Transactions on Information Systems*, vol. 10, no. 1, pp. 91–102, 1992.

40. A. Ward, A. Jones, and A. Hopper, "A New Location Technique for the Active Office," *IEEE Personal Communications*, vol. 4, no. 5, October 1997.
41. J. S. Warner and R. G. Johnston, "Think GPS Cargo Tracking = High Security? Think Again," *Technical report, Los Alamos National Laboratory*, 2003.

Distance Bounding Protocols: Authentication Logic Analysis and Collusion Attacks

Catherine Meadows[1], Radha Poovendran[2], Dusko Pavlovic[3], LiWu Chang[1], and Paul Syverson[1]

[1] Naval Research Laboratory, Code 5543, Washington, DC 20375 { meadows, lchang, syverson } @itd.nrl.navy.mil
[2] University of Washington, Department of Electrical Engineering, Seattle Washington, 98195 rp3@u.washington.edu
[3] Kestrel Institute, 3260 Hillview Avenue, Palo Alto, CA 94304 dusko@kestrel.edu

Summary. In this paper we consider the problem of securely measuring distance between two nodes in a wireless sensor network. The problem of measuring distance has fundamental applications in both localization and time synchronization, and thus would be a prime candidate for subversion by hostile attackers. We give a brief overview and history of protocols for secure distance bounding. We also give the first full-scale formal analysis of a distance bounding protocol, and we also show how this analysis helps us to reduce message and cryptographic complexity without reducing security. Finally, we address the important open problem of collusion. We analyze existing techniques for collusion prevention, and show how they are inadequate for addressing the collusion problems in sensor networks. We conclude with some suggestions for further research.

1 Introduction

Distance estimation, that is the estimate of the distance between two nodes, plays of a fundamental part in the setting up and maintenance of sensor networks. For example, a node trying to localize itself, can, if it learns its distance from three or more nodes with known locations, use multilateration to determine where it sits. This computation is a major part of many localization algorithms. Distance estimation can also be useful in synchronization: if node A knows its distance from node B, it can request a timestamp from node B and compute the clock skew by factoring in the round trip time of the request and the response.

One of the most accurate means of distance estimation is to use the time of flight of a signal. For example, one can send a signal to a seated node, have it respond, and then use the time of the round trip to measure the distance. For example, Multispectral Solutions [1] has recently developed an ultra wide band ranging radio based on such technology that measures round trip times of packets to provide range resolution of better than one foot.

Although such a technique can provide accurate measurements, it is not easy to figure out how to make use of it when a node (from now on referred to as the *verifier*) is attempting to find its distance form another node (from now on referred to as the *prover*) in the face of hostile attackers. If the prover is dishonest, it can pretend to be closer to or further away from the verifier than it actually is by either jumping the gun and sending a response before the request, or pretend to be further away than it is by delaying its response. Even if the prover is honest, a hostile attacker could attach its own identity to the prover's response, and pass off honest verifier's location as its own. Finally, dishonest provers can conspire to mislead the verifier, one prover lending the other prover its identity so that the second prover can make the first prover look closer than it is.

Probably the simplest secure distance measurement protocol is Sastry et al.'s Echo protocol [14], in which the verifier sends a nonce to the prover, and the prover returns it to the verifier. The use of a random nonce means that the prover can't respond until it has heard from the verifier, thus preventing the prover from jumping the gun. However, without any kind of authentication, it is possible for an attacker to usurp an honest prover's response and attach its own identity.

The obvious defense is to have the prover authenticate its response, and indeed, a variant of Echo protocol offers this capability. However, the time involved in computing the authentication function can be so large with respect to the travel time as to make it difficult to compute the distance except for relatively slow (and less accurate) sound frequencies.

An approach that gets around this problem is to have the prover send a *rapid*, unauthenticated, response and then send the *authenticated* response later. However, if this is not done carefully, it is again possible for an attacker to usurp an honest prover; he simply prevents the authenticated response from reaching the verifier, and substitutes his own authenticated response.

Fortunately, a solution to this problem already exists. This is the notion of a *secure distance bounding protocol*. This idea was first introduced by Brands and Chaum [2] to defend against Desmedt's Mafia attack [5] on zero knowledge protocols. The idea is that the prover first commits to a nonce using a one-way function, the verifier sends a challenge consisting of another nonce, the prover responds with the exclusive-or of its and the verifier's nonces, and then follows up with the authentication information. The verifier uses the time elapsed between sending its nonce and receiving the prover's rapid response to compute its distance from the prover, and then verifies the authenticated response when it receives it. In the Brands and Chaum protocol, the challenge and the response are done as a bit-by-bit exchange, and the time of flight is taken as the average of the time of flight of each pair of bits. Other protocols that take a similar approach, such as the Čapkun-Hubaux protocol [16], rely on a single exchange of packets. It is also possible to consider other variants, in which a single nonce is broken into k-bit chunks, and multiple packets are used.

Another, but related, approach is taken by Hancke and Kuhn in [6]. In this protocol the verifier sends the prover a nonce, and the prover computes the a collision-free one-way hash function over the nonce and a key shared between the prover and the verifier. The principals then perform a rapid bit-by-bit exchange in which the verifier sends random challenges and the prover responds with a response based on the challenge and the hash.. In this case the authentication takes place previously to the rapid exchange.

Assuming that there is no collusion between provers, the Čapkun-Hubaux protocol, like the Brands-Chaum protocol, prevents hijacking because of the commitment step, and also prevents the prover from lying about being any closer to the verifier than it is, although it can lie about being farther away simply by delaying its response. Likewise, the authentication used in the Hancke-Kuhn protocol prevents hijacking, and the verifier's random challenge prevents a premature reply.

The problem of a delayed response can be dealt with in certain instances using multiple provers or verifiers. For example, in Čapkun and Hubaux's SPINE protocol [16] three verifiers forming a triangle around a prover use a distance bounding protocol to localize it. A prover who wants to lie about its location must pretend to be closer to one of the verifiers than it is; the distance bounding protocol makes this impossible.

Running all through this is the issue of guaranteeing correctness of distance bounding protocols. The presence of time as a factor puts an extra security requirement on the protocol: not only must messages have come from the indicated principal, but in the indicated amount of time. On the other hand, time may work for us as well. Certain types of message modification attacks will not be useful if a node is trying appear closer than it is, since intercepting and modifying the message will delay its arrival.

In spite of this, very little work exists in the formal and mathematical analysis of distance bounding protocols. Sastry et al. include a security proof for the Echo protocol, but, since no authentication is involved, the proof is limited to showing that a prover cannot respond before receiving the verifier's nonce. Brands and Chaum provide a proof that their protocol is zero-knowledge but do not provide any extended analysis of the timing properties. Thus there appear to be no analyzes of distance bounding protocols available that take into account the subtle interplay between authentication and timing.

In this paper we address all of these above issues. We first give an outline of requirements that distance bounding protocols should satisfy. We then describe a new distance bounding protocol, similar in structure to Brands-Chaum, and a generalized version of the protocol we presented in [12]. We then extend the authentication logic we used in [12] to so that it can be used to reason directly about distance bounding protocols, and use the logic to give a formal analysis of the protocol's security. This formal analysis allows us to simplify greatly the type of commitment used, and to omit one cryptographic operation.

We then address the issue of collusion. The problem of collusion in distance bounding, in which two dishonest verifiers pool information to make one of the verifiers look closer than it is, was first noticed by Desmedt [5] who dubbed it the "terrorist attack." The Brands-Chaum protocol is vulnerable to a collusion attack, in which one prover sends the rapid response and then passes the information in its commitment over to another, who sends the authenticated response. Brands and Chaum were aware of this attack and left it as a an open problem. Since then, others have tackled it [3], but their solutions require colluders to share long term secrets. This approach, while possibly appropriate for the types of applications envisaged by Brands and Chaum, does not provide much help for sensor networks, in which colluding nodes are likely to be under control of the same attacker, and so would not be likely to have any objection to sharing any secret information. We study this problem in detail, showing how attacks are possible even on protocols such as ROPE [8] or SPINE that use multiple verifiers to detect cheating nodes, and make some recommendations.

2 Requirements for Distance Bounding Protocols

In [14] Sastry et al. give a set of requirements for distance bounding protocols. They are:

1. **Make few resource demands on the prover and verifier**. This means keeping the number of cryptographic operations and messages low.
2. **No previous setup required**. In particular, there should be no need for principals to share keys beforehand.
3. **Guarantees should be quantifiable.**

Although the use of authentication means that we must use some form of cryptography in the authenticated response, we can still keep costs down by minimizing its use in the rapid response. Note that hash functions and nonce generation will both count as cryptographic operations.

The second requirement seems completely at odds with any form of authenticated distance bounding protocol, but it can be partially satisfied by having the rapid exchange take place without the use of any cryptographic keys. A verifier could then request to have a key distributed to it and the prover, which the prover could then use to authenticate its authenticated response. This would be helpful, example, if a verifier was only interested in finding its nearest neighbor. This feature is present in the Brands-Chaum and Čapkun-Hubaux protocol, as well as the protocol that we present in this paper. However, it is not true of some other protocols, such as the Hanck-Kuhn protocol, which requires the prover's response to include a hash of its nonce with a key shared with the verifier.

As for quantitative guarantees, at present the Echo protocol is the only one of which we know that satisfies such quantitative guarantees, and even qualitative guarantees in the form of formal analyzes seem rare. However, we provide qualitative guarantees in this paper that we believe could ultimately be extended to quantitative guarantees in the manner of [14].

We now consider the main security requirements that have been identified in the literature.

1. A prover should be able to correctly determine its distance from an honest verifier, even when hostile attackers are present.

2. A prover should be able to determine an upper bound for its distance from even a dishonest verifier, as long as the verifier does not collude with other verifiers.
3. A prover should be able to determine an upper bound for its distance from a dishonest verifier even if it does collude.

In this paper, we prove the somewhat weaker goal that, if the prover is honest in the sense that it follows the rules of the protocol but may either delay its response (either due to dishonesty or processing time), or attempt to respond early, the verifier can compute an upper bound on the distance. Finally for (3) we argue that the known techniques for detecting or preventing fraud in distance bounding protocols are either insecure against collusion or are not applicable in sensor networks.

3 Distance Bounding Protocol and its Analysis

3.1 Assumptions

We assume that nodes have the ability to generate random or pseudorandom nonces and compute collision-free one-way hash functions. We also assume that provers have a means of authenticating themselves to verifiers, e.g. by shared keys or digital signatures. In this paper we use shared keys and message authentication codes (MACs), but the same analysis will work for digital signatures.

We also assume that principals have the ability to compute the time that an event occurs with respect to their local clocks. The unit of time may or may not be of a finer granularity then the sending or receipt of a message. If the time granularity is finer, we let the time of a message denote the beginning of a send or receive. We will be particularly interested in timed sends and receives of individual packets. In this case we will assume that it is possible to predict the time of any subevent of a send or receive (such as the end) from the time of the beginning, and vice versa. We will also assume that all subevents of the send of a given packet are engaged in by a unique principal. That is, A cannot send part of a packet and B send another, and have them both accepted as part of the same packet. Our reason for doing so is the belief that this would need a degree of synchronization that would require, at very least, cooperation between A and B, and in our analysis we are not trying to rule out collusion attacks.

3.2 The Protocol in Detail

We fix an interval I_0 that is the expected turnaround time between receiving a challenge and sending a response. Our protocol proceeds in five steps, four of which involve the sending of messages.

1. The prover P generates a nonce N_P. This, and any other computations that do not involve information from the verifier, can be done in advance of P's participating in the protocol.

2. The verifier V requests a distance measurement. This is mainly to warn P that a challenge is on the way, and to let P know V's identity.

   V **sends** V, request

3. The verifier V sends a nonce as a challenge:

   V **sends** N_V

4. The prover P sends a response, of the application of a function F to N_P, P, and N_V. We refer to this message as the *rapid response*. The only condition that we put on F is that the verifier be able to verify that $F(N_V, P, N_P)$ was constructed using N_V, P, and N_P. Examples of such functions include N_V, P, N_P, where , denotes concatenation, $N_V, (P \oplus N_P)$, assuming that names are a distinct recognizable type, and $N_V \oplus h(P, N_P)$, where h is a collision-free hash function.

   P **sends** $F(N_V, P, N_P)$

   The verifier, on receiving this message, calculates the time elapsed between sending the challenge and receiving the rapid response.

5. The prover sends a message authenticated with a key shared between it and the verifier. We refer to this message as the *authenticated response*.

   P **sends** $P, Pos_P, N_P, N_V, MAC_{K_{PV}}(P, Pos_P, N_P, N_V)$

   where Pos_P is P's position. V, on receiving the message, verifies the MAC. It also computes F from the values it receives in the authenticated response, and compares it with the value it received in the rapid response. If the two are the same, and the MAC checks out, it accepts P's response as valid. V then subtracts I_0 from the time elapsed between sending the challenge and receiving the rapid response and uses the result to calculate its distance from the prover, that is, the distance is calculated to be $v \cdot (t_2 - t_1 - I_0)/2$.

An overview of the protocol is given in Figure 1.

4 Security Analysis

4.1 Overview

In this section we give a formal analysis of the distance bounding protocol using the combined authentication and secrecy logics of [4] and [13]. Although we are interested in authentication, not secrecy, we will use some of the concepts introduced in the secrecy logic, and so we will refer to that as well. We will use the logic to show what a verifier can conclude from interacting with an honest prover. We then show how the proof breaks down if the prover is dishonest, in particular if it is in collusion with another node.

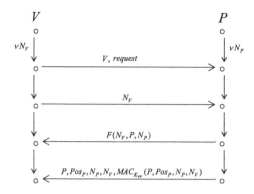

Fig. 1. Distance Bounding Protocol

4.2 The Authentication Logic

Basic Ideas and Notation

We begin by setting the stage for the logic, and introducing a little notation. The interested reader can find a more complete discussion in [4,13]. We consider a protocol as a partially ordered set of actions, as in Lamport [7], in which $a < b$ means that action a occurs before action b. We let $(t)_A$ denote t being received by A, $\langle t \rangle_A$ denote a message being received by A. We let $x \prec y$ denote the statement "if an action of the form y occurs, then an action of the form x must have occurred previously," and we let νn denote the generation of a fresh, unpredictable nonce, n.

For the purpose of the derivations in this paper, we will use a term algebra **T** consisting of constants, variables, and the following operations:available to principals: concatenation , denoted by ',' , deconcatenation, computation of message authentication codes, denoted by $MAC_{K_{XY}}(Z)$, collision-free hash functions, denoted by $h(Z)$, and exclusive-or, denoted by $\oplus$. **T** is provided with the following equational theory:

1. $MAC_{K_{XY}}(Z) = MAC_{K_{YX}}(Z)$
2. $x \oplus x = 0$
3. $0 \oplus x = x$
4. $x \oplus y = y \oplus x$
5. $(x \oplus y) \oplus z = x \oplus (y \oplus z)$

for a distinguished term 0.

We say that $g \equiv t$ if g and t can be made equal by applying the equational theory of **T**. We refer to rules (2) and (3) as the *cancellation rules*. We say that a term g is *irreducible* if no cancellation rules can be applied to g. We say that $s \sqsubseteq t$ if s is a subterm of t.

We supply **T** with a simple type theory. There is a general type "term" and one subtype,"name". A variable not of type name will often be referred to as untyped.

Free untyped variables are used to refer to terms about which the recipient knows nothing. We say that a map from variables from to terms is a *substitution* if it is the identity on all but a finite number of variables and preserves types. If σ is a substitution and t is a term, we let $t\sigma$ denote the image of t under σ.

The logic also includes a number of predicates describing states of principals. A : stands for "A knows", and HP means that P is an honest principal who follows the rules of the protocol.

Informally, a *role* is the set of actions that a principal performs to engage in a particular protocol. In the distance bounding protocol, we have two roles: the verifier and the prover. A *run* is the trace of a (possibly) partial execution of a protocol, i.e., the set of actions executed by the principals and their partial ordering. A *state* is a cut of the directed graph induced by the run; each action in the run should occur either before or after the state.

Stable Subterms

In this section we will formalize the notion that a subterm s of a term t must have been used in computing t. In our earlier work, where no cancellation was involved, this could be guaranteed by requiring that s was a subterm of t, since the term structure of t gave a unique history of the way in which t was built. However, when we allow cancellation rules, things become more complicated. For example, suppose that a principal receives a term $x \oplus s$ where x is a free untyped variable. If x were further instantiated to $y \oplus s$, then s would vanish after the cancellation rules were applied.

In order to avoid this we make the following definition.

Definition 1. *Let s be a subterm of t. We say that s is a stable subterm of t, denoted by $ss(s, t)$, if for all possible substitutions σ to the free variables in t and after all possible applications of the equations governing the term algebra* **T**, *we have $s\sigma \sqsubseteq t\sigma$. We say that a term t is simply stable, denoted by $ss(t)$, if every subterm of t is stable.*

The motivation behind the use of stable subterms is that it makes it possible to ascertain that, whatever values the free variables in t turn out to have, the stable subterm must have been used in the computation of t.

We use this insight to make the following definition:

Definition 2. *We use the notation $((s))_A$ (respectively, $\langle\langle s\rangle\rangle_A$) to denote A's receipt (respectively sending) of a message m containing s as a stable subterm.*

We note that our definition subsumes the definition used in [4, 11, 13], which required s merely to be a subterm of t. For the term algebras used in those papers, subterm implies stable subterm.

When we want to make it clear that A is receiving (respectively sending a term t with stable subterm s, we will use the notation $((s \sqsubseteq t))_A$ (respectively $\langle\langle s \sqsubseteq t\rangle\rangle_A$).

We can now formally describe the conditions on the term F used in our distance bounding protocol. We present this as an axiom that must be verified for particular choices of F.

$$ss((F(x_1, x_2, x_3)) \qquad \text{(st)}$$

We now consider the sorts of functions F that can be proved to satisfy st. For example $(N_P \oplus N_V), P)$ where N_P is a free untyped variable and P is a variable of type name, and N_V is a nonce generated by V does *not* satisfy st. Since N_P is free and untyped, any substitution may be made to it. Thus, if $N_P \sigma = X \oplus N_V$, then $((N_P \oplus N_V), P) = (X \oplus N_V \oplus N_V), P) = (X, P)$ which can be computed without N_V. However, $(N_V \oplus P), N_P)$ does satisfy st. The term N_V a random value, so we cannot make arbitrary substitutions to it. The same goes for P. We can made arbitrary substitutions to N_P, but none of them will result in canceling out N_V, P, or N_P,

Lemma 1. *Let* **T** *be the term algebra described in section 4.2, and let m be an irreducible term from* **T**. *Every subterm of m is stable if for every irreducible substitution σ to the variables of m, $m\sigma$ is also irreducible.*

Proof. (Sketch) Let m be a term satisfying the hypothesis of the lemma. We want to show that after any possible substitution σ to the variables in m, $t\sigma$ is still a subterm of $m\sigma$ after all possible reductions have been made. For the case of an irreducible σ this follows directly from the hypothesis, since no reductions are possible. For the case of a reducible σ it follows from the fact that the rewrite theory associated with our term system is Church-Rosser, modulo the associative commutativity axioms for exclusve-or, which, in our case means, that when several applications of the cancellation rule are possible, it does not matter in what order they are taken. Thus, we can apply the cancellation rules to the cancellations induced on the variables by σ first. Once that is done, then σ becomes an irreducible substitution, and we are back to the first case.

We also give as a corollary the following procedure for stable subterms:

Corollary 1. *Suppose that t contains no subterm of the form $X \oplus Y$, where one of X or Y is a free untyped variable. Then t is simply stable.*

Proof. The proof follows from Proposition 1 and the fact that the only irreducible terms that do not necessarily remain irreducible after irreducible substitutions are those that contain $Z \oplus Y$, where either Z or Y is an untyped free variable.

The corollaries below follow directly from the fact that none of the terms in question contain subterms of the form $X \oplus Y$, where one of X or Y is an untyped free variable.

Corollary 2. *Suppose that P is a variable of type name, N_V is a nonce, and N_P is an untyped free variable. Then $N_V, P, N_P, N_V, (N_P \oplus P)$, and $N_V \oplus h(P, N_P)$ are simply stable.*

Corollary 3. *Any variable or constant is simply stable. $MAC_{XY}(Z)$ is simply stable as long as Z is. $X\|Y$ is simply stable as long as X and Y are. $h(X)$ is simply stable as long as X is.*

Basic Axioms

We are now ready to describe the basic axioms of the logic as given in [4]. The logic describes what a principal can conclude from interacting via the protocol with another principal. Two basic axioms of the logic are the *receive* axiom rcv and the *freshness axiom* new, which we describe below.

The receive axiom says that everything that is received must have been originated by someone:

$$A : ((m))_A \Rightarrow \exists X. \langle\!\langle m \rangle\!\rangle_{X<} < ((m))_A \tag{rcv}$$

The freshness axiom describes the behavior of the ν operator.

$$(\nu n)_B \wedge a_A \Rightarrow (n \in FV(a) \Rightarrow (\nu n)_B < a_A \tag{new}$$
$$\wedge \ (A \neq B \Rightarrow (\nu n)_B < \langle\!\langle n \rangle\!\rangle_B < ((n))_A \leq a_A))$$

where $FV(a)$ denotes the free variables of a

The first part says that ν is a binder, that is, any event a mentioning n necessarily occurs *after* (νn). The second line requires that if the agent B executing (νn) and the principal A executing a are different, then B must have used a send action to transmit n and A must have acquired it by means of a receive action.

The fact that we can use ν as a binder means that it is possible to apply ν outside of a sequence of events S, e.g. as $(\nu n)_A(S)$. This will be convenient, since we often will not care exactly when νn occurs, as long as it occurs before n is sent in a message.

Axioms Governing Message Authentication Codes

The message authentication code has the property that it is possible to tell who created it. This property is formally derived in [4] for similar functions using their non-invertibility and assumptions about the secrecy of keys. Since we will not need the machinery of [4] for anything other than this result, we state it as an axiom here.

$$\langle\!\langle MAC_{K_{AB}}t \rangle\!\rangle_{X<} \implies X = A \vee X = B \tag{mac}$$

Timestamps, Distance, and the Axioms Governing Them

Up to now we have considered only axioms that cover the ordering of messages. Now we will extend our logic to reasoning about distance. To do this will make use of the notion of a timestamp, which was already introduced in [13], although to reason different types of properties.

A *timestamp* represents an entity's recording of its local time, For this we use the expression τt, where $(\tau t)_A$ denotes A's reading its local time and storing it a local variable. We use $a_A^{[\tau t_1, \tau t_2]}$ to denote A's engaging in event a some time between times t_1 and t_2. Where appropriate, we can use the shorthand a_A^t for $a_A^{[t-\epsilon, t+\epsilon]}$.

We note that in some cases the granularity of time measurement may actually be less than the time it takes to engage in an event. Thus, the time it takes for a principal to receive or send a message may take more than one time interval. In that case, we take a_A^t to mean the time at which A begins to engage in the action. In this case, we will need to attach a stronger meaning to $\langle\langle x \rangle\rangle_A^t$ and $((x))_A^t$ as well. They will mean, not only that x must have been used in the construction of the message, but that either x or some term each of whose bits depends on x appears at the beginning of the message as well. Our analysis will hold for either definition of timed event.

For the purposes of reasoning about time and distance, we introduce the function $d(A, B)$ where A and B are two principals (we ignore the possibility of node mobility at this point). We define $d(A, B)$ as follows:

Definition 3. *Let A and B be two principals. We define the distance between A and B or $d(A, B)$ to be $v \cdot t$, where v is the velocity at which a signal travels, and t is the minimum of all possible $(t_1 - t_2 - I)/2$ such that the following occurs:*

$$(\nu n)_A \; (\nu m)_B \; (\langle\langle n \rangle\rangle_{A<}^{t_1} < ((n))_B < \langle\langle m \rangle\rangle_{B<} < ((m))_A^{t_2})$$

and I is the turnaround time at B.

The idea is, if that B receives a nonce created by A, or vice versa, either directly or indirectly, then the time it took must be bounded below by their distance times the velocity. If one pair of send and receive events occurs after another than the total time for the whole sequence of events to occur is bounded below by twice the distance times the velocity plus the turnaround time. The remainder of this section will be devoted to the construction and analysis of authentication techniques for proving that this sequence of events has taken place.

This leads us to the following simple proposition, whose proof follows directly from the above definition.

Proposition 1. *Suppose that $A : (\nu n)_A \; (\nu m)_B \; (\langle\langle n \rangle\rangle_{A<}^{t_1} < ((n))_B < \langle\langle m \rangle\rangle_{B<} < ((m))_A^{t_2})$. Then, the distance between A and B is less than or equal to $v(t_2 - t_1 - I_0)/2$, where v is the velocity at which a signal travels and I_0 is the minimum turnaround time at B.*

The point of our analysis will be to get a verifier to the point at which she can apply Proposition 1 to calculate her distance from a prover.

Challenge-Response and Distance Bounding Templates

A key feature of the logic is the *challenge response template*, which is as follows

$$A : \Phi' \wedge (\nu n)_A < \langle\!\langle c^{AX} n \rangle\!\rangle_{A<} < ((r^{AX} n))_A$$
$$\Rightarrow (\nu n)_A < \langle\!\langle c^{AX} n \rangle\!\rangle_{A<} < ((c^{AX} n))_X < \langle\!\langle r^{AX} n \rangle\!\rangle_{X<} < ((r^{AX} n))_A$$

where c^{AX} is the challenge structure issued by A, r^{AX} is the corresponding response originated by X, and Φ' represents some additional precondition, such as an honesty assumption. For example, the challenge could be a nonce, and the response could be a MAC applied to the nonce using a key shared between A and X.

The challenge-response template is the basic building block of authentication protocols. Most authentication protocols can be built up by combining and extending various challenge-response protocols. However, the challenge-response template cannot be used in its basic form for distance bounding protocols. That is because the computational requirements on the response are so strict that much of the job of the challenge and response must be accomplished by auxiliary protocols occurring before the challenge and after the response. We refer to these auxiliary protocols as C_A and R_A, as below.

We describe the distance bounding template below:

$$A : \Phi' \wedge (\nu n)_A < C_A(n) < \langle\!\langle c^{AX} n \rangle\!\rangle_{A<}^{t_1} < ((r^{AX} n, m))_A^{t_2} < R_A(n, m)$$
$$\Rightarrow (\nu n)_A < \langle\!\langle c^{AX} n \rangle\!\rangle_{A<}^{t_1} < ((c^{AX} n))_X < \langle\!\langle r^{AX} n, m \rangle\!\rangle_{X<} < ((r^{AX} n, m))_A^{t_2}$$

There are a number of ways of constructing $C_A(n)$ and $R_A(n, m)$. In Brands-Chaum and Čapkun-Hubeaux $C_A(n)$ is a commitment, and $R_A(n, m)$ is an authentication of the rapid response, plus an opening of the commitment. In our protocol, $C_A(n)$ is empty, and $R_A(n, m)$ is the authentication of the rapid response. In the Hancke-Kuhn protocol, the $C_A(n)$ is an exchange of nonrepeatable bitstrings, the rapid exchange is the exchange of a one-way collision-free hashes of the bitstrings with a shared key, while $R_A(n, m)$ is empty.

4.3 Analysis of the Distance Bounding Protocol

Proof of Security for Honest Prover

Our logic is designed to be used in for success refinement of a protocol. Normally, this involves either increasing the functionality of the principals involved, or making assertions about their behavior that is implemented in successive refinements. We have found in it this case it makes sense to do our refinements on the honesty of the prover. Refining our analysis on different assumptions about the prover's honesty allows us to see what the different kinds of guarantees are in the different cases. We use three types of prover, first one about which we make no assumptions, then a "semi-honest" prover who sends messages in the correct order and does not reveal secrets, but who does not necessarily reply with a nonce when expected to, and who may attempt to cheat by sending the response before getting the challenge. Finally, we specify the honest prover.

The semi-honest prover is specified as follows.

$$V : \mathsf{SHP} \implies \langle F(N_V, P, N_P) \rangle_P \prec \langle P, Pos_P, N_P, N_V, MAC_{K_{PV}}(P, Pos_P, N_P, N_V) \rangle_P$$
$$\wedge \langle F(X, P, Y) \rangle_P \implies X = P \qquad \text{(shpr)}$$

We can also specify the honest prover, who follows the protocol to the letter:

$$V : \mathsf{HP} \implies (\nu N_P)_P((V, \text{request})_P < (N_V)_P < \langle\langle N_P \sqsubseteq F(N_V, P, N_P) \rangle\rangle_{P<} \prec$$
$$\langle P, Pos_P, N_P, N_V, MAC_{K_{PV}}(P, Pos_P, N_P, N_V) \rangle_P)$$
$$\wedge \langle F(X, P, Y) \rangle_P \implies X = P \qquad \text{(hpr)}$$

Finally, we specify a necessary piece of information about the honest verifier's behavior.

$$V : \langle MAC_{K_{XV}}(Y, Pos, N, M) \rangle_V) \implies Y = V \qquad \text{(hv)}$$

This prevents V from concluding the a message sent by herself is from P.

Our proof will proceed incrementally, using stronger and stronger assumptions about P. We will start with proving what can the verifier can conclude when nothing at all is known about the principal or principals with which she is interacting. We will then progress to the case of the semi-honest prover, and conclude with the honest prover.

We start with what the verifier observes:

$$V\text{ sees} : \langle V, \text{request} \rangle < (\nu N_V)_V < \langle N_V \rangle_V^{t_1} < (F(P, N_V, N_P))_V = ((N_V))_V = ((N_V))_V^{t_2}$$
$$< (P, Pos_P, N_P, N_V, MAC_{K_{PV}}(P, Pos_P, N_P, N_V))_V \qquad \text{(vfr)}$$

where P is a variable of type name.

By applying the rcv axiom twice, we obtain from it together with the st axiom governing $F(P, N_V, N_P)$ and the simple stability of the MAC expression that:

$$V : (F(P, N_V, N_P))_V^{t_2} \implies \langle F(P, N_V, N_P) \rangle_X < (F(P, N_V, N_P))_V^{t_2}$$
$$\text{(a1)}$$
$$V : (MAC_{K_{PV}}(P, Pos_P, N_P, N_V))_V \implies \langle MAC_{K_{PV}}(P, Pos_P, N_P, N_V) \rangle_Y < \quad (1)$$
$$(MAC_{K_{PV}}(P, Pos_P, N_P, N_V))_V \qquad \text{(a2)}$$

From the st axiom governing $F(P, N_V, N_P)$, the new axiom, and a1 we obtain

$$V : (\nu N_V)_V < \langle N_V \rangle_V^{t_1} < ((N_V))_X < \langle F(P, N_V, N_P) \rangle_X < (F(P, N_V, N_P))_V^{t_2} \quad \text{(a3)}$$

This is as far as we can go without making some assumptions about honesty. Since we have two principals now, X and Y, we will need to make honesty assumptions about them both. The condition Φ that we assume will be $\mathsf{SH}X \vee \mathsf{SH}Y$.

Proposition 2. *Suppose that* a2 *and* a3 *hold. From* vfr *and* $\mathsf{SH}X \vee \mathsf{SH}Y$ *we can further conclude that*

$$V : (\nu N_V)_V < \langle N_V \rangle_V^{t_1} < ((N_V))_P < \langle F(P, N_V, N_P) \rangle_P <$$
$$\langle P, Pos_P, N_P, N_V, MAC_{K_{PV}}(P, Pos_P, N_P, N_V) \rangle_P <$$
$$(P, Pos_P, N_P, N_V, MAC_{K_{PV}}(P, Pos_P, N_P, N_V))_P \qquad \text{(b1)}$$

Proof. Suppose that X is semi-honest. From the shpr axiom and a3, we obtain that $X = P$. From the mac axiom, we get that $\langle P, Pos_P, N_P, N_V, MAC_{K_{PV}}(P, Pos_P, N_P, N_V) \rangle_P$ as well. The semihonesty of $X = P$ gives us $\langle F(P, N_V, N_P) \rangle_P < \langle P, Pos_P, N_P, N_V, MAC_{K_{PV}}(P, Pos_P, N_P, N_V) \rangle_P$, and the remainder follows from a2 and a3.

Suppose now that Y is semi-honest. Then, by the mac axiom, we get that $Y = P$. From the shpr axiom, we get that $\langle F(P, N_V, N_P) \rangle_P < \langle P, Pos_P, N_P, N_V, MAC_{K_{PV}}(P, Pos_P, N_P, N_V) \rangle_P$ as well, and from this and a2 and a3 we get the result.

We are now left with two things to prove, first that P was the first to send the rapid response, and secondly that V receives P's response after P sends it. The first is necessary in order for V to be able to conclude the second. We get both from the honest of P.

Proposition 3. *Suppose that* HP *and* vfr *hold. Then*

$$V : (\nu N_V)_V \wedge (\nu N_P)_P \, (\langle N_V \rangle_V^{t_1} < ((N_V))_P < \langle\langle N_P \sqsubseteq F(P, N_V, N_P) \rangle\rangle_{P<}$$
$$< ((N_V \sqsubseteq F(P, N_V, N_P)))_V^{t_2} < \langle MAC_{K_{PV}}(P, Pos_P, N_P, N_V) \rangle_P$$
$$< (P, Pos_P, N_P, N_V, MAC_{K_{PV}}(P, Pos_P, N_P, N_V))_V) \quad \text{(c1)}$$

Proof. From HP we get SHP, and from that and vfr we get b1. From HP we also get that P was the first to sent a message constructed with N_P. From the new axiom, we get that $(F(P, N_V, N_P))_V$ must occur after $\langle\langle N_P \rangle\rangle_{P<}$. This, together with c2, gives us the result we need.

We are now able to conclude that V knows

$$\nu N_V)_V \wedge (\nu N_P)_P \, (\langle N_V \rangle_V^{t_1} < ((N_V))_P < \langle\langle N_P \rangle\rangle_P < ((N_P))_V$$

and we are thus able to conclude that $v((t_2 - t_1 - I_0)/2 \geq d(V, P)$.

Case of the Dishonest Prover

We first note that, although in our definition of an honest prover the prover responds after receiving the verifier's nonce, our result would hold even if it attempted to respond before receiving it, thanks to the new axiom. Thus our result holds for a prover who follows all the rules of the protocol but may attempt to respond early or late, as well as a completely honest prover.

However, we are not able to prove any such results about the semi-honest prover. The reason is that proving such result would require strong assumptions about the behavior of other dishonest nodes as well. Suppose that a dishonest (or badly implemented) prover P sends out a predictable value instead of generating a nonce. Then an attacker A who is closer to the verifier than P is, could if it is aware of this, anticipate P's rapid response before P does, thus making P looking closer than it is. In order to rule out this kind of attack, we would need to make the assumption that A could not anticipate P's response, which is so close to the assumption of the behavior of the honest P who sends an unpredictable nonce as to make no difference.

This problem is closely related to Desmedt's "terrorist attack" involving colluding verifiers. Consider the case in which both Q, the sender of the rapid response and P, the sender of the authenticated response, disobey pr. If Q and P share N_P, then Q could send $h(N_P, P) \oplus N_V$ in P's stead. If Q was closer to V than P, then P could use Q's response to pretend to be closer to V than it was. Of course, there is no reason for Q to cooperate with P in this way unless they are actively colluding, which is why we say that the protocol is vulnerable to collusion attacks.

We note that the Čapkun-Hubaux, Brands-Chaum, and Hancke-Kuhn protocols are vulnerable to the same type of collusion attacks, as are most other distance bounding protocols. Indeed, Brands and Chaum [2] pointed out in their original paper that their protocol was subject to this type of attack. Existing schemes for avoiding the terrorist attack rely either on tamper-proof hardware [15, 17] or on forcing the conspirators to reveal long-term keys to each other [3]. However, we would expect both of these types of solutions, although they may be useful for certain kinds of wireless networks, to find only limited applications in sensor network security. Forcing potential cheaters to share long-term secrets if they want to collude only makes sense when the parties are mutually distrusting. If, as in the case of a sensor network, they are more likely to have been compromised by the same attacker, it is not likely to provide much deterrence. Likewise, tamperproof hardware may not be the optimal solution in a sensor network in which one is highly motivated to keep hardware costs low because nodes may be lost, stolen, destroyed, or power-depleted. In the next section, we consider the problem of detecting, rather than preventing or deterring, collusion attacks. We show that colluding verifiers who are capable of implementing wormhole attacks can defeat even protocols such as SPINE that use triangulation to detect cheating.

5 Analysis of Distance Bounding Under User Collusion

We now address the question of the impact of collusion on the distance bounding protocols that try to incorporate security. Distance bounding protocols are subject to collusion because they rely upon the keeping secrets and/or the delayed release of information to achieve security. If that information is shared, then collusion is possible. Using this shared information, the adversary then tries to make the verifier believe it is indeed executing all steps of the protocol. The desired outcome for the colluders is to make an adversary or cheater appear closer to the verifier than it really is. The end effect is that the relative location is artificially enlarged to include the colluding node that is far away. We will illustrate this using (a) standard collusion and also (b) wormhole in Figure 2. A wormhole attack is one in which a fast link is set up between the victims and an attacker who is outside of the normal range. The wormhole attack may appear to be an overkill for this problem since even without the resources to establish wormhole, adversaries can collude and create damage. However, wormholes can increase the range of the colluder who is farther away, thus increasing the amount of error that can be induced by collusion.

The attack for Figure 2 proceeds as follows: Colluding node P receives a nonce from colluder node A (This step can be removed if the reduction in communication is to be minimized, but the nodes P and A must know the nonce N_P at some point for executing the MAC. Node P would then declare its distance from node V to be that between nodes V and A, denoted d_{AV}. The third step in the protocol is now changed to $F(N_V, A, N_A)$. The node P then transmits the nonce N_V to node A, which computes $MAC_{K_{AV}}(A, Pos_A, N_A, N_V)$ and transmits it to node P. The final step is executed by node P by transmitting the message to verifier node V. Note that node A is assumed not to be able to communicate the information directly to the verifier node V in this version of the protocol. Thus the node P must execute the last step to complete the protocol.

In the case that there is a wormhole link such that the node A is able to transit the data without having a terminal node at both ends of the wormhole (say for example using a directional antenna) then the last step of the protocol termination does not have to involve node P and is modified as shown in figure2.

The next question then is: how does one recognize the existence of such collusion in the distance bounding protocol. We claim that if the nodes A, and P do form a collusion, and behave consistently with respect to relative distance measures and compute the MAC and terminate the last step of the protocol within a "reasonable" time interval, there is no mechanism to detect the user collusion since it will create no inconsistency, and hence the protocol will exist with faulty measurements only. In making this claim, we have tried to stay away from showing how the detection can be made under certain assumptions about the existence of honest nodes since, even with that assumption, one can show that for infinitely many cases that the collusion cannot be detected. Our claims hold even if there are more than three independent verifiers as in the case of multilateration, as long as the power levels of the transmissions from nodes A and P are consistent. Figure3 illustrates a successful collusion with three

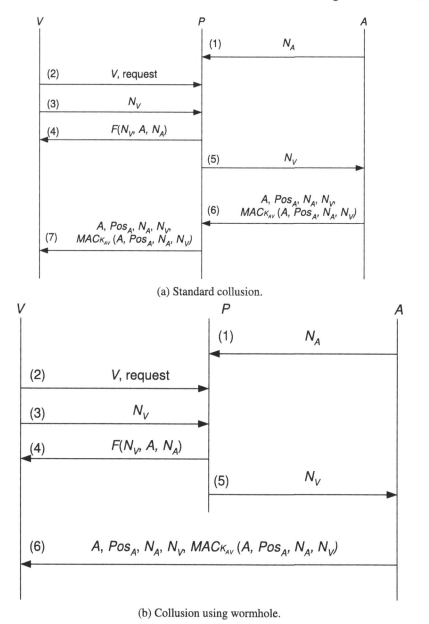

(a) Standard collusion.

(b) Collusion using wormhole.

Fig. 2. Protocols under Collusion of nodes A and P

verifiers. Hence our conclusion is that the secure distance bounding is vulnerable to collusion and in general the collusion cannot be detected.

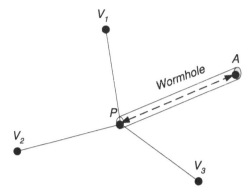

Fig. 3. Illustration of a successful collusion of nodes P and A through wormhole in presence of three verifiers V_1, V_2, and V_3.

6 Conclusion

We have presented a new protocol for distance bounding that requires less message and cryptographic overhead than similar protocols, while still possessing the property of delaying authentication, which can be desirable in a number of applications. More importantly, we have provided a qualitative logical analysis that makes the relationship between authentication and distance measurement clear. Moreover, in doing so we have extended our logic to cover exclusive-or in a way that we think will be applicable to many other equational theories as well. Furthermore, we provide a framework which can be used to in the evaluation of other distance bounding protocols as well. Finally, we point out some fundamental limitations in current distance bounding technologies; the use of cryptographic authentication means that even the ones that are designed to resist collusion are subject to attacks in which attacks in which one dishonest verifier shares its keying material with another. Moreover, these attacks are not detectable by protocols such as SPINE that use triangulation to detect dishonest non-colluding verifiers.

There is still of course, much to be done. Although our logic gives a framework for analyzing distance bounding protocols, it is still only a qualitative framework. What is really needs is a method for analyzing distance bounding protocols that combines both the logical method and the analytical approach of Sastry et al. This will also help us to derive tight bounds on the errors involved in communicating with an honest or isolated dishonest prover. Our logic is intended to be extensible to quantitative as well as qualitative theories, and we believe that the relatively simple distance bounding protocols would make a good test case, as will as providing a unified theory within which distance bounding can be analyzed.

Even if a fully worked our formal theory that covers both qualitative and quantitative aspects is developed, however, the problem of collusion remains. At this point, it seems to make sense to regard distance bounding as a tool which can be used to verify distance from an honest prover and provide a lower bound on the distance of an isolated cheating prover. Incorrect locations calculated from colluding verifiers may show up as inconsistencies when compared against locations computed from honest verifiers. These inconsistencies could then be exploited to detect the dishonest verifiers, as long as certain assumptions about the distribution of the verifiers hold (e.g. that they are in the minority). We note that such techniques, e.g. SERLOC [9]and HIRLOC [10] have been developed to detect wormhole attacks on range-free location, and would expect a similar approach to work here.

References

1. MSSI Completes Phase II SBIR Contract for UWB-Based Urban Positioning System ('UPS') - Awarded $1M Phase II Plus, June 27 2006.
2. S. Brands and D. Chaum. Distance-bounding protocols. In *Advances in Cryptology - Eurocrypt '93*. LNCS 765, Springer-Verlag, 1995.
3. L. Bussard. *Trust Establishment Protocols for Communicating Devices*. PhD thesis, ESNT Paris, October 2004.
4. Iliano Cervesato, Catherine Meadows, and Dusko Pavlovic. An encapsulated authentication logic for reasoning about key distribution protocols. In Joshua Guttman, editor, *Proceedings of CSFW 2005*, pages 48–61. IEEE, 2005.
5. Y. Desmedt. Major security problems with the 'unforgeable' Feige-Shamir proofs of identity and how to overcome them. In *Proceedings of Securicom '88*, 1988.
6. G. Hancke and M. Kuhn. An RFID distance bounding protocol. In *Proc. of Securecomm 2005*, 2005.
7. Leslie Lamport. Time, clocks, and the ordering of events in a distributed system. *Commun. ACM*, 21(7):558–565, 1978.
8. L. Lazos, S. Čapkun, and R. Poovendran. ROPE: Robust position estimation in wireless sensor networs. In *The Fourth International Conference on Information Processing in Sensor Networks (ISPN '05)*, April 2005.
9. L. Lazos and R. Poovendran. SerLoc: Robust localization for wireless sensor network protocols. *ACM Transactions on Sensor Networks*, 2005.
10. Loukas Lazos and Radha Poovendran. HiRLoc: Hi-resolution robust localization for wireless sensor networks. *IEEE Journal on Selected Areas in Communication*, 24(2):993–999, February 2006.
11. Catherine Meadows and Dusko Pavlovic. Deriving, attacking and defending the GDOI protocol. In Peter Ryan, Pierangela Samarati, Dieter Gollmann, and Refik Molva, editors, *Proc. ESORICS 2004*, volume 3193 of *Lecture Notes in Computer Science*, pages 53–72. Springer Verlag, 2004.
12. Catherine Meadows, Paul Syverson, and LiWu Chang. Towards more efficient distance bounding protocols. In *SecureComm 2006*, August 2006.
13. Dusko Pavlovic and Catherine Meadows. Deriving secrecy properties in key establishment protocols. In Dieter Gollmann and Andrei Sabelfeld, editors, *Proceedings of ESORICS 2006*, Lecture Notes in Computer Science. Springer Verlag, 2006. to appear.
14. N. Sastry, U. Shankar, and D. Wagner. Secure verification of location claims. In *ACM Workshop on Wireless Security (WiSe 2003)*, pages 48–61. ACM, September 19 2003.
15. D. Singleé and B. Preneel. Location verification using secure distance bounding protocols. In *International Workshop on Wireless and Sensor Network Security*. IEEE Computer Society Press, 2005.

16. S. Čapkun and J. P. Hubaux. Secure positioning in wireless networks. *IEEE Journal on Selected Areas in Communication*, 24(2), February 2006.
17. B. Waters and E. Felten. Secure, private proofs of location. Technical Report TR-667-03, Princeton, 2003.

Location Privacy in Wireless LAN

Leping Huang[1,2], Hiroshi Yamane[2], Kanta Matsuura[2], and Kaoru Sezaki[2]

[1] Nokia Research Center Japan,1-8-1,Shimomeguro, Meguro-ku, Tokyo, Japan
   leping.huang@nokia.com
[2] University of Tokyo, 4-6-1 Komaba, Meguro-ku, Tokyo, Japan
   yamane@mcl.iis.u-tokyo.ac.jp,
   {kanta,sezaki}@iis.u-tokyo.ac.jp

Summary. Current Wireless LAN(WLAN) specification require all stations to use globally unique fixed MAC addresses. The MAC address is visible in all WLAN packets. Globally unique and fixed addresses enable an observer to collect history and profile data of wireless users. This cause a serious location privacy breach especially in public access networks. Several protection methods based on temporal address and periodical address updates have already been proposed. In this paper, we identify a new wireless location privacy attack–correlation attack–in the context of wireless LAN and high accuracy localization technique. Correlation attack is a method of utilizing the temporal and spatial correlation between the old and new pseudonym of nodes. we identified that solutions based on periodical address update cannot protect users from advanced tracking methods including correlation attack under such context. To combat such attacks, we propose the concept of a silent period. A silent period is defined as a transition period between the use of new and old pseudonyms, when a node is not allowed to disclose either the old or the new address. This makes it more difficult to associate two separately received pseudonyms with the same station, because the silent period disrupts the temporal and/or spatial correlation between two separately received pseudonyms, and obscures the time and place where a pseudonym changed. we generalizes a wireless LAN system into a MIX based anonymity model. The model offers two insights: a way of evaluating location privacy protection systems; and serving as a bridge between the new location privacy protection problem and existing defense and attack approaches in the MIX related research.

1 Introduction

Recent technological advances in wireless location tracking have presented unprecedented opportunities for monitoring the movements of individuals. While such technology may support many useful location-based services (LBSs), which tailor their functionality to a user's current location, privacy concerns might seriously hamper user acceptance. Within those technologies, the positioning systems that utilized short-range radio such as Bluetooth and Wireless local area network (WLAN) have been receiving great attention recently. However, due to the advances in localization and tracking techniques [1–3] that can accurately estimate location of a wireless node based on its transmitted signal properties, any wireless communication can be utilized by an adversary to locate and/or track the user of wireless device. Some of those systems track users' movement by eavesdropping on their communication. Eavesdropping is invasive and does not require the cooperation or approval of the users. This leads to breach the *location privacy* of users, i.e. the adversary can identify the locations visited by a user.

In this paper, we concentrate on one location privacy threats *trajectory privacy*. *Trajectory privacy* represents a user's right to prevent other parties from collecting a set of locations, visited by the user over a period of time. If an adversary is able to link two or more locations of an entity over time to build its trajectory, then it is able to profile the entity's behavior over time. This presents a threat to the user's privacy. This definition is similar to the definition of location privacy used by Beresford in [4]. The only difference is that we more concentrate on one's continuous movement than one's position at discrete time instant.

Our motivation of this research is to provide better protection on user's location privacy to those eavesdropping-based location tracking attack. Some solutions to this location privacy threat have already been proposed for both Bluetooth and WLAN [3, 5], which are based on the idea of periodic address updates. Our major contributions in this paper are as follows:

- we identify a new wireless location privacy attack– correlation attack–in the context of wireless LAN and high accuracy localization technique. Correlation attack is a method of utilizing the temporal and spatial correlation between the old and new pseudonym of nodes. we identified that solutions based on periodical address update cannot protect users from advanced tracking methods including correlation attack under such context.
- we provide a solution–silent period–to circumvent the correlation attack. A silent period is defined as a transition period between the use of new and old pseudonyms, when a node is not allowed to disclose either of them. This makes it more difficult to associate two separately received pseudonyms with the same station, because the silent period disrupts the temporal and/or spatial correlation between two separately received pseudonyms, and obscures the time and place where a pseudonym changed.
- we generalizes a WLP^2S into a MIX based anonymity model. The model offers two insights: a way of evaluating location privacy protection systems; and serving as a bridge between the new location privacy protection problem and existing defense and attack approaches in the MIX related research.

The rest of the paper is organized as follows. Section 2 describes the wireless location privacy protection system (WLP^2S) model to be used throughout this paper, as well as location privacy threats and adversary models considered. In Section 3, we propose the privacy protection algorithm: silent period and its effect to alleviate address's temporal and spatial relations. Section 4 proposes a general model for WLP^2S which includes an abstracted framework of system, information-theory based definitions of anonymity and measures, as well as detailed description about the impact of WLAN radio on the formal model. Section 5 describes two wireless location privacy protection systems with different tracking algorithms, and evaluate the performance of these two systems by using proposed measures. Simulation results reveal the effect of many parameters of a WLP^2S on users' privacy level. Section 6 utilizes our formal model to identify potential protocol threats and suggests areas for improving silent period protocol. Background and prior arts are presented in Section 7. Finally, we conclude the paper in Section 8 with a discussion of proposed future work.

2 System Assumption and Problem Analysis

In this section, we first define the WLAN system model to be used in this paper, and then analyze the location privacy problems and adversary model in this model. Finally, we discuss requirements and constraints of applications used in this model. The notations and abbreviations used throughout this paper, is provided in Table 3 in the Appendix.

2.1 System Model and Assumptions

Figure 1 illustrates a typical WLAN location privacy protection system (WLP^2S). A WLP^2S is conceptualized to comprise mobile station(MS), operator network, eavesdropper(E), and service providers on Internet. Operator network is composed of Access Point (AP) to provide wireless access service to MSs, gateway as the interface to service provider, access routers to forward IP packets between MS and service provider and several management servers. AP, MS, and E are incorporated with the WLAN radio interfaces operating at identical frequencies. A user carrying a MS moves on streets, and use wireless service provided by operator network to access service provider.

We assume that a public key infrastructure is employed in the WLP^2S. Before joining a WLP^2S, a MS i first authenticates with operators' Authentication, Authorization, Accounting (AAA) server, and then obtains a public/private key pair (PK_i, PK_i^{-1}), a symmetric key SK_i. Those keys are used to protect the confidentiality of the wireless communication between AP and MS. The whole operator network is a trusted entity in our model; service provider may be compromised by adversary; eavesdroppers belong to the adversary.

Figure 1 further explains the network architecture from layered communication point of view. In this figure, we illustrated three communication layers: MAC, IP and Application layer. In MAC layer, MS and AP are connected by WLAN radio, access router and AP are connected by Ethernet link. Eavesdropper monitors MAC frames between MS and AP. In addition. AP serves as a MAC layer bridge/switch to forward MAC layer frames between access router and MS. From IP layer point of view, MS is directly connected to access router. IP frames from service provider are forwarded to MS via access routers and gateway in operator's network. In application layer, MS is directed connected to its service provider.

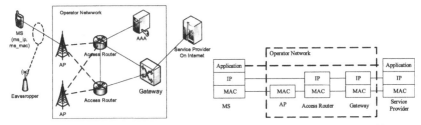

Fig. 1. (left) Illustration of a WLAN Communication System, (right) Layered architecture of WLP^2S

2.2 Location Privacy Threat Analysis in WLAN

Generally, a location privacy threat occurs when an untrusted party can locate a transmitting device and identify the subject using the device. A wireless LAN network poses very serious location privacy threats for the following reasons.

802.11 WLAN standards adopt Carrier Sense Multiple Access with Collision Detection (CSMA/CD) as its MAC layer protocol [6]. CSMA/CD is a network control protocol in which (a) a carrier sensing scheme is used and (b) a transmitting data station that detects another signal while transmitting a frame, stops transmitting that frame, transmits a jam signal, and then waits for a random time interval (known as "backoff delay" and determined using the truncated binary exponential backoff algorithm) before trying to send that frame again.

This protocol has two impacts related to transmitting node's location privacy. First, *carrier sensing scheme* requires all nodes in its proximity to share the same communication channel. Any node including eavesdroppers can overhear all frames sent by others within its proximity. Secondly, *random access scheme* requires each transmitting node to include its identifier (MAC address) in all its frames to distinguish itself with others for correct data reception.

The format of the 802.11 MAC frame is shown in Table 1. The encryption algorithms of 802.11 (AES and WEP) provide data confidentiality only for the frame body field of the frame; other fields are sent in plain text. Considering the shared channel characteristics of 802.11 radios, it is obvious that the MS understands the sender address (field A1) and receiver address (field A2) of all frames sent by nodes within its proximity. This opens the possibility for an adversary to monitor the movement of a MS without its cooperation. If the sender transmits frames continuously, an eavesdropper can obtain the identity of the sender regularly. Continuous reception of the frame identity improves the accuracy of the tracking system. Maximum MAC address broadcast frequency can be derived by calculating the time to transmit one IP frame (containing only an IP header) from one station to the AP. The operation rate of 802.11 used here is 11Mbps DSSS/CCK in short-preamble mode. The procedure of frame transmission is as follows. MS first sends an IP frame with T_{data} μs, and the AP sends back an acknowledgment with T_{ack} μs, after waiting for the SIFS (short inter-frame space) to avoid frame collision. The station then waits for the SIFS again before being ready for the next transmission. SIFS is 10 μs in 802.11b. The round trip time(RTT) of transmitting one frame is shown in Eq. 1. Therefore, the eavesdropper will receive the identity of sender at a frequency up to 5KHz. In general, more frequent identity broadcasting improves the location-tracking accuracy.

$$RTT = T_{data} + SIFS + T_{ack} + SIFS$$
$$= (96\mu s + 54 \times 8/11\mu s) + 10\mu s + (96\mu s + 14 \times 8/11\mu s) + 10\mu s$$
$$\cong 262\mu s \tag{1}$$

Table 1. 802.11 MAC Frame Format

2	2	2	6	6	2	6	0-2312	4
Frame Control	Dur. /ID	A1	A2	A3	Sequence Control	A4	Frame Body	FCS

Besides, the radio signal properties of a WLAN system allow relatively precise determination of a client's position. When an access point receives a signal from a client, it is highly probable that the client's position will be within a typical range of the AP (say, 100 meters). As discussed in Section 7, tracking accuracy can be further improved to the order of 1 meter by triangulation or pattern matching methods.

Finally, it is relatively inexpensive to deploy enough wireless LAN nodes to cover large areas for location tracking, compared with covering the same area with a cellular-network based tracking system. This non-technical reason also greatly increases the risk to location privacy in wireless LANs.

From the analysis above, we can summarize as that although the communication between two wireless LAN nodes are protected by encryption, a transmitting node still needs to disclose its identifier frequently to others in its proximity.

2.3 Adversary model and Correlation Attack

In this paper, we assume that an eavesdropper is capable of capturing all frames with some radio metric transmitted in the channel within its proximity, and estimating the position of frame sender based on measured radio metric (e.g. Radio signal Strength Indication(RSSI)). We also assume a global passive adversary model(GPA), which means that all regions that the MS may visit are covered by adequate number of eavesdroppers for position tracking. From the analysis above, we know that although the communication between two wireless LAN nodes are protected by encryption, a transmitting node still needs to disclose its identifier frequently to other nodes in its proximity. If a user use a same identifier continuously, the adversary can records a series of locations with the same identifier by capturing frames continuously. An adversary can use these records to *link two or more locations of an entity over time* to build its trajectory, then it is able to profile the entity's behavior over time. This greatly violates the user's location privacy.

Some solutions to this location privacy threat have already been proposed for both Bluetooth and WLAN [3, 5], which are based on the idea of periodic address updates. However, those solutions may not be effective in the WLP^2S. Frequent broadcasts of the MAC address provide adequate temporal and spatial tracking accuracy. With enough temporal and spatial precision, it may be possible for an adversary to correlate two pseudonyms that are sent separately from the same device moving through space. Temporal correlation may be used because the period with which stations change their pseudonym may be small. Spatial correlation may be used if it is assumed that a station will generally continue in the same direction, with the same speed as it traveled in the past. Such *correlation attack* becomes easier as the distance between devices increases, as the speed of a device decreases, and as the accuracy with which a device may be located increases.

An example is shown in Figure 2. The definitions of disappearing time (DT), disappearing place (DP), emerging time (ET), and emerging place (EP) can be found in Table 3. Assuming that the system provides enough temporal and spatial precision, a node coming from the right with identity A updates its identity to A' at some time within $[DT_1, ET_1]$. The last trail monitored by the system for A is recorded at time DT_1 and position DP_1; the first trail monitored by the system for A' is recorded at time ET_1 and EP_1. The objective of an adversary is to link A with A' with high probability by using knowledge such as the tracking history for A, a user movement model, or a building layout. From the similarity between both the times and locations of the two records, the tracking system can infer that node A changed its address to A' sometime between $[DT_1, ET_1]$ and near the position $[DP_1, EP_1]$.

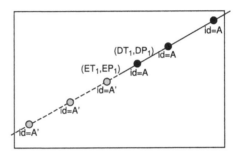

Fig. 2. The correlation attack on periodical pseudonym updates

In summary, in this paper we assume an adversary is capable of (1) capturing the identifiers in all frames transmitted by MS, (2) estimating the position of frame sender based on measured radio metric, and (3) using temporal and spatial correlation of frames from same device to associate two or more locations of an entity over time.

3 Silent Period For Location Privacy

To combat the correlation attack discussed in the previous section, we propose the new concept of a silent period. The silent period is defined as a transition period between using new and old pseudonyms in which a station is not allowed to disclose either the old or the new pseudonym. As a result, the silent period introduces ambiguity into the determinations of the time and/or place at which a change of anonymous address occurred. This makes it more difficult to associate two separately received pseudonyms with the same station, because the silent period disrupts the temporal and/or spatial correlation between two separately received pseudonyms and obscures the time and place where a pseudonym changed.

When multiple stations within the same region follow the rule of the silent period by ceasing transmission after updating their MAC address (pseudonym), the effect is the same as if those nodes had entered a mix zone [4] (also see section 7) in which no user's movement could be monitored by the system. Therefore, the silent period can be seen as an extension or implementation of the mix zone concept. It creates virtual mix zones by controlling the transmission of frames.

An example using the silent period is shown in Figure 3. In the figure, node 1 moves along a path from the upper right corner to the lower left corner. Meanwhile, node 2 moves along a path from the lower right corner to the upper left corner. Both nodes update their addresses and then enter the silent period. The effects of the proposal are illustrated near the intersection of the path of both nodes. The silent period is illustrated as a rectangle in the center of this graph. Node 1 arrives at the border of the silent period at time DT_1 and position DP_1, and node 2 arrives at time DT_2 and position DP_2. Here we assume that two nodes arrive at the border simultaneously ($DT_1 = DT_2$). Both node 1 and node 2 disable frame transmission for a silent period. After the silent period, nodes 1 and 2 restart frame transmission. The tracking system monitors a new trail with address A' emerging at position EP_1 and time ET_1, and another new trail with address B' emerging at time ET_2 and position EP_2. The tracking system knows that node 1 changed its address within the silent period. However, as it detects that two new address A' and B' emerge after the silent period, the tracking system cannot determine whether node 1 changed its address to A' or B', because both position EP1 and EP2 are reachable by node 1 from position DP_1, due to the silent period introduced during the moving time. This method obscures the temporal and spatial correlation between new and old pseudonyms by "mixing" the pseudonyms of nodes.

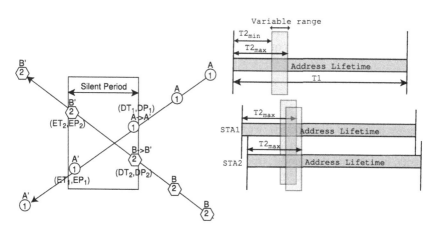

Fig. 3. (left)Synchronization issues in silent networks. Anonymity is guaranteed when the variable periods of Silent Period overlap, (right)Illustration of the movement of two stations, node 1 and node 2, which both use a silent period

The discussion above assumes that nodes 1 and 2 switch their addresses at the same time for the sake of simplicity. The length of the silent period is assumed implicitly constant and identical for all nodes. However, when the address update time is not synchronized between nodes, the constant length of the silent period may not effectively mix the pseudonyms of nodes. This is because when the length of the silent period is constant, the tracking system can link the old and new pseudonyms to the same node. It does this by comparing the order of the emerging times of the two nodes (with their new pseudonyms) with the disappearing time of those with the old pseudonyms. As shown in Figure 3, if the tracking system knows that node 1 with identity A enters the silent period at a time DT_1 (earlier than node 2), it can infer that the emerging time of node 1 is earlier than that of node 2. When the system detects that a node with identity A' emerges earlier than another node with identity B', the system can easily infer that node 1 updates its address to A'.

To solve this problem, we propose the use of a variable length silent period. If the range of the emerging time of a node overlaps with that of another node, the tracking system cannot link the detected pseudonym to the correct node. We illustrate the idea of a variable-length silent period in Figure 3. The silent period is determined by a random variable varying within the range $[T2_{min}, T2_{max}]$. When the ranges of the emerging times of two nodes overlap, the temporal relation between the emerging time and the disappearing time of the two nodes is obscured. The tracking system cannot correctly link the new pseudonym of a node to its old pseudonym. This approach is motivated by the CSMA/CD medium-access control algorithm used in collision-avoidance schemes.

In summary, the silent period should contain constant and variable periods. The effect of the constant period is to mix the spatial relation between a node's disappearing points and its emerging points, while the effect of the variable period is to mix the temporal relation between the node's disappearing times and its emerging times. For clarity, we also define *lifetime* as the duration the user use one identifier continuously. The ratio of silent period to lifetime, called *silent period ratio* indicates how much time user spends on privacy protection. In general, larger silent period ratio results lower communication quality.

4 Modeling of Wireless Location Privacy Protection System

After we proposed silent period, we noticed that the privacy level that our silent period proposal can achieve in a WLP^2S is affected by many parameters such as node density, nodes' mobility model, tracking algorithm, length of silent period and lifetime etc. Unless we have a generalized model for such a system, it will be difficult to evaluate the effect of those system's parameters on users' privacy level and enhance the protocol Our aim therefore is to generalize wireless location privacy protection system into a formal anonymity model. In this section, we first we describe the formal anonymity model for a WLP^2S, and then analyze the relevant features and constraints of a WLP^2S caused by WLAN radio and applications.

4.1 Formal description of WLP^2S

We describe this model in three steps. First, we map different roles of nodes of a WLP^2S into a MIX based system $\mathcal{S}$, including a MIX $\mathcal{M}$, a set of MIX's users $\mathcal{U}$ and MIX's intruder $\mathcal{I}$. Concisely, $\mathcal{S} = \{\mathcal{U}, \mathcal{I}, \mathcal{M}\}$. Then, we use equivalence relation to define anonymity set and measures for the generalized WLP^2S. Finally, we analyze the features–mainly accuracy–of a intruder's observation process in detail. Process algebra-based approach defined in [7] is used in step one and three; while information-theory based approach proposed in [8] is used in step two. Our model does not include a formal protocol validation module because we only addressed global passive attacks(GPA) now. Future work may consider adding protocol validations against active attacks if deemed necessary.

To help readers understand our formal model, we first give a visual overview of the model in Figure 4. A WLP^2S is abstracted into three entities (a set of MIX users $\mathcal{U}$, MIX Intruder $\mathcal{I}$, MIX $\mathcal{M}$), functions between those entities(selection function σ, action $\mathcal{A}$, observation function ω). A MIX user u participates in MIX by executing a series of actions $\mathcal{A}$; while a MIX intruder $\mathcal{I}$ observes each of those actions through an observation function ω. This observation function causes loss and bias on the observed actions. Intruder uses selection function σ and a prior knowledge about MIX $k_\mathcal{M}$ to find out the correlation between observed actions. This can be seen as a passive attack on the MIX. The *analysis and measurement* blocks in Figure 4 indicate a set of tools for analyzing the generalized system. In current system, we only use size $\mathcal{S}$ and entropy $\mathcal{H}$ of MIX as measures. Explanation of notations used in this paper are listed in table 3 in Appendix.

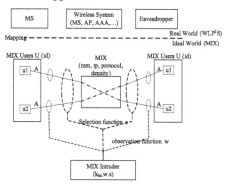

Fig. 4. Formal model of a wireless location privacy protection system.

After preliminary analysis, we noticed that every MS has two roles in the system. Firstly, a user of the MIX u executes a series of actions $\mathcal{A}$(i.e. send a frame). Each MS has one identifier id uniquely identified by others within a specific period of time, and use this identifier as sender id when transmitting a frame. Secondly, a set of MSs and other entities(i.e., AP, AS) in the system forms a MIX between MSs by utilizing their features such as MSs' mobility pattern mm, traffic pattern tp, privacy protection protocol *protocol*. The limits of an intruder's knowledge $k_\mathcal{M}$ about some of those features helps creating this MIX. Here, we generalize the first role of a

MS to a MIX user $u \in \mathcal{U}$, which contains only one parameter: user identifier id, or concisely: $u = \{id\}$. In WLAN, identifier id is a 48bits MAC address. We also define a set of identifiers $\mathcal{ID}$. Reader should notice that although a set of users $\mathcal{U}$ and a set of identifiers $\mathcal{ID}$ are related with each other, but they are not identical. In a WLP^2S, it is allowed that a user utilizes different identifiers at different period of time; while a identifier is used by different user at different period of time. To avoid address collision–which will seriously disturb normal communication–we assume that one address is not used by multiple users simultaneously. This property can be guaranteed by some form of Duplicate Address Detection(DAD) protocol such as [9].

In a MIX system, users normally execute a series of actions $\mathcal{A}$. Some of those actions are used by intruder $\mathcal{I}$ to attack the system. In a WLP^2S, we think there are three types of actions: node movement $\mathcal{A}_m$, frame transmission $\mathcal{A}_t$ and frame reception $\mathcal{A}_r$. Or concisely $\mathcal{A} := \mathcal{A}_m \| \mathcal{A}_t \| \mathcal{A}_r$. We classify those actions into observable actions $\mathcal{A}_{obs}$ and invisible actions $\mathcal{A}_{inv}$. $\mathcal{A}_m$ represents the actions that a node roams within a geographical area. As an intruder utilizes only the radio metrics in captured frames to estimate a node's position, the movement of the node itself is considered invisible to the intruder. $\mathcal{A}_r$ represents the actions that a node receives a frame targeted to itself. According to WLAN specification, nodes do not need to transmit any signals for frame reception. Consequently, $\mathcal{A}_r$ is also invisible to the intruder. Only $\mathcal{A}_t$ action is observable to the intruder because sender should explicitly send radio signal for each frame transmission. Considering the feature of a WLAN tracking system, we generalize a frame transmission action $a_t = id_s \| id_r \| msg \| t \| sm \| pos$. It means that one message msg is sent from user with id_s to user with id_r with signal metric sm at position pos and time t, while $id_s, id_r \in \mathcal{ID}$. In addition, we also define a function $id(a)$ on $\mathcal{A}_t$, which returns sender id id_s of an action a_t, and another function $time(a_t)$, which returns an action a_t's time instant t when it is generated. Most of tracking algorithms analyze the correlations between groups of actions with same identifiers to track node's movement. For clarity, we define trajectory T_i as set of all $a_t \in \mathcal{A}_t$ with same sender id i, $T_i = \{a_t | id(a_t) = i\}$. One thing we want to emphasize here is that we define trajectory not for users , but for identifiers. $\mathcal{T}$ is defined as set of all trajectories $\mathcal{T} = \{t_i\}$. we notice that set of trajectory is a subset of observable actions: $\mathcal{T} \subseteq \mathcal{A}_{obs}$.

Besides, a group of MS and other entities(i.e., AS, AP) in a WLP^2S forms the MIX. Because intruder lacks knowledges about some features of MIX, it will have some ambiguity(anonymity) when associating users and its actions. The knowledge $k_{\mathcal{M}}$ that may affect the ambiguity of intruder includes randomness of user's behaviors such as mobility pattern mm, traffic pattern tp, privacy protection protocol $protocol$ and number of MIX users $density$.

On the other hand, eavesdroppers use observed actions to attack the unlinkability of MIX. We generalize the behaviors of eavesdroppers into a intruder $\mathcal{I} := \{k_{\mathcal{M}}, \omega, \sigma\}$, while $k_{\mathcal{M}}$ is an abstraction of intruder's prior knowledge about $\mathcal{M}$, ω is an observation function, and σ is a selection function. The intruder $\mathcal{I}$ first observes a set of actions $\mathcal{A}$ generated by a set of users $\mathcal{U}$ by an observation function $\omega : \mathcal{A} \rightarrow \mathcal{A}_{obs}$. Only part of the actions $\mathcal{A}_{obs}$ are observable to intruder $\mathcal{I}$, and only fraction of each observable action are understood by intruder. Then, based on the a prior knowledge $k_{\mathcal{M}}$ and observed actions $\mathcal{A}_{obs}$, the intruder uses some selection functions σ to link actions with different ids.

The objective of selection function σ is to find out the relationship between trajectories $\mathcal{T}$. Following the approach described in [8], the relationship between trajectories $\mathcal{T}$ can be expressed as equivalence relation $\sim_\sigma$ on $\mathcal{T}$. A prior intruder should not know anything about the structure of $\mathcal{T}$, but by utilizing a prior knowledge about MIX $k_{\mathcal{M}}$ and observed action, intruder may know more about it. For a random variable X, let $P(T_i \sim_\sigma T_j) = P(X = (T_i \sim_\sigma T_j))$ denotes the attacker's a posteriori probability that given two trajectories T_i and T_j, X takes the value $(T_i \sim T_j)$. Following the approach used in [8, 10], we define our anonymity model.

Definition 1. *Given an intruder I, a finite set of trajectory $\mathcal{T}$, and a set of identifiers $\mathcal{ID}$'s equivalence relation $\sim_\sigma$, we define a set of discrete probability distribution function $\mathcal{P}$. Let $\mathcal{P}_i \in \mathcal{P}$ be the attacker's a-posteriori probability distribution for a trajectory T_i that T_i is equivalent to trajectories $T_j \in \mathcal{T}$ with respect to equivalence relation $\sim_\sigma$. Each value $p_{i,j} \in \mathcal{P}_i$ is defined as $p_{i,j} = P(T_i \sim_\sigma T_j)$*

In other words, this model defines a function $p : \mathcal{T} \times \mathcal{T} \to [0,1]$, each value $p_{i,j}$ indicates the linkability between two trajectories T_i and T_j. And mathematically this function satisfy following condition. $\forall i \in I, \sum_{j \in I} p(i,j) = 1$.

The role of a location tracking algorithm is to track nodes' movement by associating trajectories with different identifiers to the same node. Considering with the discussion about the objective of selection function several paragraphs above, we know that a selection function σ is the abstraction of a location tracking algorithm. Two tracking algorithms (implementations of selection function) are discussed in detail in next section.

We can find more features about function $\mathcal{P}_i$ by considering the features of a WLP^2S. In the system, we assume that nodes do not use two identifiers simultaneously, and one identifier is used by at most one user at a time. If the address update time of a node we are interested is t_0, we can easily derive following features of function $p(i,j)$. First, if one trajectory T_j contains actions both before and after time t_0, it means that node with id id_j does not change its identifier at time t_0. Consequently, its linkability with other nodes are zero. Or mathematically,

$$\exists a_1, a_2 \in T_j, time(a_1)\langle t_0 \wedge time(a_2)\rangle t_0 \to \forall T_i \in \mathcal{T} \wedge T_i \neq T_j, p(i,j) = 0.$$

Besides, if two trajectories T_i, T_j contains actions either before or after address update time t_0 at the same time, the probability between this two trajectories $p(i,j)$ is zero.

$$\exists a_i \in T_i, a_j \in T_j, time(a_i)\rangle t_0 \wedge time(a_j)\rangle t_0 \to p(i,j) = 0.$$

$$\exists a_i \in T_i, a_j \in T_j, time(a_i)\langle t_0 \wedge time(a_j)\langle t_0 \to p(i,j) = 0.$$

With this probability distribution $p(i,j)$ in mind, we are ready for the definition of the anonymity set. Instead of defining the anonymity group of users $\mathcal{U}$, we define the anonymity set of identifier set $\mathcal{ID}$.

Definition 2. *Geographical Anonymity Set(GAS) of a identifier: Given an identifier* $i \in \mathcal{ID}$ *and its trajectory* T_i, *the anonymity set–which is called as Geographical Anonymity Set–is defined as a subset of* $\mathcal{ID}$, *which satisfy following conditions.*

$$GAS(i) = \{j | j \in \mathcal{ID}, \exists T_i, T_j \in T, p(i,j) \neq 0\} \tag{2}$$

It means that GAS includes all identifiers whose trajectory may be equivalent to T_i. We define the size of a identifier i's GAS $\mathcal{S}_i$ and entropy of identifier i's GAS $\mathcal{H}_i$ as two measures of identifier i's location privacy.

$$\mathcal{S}_i = |GAS(i)|, \mathcal{H}_i = - \sum_{j \in \mathcal{ID}} p_{i,j} \times \log_2(p_{i,j}) \tag{3}$$

Based on the measures for a specific user, we can also measure the privacy level of the whole system by using some statistical tools. In this paper, we define two system-wide measures: GAS size $\mathcal{S}$ and GAS entropy $\mathcal{H}$ for all observed identifiers as equations below. Other statistical measures such as minimum, maximum of all identifiers' measure (size, entropy) can also be used here depending on different applications.

$$\mathcal{S} = \frac{\sum_{i \in \mathcal{ID}}(\mathcal{S}_i)}{|\mathcal{ID}|}, \mathcal{H} = \frac{\sum_{i \in \mathcal{ID}}(\mathcal{H}_i)}{|\mathcal{ID}|} \tag{4}$$

Where $|\mathcal{ID}|$ is the size of identifier set $\mathcal{ID}$.

The relationships between those abstracted entities in a MIX system are also summarized in Figure 4 at the beginning of this sub-section.

4.2 Relevant Features and Constraints of WLP^2S

One major difference between a communications system and a WLP^2S is on their observation functions. In a communications network such as the Internet, although an intruder may observe only a fraction of the actions (e.g., communication of frames between routers) and understand only part of each action (e.g., frame header only), the part of each action (e.g., frame) understood by an intruder is exactly the same as what a user takes. However, in a WLP^2S, the action observed by an intruder is not exactly the same as what a user executes due to the errors introduced in the location observation process. Normally, a location estimation process is composed of two steps. The first step is to measure the set of frame transmission actions $\mathcal{A}_t$. In the measurement process, position pos of each transmission action a_t is invisible to intruders; signal metric sm' received by intruder is smaller than that sent by sender due to the radio attenuation; but other fields are captured correctly. The second step is to estimate node's position pos' based on the observed actions with the same identifier and at same time t. The output of this step a'_t includes an estimation of position pos' of where this action is executed. We notice that set of all estimation results a'_t is not a subset of original transmission actions, or mathematically $\{a'_t\} \nsubseteq \mathcal{A}_t$. Especially, signal metric sm and position pos of each a_t is modified by estimation process because of the accuracy of tracking system.

In addition, similar to the process in [7], only part of the action a_t can be understood by the intruder due to communication mechanisms such as encryption. In a WLP^2S, we assume that the intruder cannot understand the whole message body msg, because it is protected by some encryption mechanism. This process is called tagging function here. Another feature of WLP^2S is that intruders cannot record traces continuously in many cases because of hardware constraints. Instead, an intruder may record actions at a series of time instant.

In summary, the whole observation function ω of a WLP^2S requires three steps: sampling function, position estimation function, and tagging function. In WLP^2S, ω is affected by (1) accuracy of position estimation function, (2) sampling interval of sampling function. We will evaluate the effect of these two parameters in next section.

5 Simulation Study

In this section, we utilize the formal model and its measures to evaluate the privacy level of silent period protocol and formed WLP^2S.

5.1 Design of two protection systems

We specify two systems in this section for our simulation study. There are many variable parts in proposed MIX-based formal model. We need to specify following information to define a WLP^2S: (1) MIX $\mathcal{M}$'s mobility model mm, traffic pattern tp, privacy protection protocol $protocol$ (silent period, lifetime), node density $density$, (2) Intruder $\mathcal{I}$'s selection function σ (tracking algorithm), and observation function ω(including tracking accuracy, sampling interval), knowledge about MIX $k_{\mathcal{M}}$. The only difference between these two systems is on intruder's selection function σ(location tracking algorithm). A summary of the system specification is listed in Table 2.

MIX $\mathcal{M}$

We first discuss the specification of MIX $\mathcal{M}$. In our system, MIX is formed only by MSs. AP and other servers in operator network is not involved in the formation of MIX. First, regarding to traffic pattern tp, we consider following two situations: continuous broadcast and application specific traffic model. Continuous broadcast is used as an ideal traffic model. In continuous broadcast, all nodes broadcast their identifiers continuously except for silent period in simulation, and they restart communication just after the end of silent period. This implies that there is no frame collision in the system and the effect of different traffic patterns is not taken into account.

Secondly, MS use random walk as its mobility model mm when moving around the simulation area. In random walk model, the user first selects one direction between 0 and 2π, and a speed between 0 and 1 m/s, which are maintained for about 10 seconds. Afterward, the user iteratively selects its speed and direction.

Thirdly, MS uses silent period protocol as its privacy protection protocol. There are three parameters in this protocol: length of fixed silent period sp_{fixed}, length of variable silent period $sp_{variable}$, and address lifetime $lifetime$. However, as discussed in Section 4.2, because of frame collision in WLAN MAC and traffic models of WLAN application, nodes with new id may appear later than the end of silent period. in other words, the duration of time between the end of silent period and the first observation of node with new id is variable because of application's traffic model and load of channel. This variable period of time has already mixed the temporal relationship between inputs (nodes' id), hence it is not necessary to explicitly allocate a variable silent period $sp_{variable}$ in MIX's protection protocol anymore. The impact of WLAN radio and traffic model on intruder's selection function is described in next section.

Finally, we assume that all nodes update their address independently, and use density between $0.04/m^2$ and $1/m^2$ in simulation.

Intruder $\mathcal{I}$

Then, we discuss the specification of Intruder $\mathcal{I}$. In our system, Intruder $\mathcal{I}$'s knowledge about MIX $k_{\mathcal{M}}$ includes fixed silent period sp_{fixed}, and mobility model mm.

Besides, two tracking algorithms, *simple tracking* and *correlation tracking* are used as the implementation of selection function σ in our simulations. The objective of tracking algorithms is to associate old and new identifiers of nodes from observed frames. We first introduce the notations used by the tracking system. The node under measurement is called target, and others are called mixers. We use $IDP_{type,time}$ to represent a position of a node. IDP is the position of node with id ID; we use identifiers TP to represent target position, and MPn to represent mixer n's identifier. Subscript *type* indicates how the information is gathered (m, measured; e, estimated; a, actual position), and the subscript *time* indicates when this sample is captured. In compliance with tracking algorithm notations, we assume that a node enters the silent period at time t_{-1}, and leaves it at time t_0. An illustration of notations is given in Figure 5 below.

Simple tracking utilize its knowledge about nodes' mobility model and the observed target position $TP_{m,t_{-1}}$ at time t_{-1} to estimate its next position $TP_{e,t}$. As discussed in Section 4.2, length of silent period is variable due to WLAN's radio and traffic features. Instead of specifying the maximum variable silent period explicitly, we introduce a new terminology *detection window* in Intruder's selection function. Detection window is a period of time estimated by adversary. Within the detection window, adversary can observe target's first frame with new id for sure. From adversary's perspective, all nodes, whose first frame with new identifier is observed within detection window, are temporally overlapped with target. The size of the detection window depends on intruder's knowledge about system such as users' traffic model and node density. Usually, intruder can run some simulation to determine this value in advance. Meanwhile, given the length of silent period, the detection window and users' mobility model, intruder determines a geographical area *reachable area(RA)* where node may appear with new id. Within a specified time-frame, it is impossible for any nodes to drift out of such reachable area due to the constraint from mobility model; as such, the RA and *detection window* serves as a maximum geographical and temporal boundary where a member of the GAS should locate after silent period. Then it chooses a group of nodes who have sent frames with unobserved new identifiers in RA after time t_0 form the GAS of target. This group of nodes forms GAS of target under simple tracking algorithm. Finally, attackers select one node from this group of nodes randomly to associate to target. As a result, the probability that the old and new targets' identifier can be associated correctly is $1/n$ given the size of GAS is n.

We give an example in Figure 5 to further explain the refined intruder's selection algorithm. Before tracking, intruder first calculates the size of detection window and reachable area based on length of silent period, users' mobility model and node density. Adversary records the time when node T transmits the last frame with its old identifier, and then decide a period of time that T may appear after silent period. From left-side of this figure, we perceive that only node M_1 is considered as MIX users because its ET (Emerging Time) is within detection window and its position is within reachable area. Where, ET is defined as first time when intruder observe nodes' frame with new id, DT(Disappearing Time) is defined as last time when intruder observe nodes' frame with old id. On the other hand, right-side of this figure illustrates the spatial overlapping in simple tracking. From this figure, we observe that T', M_1, M_2 and M_3 are observed within reachable area. Considering both the temporal and spatial overlapping. T' and M_1 are considered as belonging to GAS. Simple tracking algorithm selects one from these two identifiers randomly as target's next identifier.

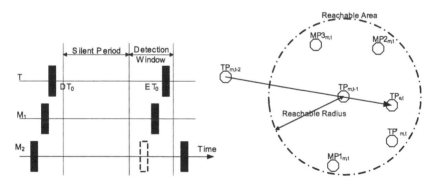

Fig. 5. (a) Illustration of detection window and temporal overlapping in refined selection function, (b) Illustration of spatial overlapping in refined selection function

Correlation tracking uses the measured position $TP_{m,t_{-2}}$ and $TP_{m,t_{-1}}$ at time t_{-1}, t_{-2}, respectively, to estimate the position TP_{e,t_0} of the target at time t_0. Correlation tracking-algorithm estimates TP_{e,t_0} by assuming that the nodes keep the same speed and direction from time t_{-1} to t_0 exactly like that from time t_{-2} to t_{-1}. Equation (5) is used to calculate TP_{e,t_0}. We also use detection window to determine temporal overlapping in correlation tracking. Based on the estimation result TP_{e,t_0}, the attacker selects one node from new identifiers measured at time t_0, which is nearest to TP_{e,t_0}, as the next identifier of the target.

$$\frac{\overrightarrow{TP_{m,t-2} - TP_{m,t-1}}}{interval} = \frac{\overrightarrow{TP_{m,t-1} - TP_{m,t_0}}}{speriod} \tag{5}$$

In addition, intruder's observation function ω contains two parameters: positioning accuracy and sampling interval. In this simulation, we assume positioning accuracy(error) of observation process follows normal distribution with standard deviation between 0.1 and 2.0. The sampling interval varies between 0.1s to 2s in the simulation.

Analysis and Measurement

Finally, we use GAS size S and GAS entropy $\mathcal{H}$ (Equation (4)) as measures of these two systems. A summary of the system's configuration is listed in Table 2.

Table 2. Specification of two WLP^2S

Entities	Parameter	Value
$\mathcal{M}$	mobility model mm	random walk(speed: $0 - 1m/s$, step time 10s) node's init position: Uniform distribution
	traffic pattern tp	continuous broadcasting, Applications: VoIP, Video Streaming, FTP, HTTP
	protection protocol $protocol$	silent period protocol, fixed silent period(s) sp_{fixed}: $0.1 - 5$ variable silent period(s) $sp_{variable}$: $1.0 - 20$ lifetime(s): $25 - 500$
	$density(/m^2)$	0.04 - 1
$\mathcal{I}$	selection function σ	simple/correlation tracking
	observation function ω	accuracy: normal distribution, std dev. 0.1-2.0 sampling interval(s): $0.1 - 2$
	knowledge $k_{\mathcal{M}}$	sp_{fixed}, maximum value of $sp_{variable}$, mm
Others	simulation area	$20m \times 20m$
	measures	GAS size S, GAS entropy $\mathcal{H}$

5.2 Result analysis

We first analyze the system performance when nodes broadcast their identifier continuously. Left side of Figure 6 illustrates the relationship between the size of GAS and the length of the silent period. In the figures, the solid lines denote simulation, while the dotted lines denote analytical calculations. From this figures, we observe that the size of GAS increases proportionately with length of the silent period. This trend can be explained. In the random walk mobility model, the nodes select one direction and speed randomly for a fixed step-time (10 sec in our simulation). The RA of the user is thus a circle centered at the nodes' current position, whose radius is determined by the product of velocity and duration. Because nodes are evenly distributed within the simulation area, the number of nodes that forms the target's GAS can be represented by (6). Equation (6) is plotted, as dotted lines, in left-side of Figure 6. The comparison between the plots in dotted and solid lines proves the consistency between simulation and theoretical results. The figure between density and size of GAS is omitted because of page limitation. From that figure, we also observe the same trend between density and size of GAS.

$$S(duration) = \pi(velocity \times duration)^2 \times density \qquad (6)$$

The right-side graph of Figure 6 illustrates the relationship between entropy of GAS and nodes accuracy, under two tracking algorithms. We observe that the entropy of the system increases proportionately with the nodes' standard deviation of accuracy. In other words, a higher accuracy level of an intruder reduces the privacy level of users. We also observe that the user receives a lower privacy level when a more accurate tracking algorithm is applied.

The left-side graph of Figure 7 illustrates the relationship between the sampling interval and entropy of GAS. We observe in this figure that the system's entropy increases proportionally with an increase in the tracking system's sampling interval. Right-side of Figure 7 shows the performance of the silent period when nodes update their addresses independently. From this figure, we perceive relationship between the silent period ratio and the normalized size of GAS. The silent period ratio is the ratio of the silent period to the lifetime, which represents the share of time a user allocates to protect its privacy. The normalized size of GAS is the ratio of the size of GAS under an independent address update to that under a synchronized address update. The normalized size therefore represents the relative privacy level a user receives when it updates its address independently relative to the synchronized address update. In this figure, we observe that users receive a higher privacy level when they allocate more time to privacy protection.

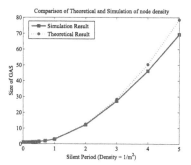 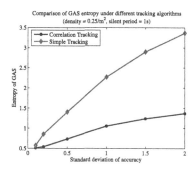

Fig. 6. (left) Comparison of Theoretical and Simulation Results of Silent Period vs. size of GAS, (right) Entropy of GAS vs. accuracy under two tracking algorithms.

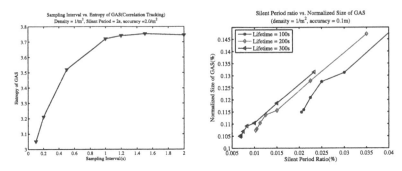

Fig. 7. (left) sampling interval vs. entropy of GAS, (right) silent period ratio vs. normalized size of GAS

As a result of the preceding analysis, we can conclude that for a higher node density, the longer silent period increases the size of GAS; while a more accurate tracking algorithm and a shorter sampling interval reduces both entropy and size of GAS. The privacy level decreases sharply when nodes update their addresses independently, but increases when nodes allocate a greater ratio of time to keeping silent. In addition, we guess that less predictable mobility model results in the higher privacy level, but this is still to be verified by simulation.

6 Discussion and Extension

In this section, we utilize the existing results in literatures [8, 11–13] regarding defense and attack methods on MIX to identify potential threats to the system, and improve the performance of current defense protocols.

Regarding the attack methods, we only consider passive attack based on traffic eavesdropping in current WLP^2S. In addition to this, there are some powerful active and passive attacks such as *blending attack* and *selective attack* introduced in the literatures [11]. Here, we analyze the performance of silent period proposal against these two attacks respectively. First, in a blending attack, intruder floods the MIX with attacker traffic or delay or drop other incoming traffic. Current WLP^2S is robust to such attack because the nodes independently go into silence, the adversary cannot stop the nodes' address update or reduce the anonymity set by adding traffic during the address update. However, this attack should be taken into consideration if variable silent period is decided in the centralized way such as our extension below. Secondly, intruder may use selective attack to reduce the entropy of anonymity group. As discussed in [8], selective attack is a method to exclude other items to be linkable to the items we are interested in. One example of selective attack in WLP^2S is as below. Intruder first tries to link mixers' trajectories before and after silent period, then it exclude all trajectories that has been successfully linked from the GAS of target. Consequently, this method may reduce the size of target's GAS. We think potential advanced tracking algorithm may use this approach to track the target.

On the other hand, we notice that our defense protocol–silent period–can be categorized as a gossip based system. Such systems fundamentally rely on probability to guarantee anonymity. The main representative of this class of anonymity systems is Crowds [14]. One direction to improve our approach is to introduce some deterministic factors in our defense protocol, as a result, defense protocol can provide non-probabilistic anonymity guarantees. The first issue is the address update timing. The simulation results in Sec.5.2 demonstrate that privacy level seriously decreases when nodes update their addresses independently. We suggest that it would be beneficial to add a coordinator to the protection system to help synchronize the address updates. Access point (AP) would be the best-suited node to take such a responsibility because it has connection with all MSs it is serving. Because of page limitation, we only introduce the basic idea of this protocol. MS first registers itself to AP whenever it requests to update its new address, AP schedules the timing of address update for all registered MS, and then notify them when the time arrives by broadcasting some control packets. To prevent compromised AP from leaking the links between old and new identifiers of MS, MS should register itself with new identifier to AS after each address update. This prevent AP from knowing the old identifier of MSs who will go silence.

The second issue is the variable part of silent period. Variable silent period is proposed to mix the temporal relationship between MIX's users. In current system, each node randomly decide the length of variable silent period from a range of value. In simulation, we also assume that two nodes are fully mixed if there are some overlapped variable period between nodes. In comparison with existing MIX research results, we notice that the length of overlapped period *do* affects anonymity level. We think silent period length is analogical to the delay a message experienced in a MIX; and the algorithm to determine silent period length is analogical to batching algorithms used in a MIX. Batch algorithm determines the delay a message should experience for passing through a MIX. Our variable silent period algorithm is similar to the timed MIX, in which all messages are flushed by the MIX in the time of flushing. As discussed in [13], timed MIX are subject to the many attacks. To utilize more robust MIX such as those proposed in [13], we suggest that it would be beneficial to assign AP the responsibility of deciding the length of variable period for all MSs,or in other words, to assign AP the responsibility of batching for all MSs. The benefit of this change is that length of silent period can be decided based on the current status of MIX (i.e., number of nodes in silent period). We expect that when this change is applied, user needs to allocate less share of time on variable silent period for the same privacy level as before. Finally, we also propose to utilize a cascade of MIX nodes in WLP^2S. There are two notable advantages for the construction, namely: avoidance of a single point of failure when one MIX is compromised; and improvement of the privacy level by connecting multiple stages of mixes, which allows repeated entry of users in the silent period. Although an intruder may keep tracking a user for one round, it will lose tracking finally if the user enters MIX repeatedly. Mixes in this case may be one or multiple APs in the same serving area.

By conducting a preliminary study based on formal model, we identified the potential attacks to WLP^2S, and extended our current defense protocol. These results demonstrates the effectiveness and promising potential of proposed formal model as a bridge between new location privacy protection problem and existing research results in anonymity research.

7 Prior Arts

Various techniques have been proposed in the literature for estimating the location of a mobile node [15]. These techniques are broadly classified by signal metric-for example, Angle of Arrival (AOA), Received Signal Strength (RSS), and Time of Arrival (TOA)-or by metric processing-for example, triangulation and pattern matching. Readers can find a survey about latest indoor location tracking technologies in [16]. Latest short-range radio based tracking systems [15] achieves accuracy up to 1 meter. Such a high precision tracking system may erode users' location privacy in the future.

To protect users from potential threat, there are several research studies in commercial radios such as WLAN. Gruteser and Grunwald [17–19] have worked extensively on protecting location privacy in WLAN. In their works, they presented a middleware architecture and algorithm to adjust the resolution of location information along spatial and temporal dimensions, and enhanced location privacy by frequently disposing of a client's interface identifier. They proposed updating the node's interface identifier whenever a station associates with a new access point (AP). On the other hand, Beresford and Stajano [4] proposed the concept of the MIX zone based on Chaum's [20] MIX. A MIX zone for a group of users is defined as the largest connected spatial region in which none of the users in the area has registered an application callback. They assumed the LBS application providers as hostile adversaries, and suggested that application users hide their own identifier from providers. Because application providers do not receive any location information when users are in a MIX zone, the users' identities are *mixed*. The defects of those proposals in the context of WLP^2S are described in [21].

Many modern anonymity systems are based on the MIX concept first proposed by Chaum in [20]. MIX is a set of servers that serially decrypt or encrypt lists of incoming messages. These messages are sent out in a random order, in such a way that an attacker cannot correlate output messages with input messages without the aid of MIX (i.e., when several messages are sent out in a different order than they are received). There are many anonymous applications based on MIX such as untraceable electronic transactions, electrical voting, anonymous mailer and anonymous web surfing.

Increasing attention has been paid to the development of a formal method to verify and evaluate privacy protocol recently. Diaz et al. [22] and Serjantov and Danezis [10] proposed an information theoretical approach to measure the level of anonymity of a system. In these papers, the authors identified that not all nodes involved in anonymous communication contribute the same degree of anonymity to the system. Neither can the size of anonymity set precisely describe the degree of anonymity a system provides. In their proposals, they take into account the entropy of users sending and/or receiving messages as a measure of the anonymous system.

In addition, Diaz and Serjantov [13] presented a simple function to generalize different types of mixes: a function that is based on the number of messages inside the MIX to the fraction of message to be flushed. By using ACP-style process algebra, Mauw et al. [7] proposed a formal definition of the notation of anonymity in the presence of an observing intruder, and validate this definition by analyzing a well-known anonymity preserving protocol *Onion Routing* [23]. On the other hand, Steinbrecher and Kopsell [8] chose information theory and equivalence class to describe unlinkability because it allows an easy probabilistic description. Hughes and Shmatikov [24] proposed a new specification framework for information hiding properties such as anonymity and privacy. This framework is based on the concept of a function view, which is a concise representation of the attacker's partial knowledge about the function.

8 Conclusion and Future Work

In this paper, we first evaluated the new risks imposed by high-accuracy location-tracking systems. A correlation attack is identified as a threat that cannot be defeated using existing periodical pseudonym update solutions. We proposed the new concept of a silent period to combat correlation attacks. Through analysis, we determined that the silent period should contain constant and variable periods. The effect of the constant period is to mix the spatial relation between a node's disappearing points and emerging points, while the variable period is used to mix the temporal relation between a node's disappearing times and emerging times. In the study, we notice that the lack of a formal model for this problem makes it difficult to evaluate any new location privacy protection proposals accurately, and difficult to utilize existing research results in anonymous communication into this new problem. Consequently, we analyzed wireless location privacy protection system formally, and then generalized it into a formal model based on MIX. Two measures, size and entropy of MIX's anonymity group are proposed. Use of these measures to analyze our system revealed that higher node density, longer silent period increases user's privacy level; while a more accurate tracking algorithm and a shorter sampling interval reduces the privacy level. Besides, user receives much lower privacy when nodes update their addresses independently. On the other hand, based on the formal model, we utilized existing research results in MIX to identify potential threat and improve efficiency of current defense protocol. We also discussed the possibility of introducing deterministic factors into the protocol to improve the protocol efficiency and provide non-probabilistic anonymity guarantee. The result of simulation and preliminary analytical study demonstrated the effectiveness and promising potential of formal model. Furthermore, we discuss the system design issues when introducing periodical address update scheme into current network, and proposed a solution based on the principle of radio and network identifier separation and centralized address management entity.

We think correlation attack is a common problem for all identity-based wireless communication system. The model and solutions described in this paper is not only applicable for location privacy problem in WLAN and Bluetooth system, but also would be applicable for the very important problem of location privacy in many other areas such as Vehicle ad-hoc networks(VANETs), ubiquitous computing. We will study the feasibility of extending current proposals to those application areas. We also noticed that current model lacks a model check and protocol verification module. Latest research results regarding model checking for probabilistic system [25] as well as deterministic system [7] is a good reference and starting point for us. In addition, we would evaluate those extensions proposed in this paper by more accurate mobility models.

References

1. Bandara, U., Hasegawa, M., Inoue, M., Morikawa, H., Aoyama, T.: Design and implementation of a bluetooth signal strength based location sensing system. In: Proc. of IEEE Radio and Wireless Conference (RAWCON 2004), Atlanta, U.S.A (2004)
2. Hitachi: Hitachi's air location (2004) http://www.hitachi.co.jp/airlocation/.
3. Bahl, P., Padmanabhan, V.: Radar: an in-building rf-based user location and tracking system. In: Proc. of IEEE INFOCOM 2000, Tel-Aviv,Israel (2000)
4. Beresford, A., Stajano, F.: Location privacy in pervasive computing. IEEE Pervasive Computing 2 (2003) 46–55
5. BluetoothSIG: Bluetooth 1.2 draft 4 (2003)
6. IEEE: 802.11-1999(reaffr 2003) (2003)
7. Mauw, S., Verschuren, J., Vink, E.d.: A formalization of anonymity and onion routing. In: ESORICS 2004, Sophia Antipolis, France (2004)
8. Steinbrecher, S., Kopsell, S.: Modelling unlinkability. In: Proc. of PET 2003, Elbflorenz Dresden, Germany (2003)
9. Yamazaki, K., Sezaki, K.: Spatio-temporal addressing scheme for mobile ad hoc networks. In: Proc. of IEEE TENCON 2004, Chiang Mai, Thailand (2004)
10. Serjantov, A., Danezis, G.: Towards an information theoretic metric for anonymity. In: Proc. of PET 2002. Volume 2482 of LNCS., Springer (2002)
11. Serjantov, A., Dingledine, R., Syverson, P.: From a trickle to a flood: Active attacks on several mix types. In: 5th International Workshop on Information Hiding(IH2002), Noordwijkerhout, The Netherlands, spinger (2002)
12. Chaum, D.: The dining cryptographers problem: unconditional sender and recipient untraceability. Journal of Cryptology 1 (1988) 65–75
13. Diaz, C., Serjantov, A.: Generalising mixes. In: Proc. of PET 2003. LNCS 2760, Dresden, Germany, Springer-Verlag (2003)
14. Reiter, M., Rubin, A.: Crowds: Anonymity for web transactions. ACM Transactions on Information and System Security 1 (1998) 66–92
15. Guvenc, I., Abdallah, C., Jordan, R., Dedeoglu, O.: Enhancements to rss based indoor tracking systems using kalman filter. In: Proc. of Intl. Signal Processing Conf.(ISPC), Dallas, TX, U.S. (2003)
16. Pahlavan, K., Li, X., Makela, J.P.: Indoor geolocation science and technology. IEEE Communications Magazine (2002)
17. Gruteser, M., Grunwald, D.: Enhancing location privacy in wireless lan through disposable interface identifiers: a quantitative analysis. In: Proc. of 1st ACM international workshop on Wireless mobile applications and services on WLAN hotspots(WMASH 2003), San Diego, CA, USA (2003)

18. Gruteser, M., Grunwald, D.: A methodological assessment of location privacy risks in wireless hotspot networks. In: Proc. of 1st Intl. Conf. on Security in Pervasive Computing(SPC 2003). Volume 2802 of LNCS., Boppard, Germany, Springer (2003)
19. Gruteser, M., Grunwald, D.: Anonymous usage of location-based services through spatial and temporal cloaking. In: Proc. of ACM MobiSys 2003, San Francisco, CA, USA, USENIX (2003) 31–42
20. Chaum, D.: Untraceable electronic mail, return addresses, and digital pseudonyms. Communications of the ACM **24** (1981) 84–88
21. Huang, L., Matsuura, K., Yamane, H., Sezaki, K.: Enhancing wireless location privacy using silent period. In: IEEE Wireless Communications and Networking Conference (WCNC 2005), NL, U.S. (2005)
22. Diaz, C., Seys, S., Claessens, J., Preneel, B.: Towards measuring anonymity. In: Proc. of PET 2002. Volume 2482 of LNCS., Springer (2002)
23. Syverson, P., Tsudik, G., Reed, M., Landwehr, C.: Towards an analysis of onion routing security. In: Proc. of Workshop on Design Issues in Anonymity and Unobservability, Berkeley, CA, USA (2000)
24. Hughes, D., Shmatikov, V.: Information hiding, anonymity and privacy: A modular approach. Journal of Computer Security **12** (2004) 3–36
25. Shmatikov, V.: Probabilistic model checking of an anonymity system. Journal of Computer Security (2004)

Appendix

Table 3. Notations used in this paper

Notation	Description
u	a MIX user
$\mathcal{U}$	set of all MIX users $\mathcal{U} = \{u\}$
a	an action executed by a MIX user
$\mathcal{A}$	set of all action $\mathcal{A} = \{a\}$
$\mathcal{I}$	MIX's intruder
$\mathcal{M}$	MIX in abstracted MIX system
$\mathcal{S}$	THE system, an abstraction of WLP^2S, $\mathcal{S} = \{\mathcal{U}, \mathcal{I}, \mathcal{M}\}$
$\mathcal{ID}$	set of all identifiers $\mathcal{ID} = \{id\}$
T_i	identifier i's trajectory, set of all a_t with same sender id i, $T_i = \{a_t \vert id(a_t) = i\}$
$\mathcal{T}$	set of all trajectories $\mathcal{T} = \{T_i\}$, $\mathcal{T} \subseteq \mathcal{A}_{obs}$
a_t	a transmission action, $a_t = id_s \Vert id_r \Vert msg \Vert time \Vert sm \Vert pos$
$\mathcal{A}_m$	set of all movement actions
$\mathcal{A}_t$	set of all transmission actions $\mathcal{A}_t = \{a_t\}$
$\mathcal{A}_r$	set of all reception actions
$\mathcal{A}_{obs}$	set of all observable actions
$\mathcal{A}_{inv}$	set of all invisible actions
$k_{\mathcal{M}}$	$\mathcal{I}$'s knowledge about $\mathcal{M}$
σ	$\mathcal{I}$'s selection function $\sigma : \mathcal{T} \times \mathcal{T} \to P$
ω	$\mathcal{I}$'s observation function $\omega : \mathcal{A} \to \mathcal{A}_{obs}$

$\sim_\sigma$	selection function σ's equivalence relation on T
$p_{i,i}$	probability that T_i and T_j are equivalent based on equivalence relation $\sim_\sigma$
$\mathcal{P}_i$	attacker's a-posteriori probability distribution function for t_i
$\mathcal{P}$	set of all identifiers' a-posteriori probability distribution function $\{\mathcal{P}_i\}$
msg	an a_t's message payload
$time$	time when a_t is executed
sm	signal metric of a_t
pos	position where a_t is executed
mm	mobility pattern of $\mathcal{M}$
tp	traffic pattern of $\mathcal{M}$
$protocol$	privacy protection protocol used in $\mathcal{M}$
$density$	number of MIX users in $\mathcal{M}$, $density = |\mathcal{U}|$
$GAS(id)$	geographical anonymity set(GAS) of identifier id
$\mathcal{S}_{id}$	identifier id's GAS size
$\mathcal{H}_{id}$	identifier id's GAS entropy
$\mathcal{S}$	GAS size of THE system $\mathcal{S}$
$\mathcal{H}$	GAS entropy of THE system $\mathcal{S}$
$id(a_t)$	identifier of a transmission action a_t
$time(a_t)$	time instant of a transmission action a_t
$\{x\}$	a set of elements
$x\|y$	x concatenated to y

Part III

Secure Time Synchronization

Time Synchronization Attacks in Sensor Networks

Tanya Roosta[1], Mike Manzo[2], and Shankar Sastry[3]

[1] University of California at Berkeley roosta@eecs.berkeley.edu
[2] University of California at Berkeley mike@manzo.org
[3] University of California at Berkeley sastry@eecs.berkeley.edu

In this chapter, we review time synchronization attacks in wireless sensor networks. We will first consider three of the main time synchronization protocols in sensor network in sections. In section we discuss applications of time synchronization in sensor networks. In section we analyze possible security attacks on the existing time synchronization protocols. In section we examine how different sensor network applications are affected by time synchronization attacks. Finally in section we propose possible countermeasures to secure the time synchronization protocols.

1 Introduction

Ad hoc networks are infrastructure-less, possibly multi-hop wireless networks where every node can be either a host or a router, forwarding packets to other nodes in the network. Some applications of sensor networks are in providing health care for elderly, surveillance, emergency disaster relief, and battlefield intelligence gathering. A sensor network consists of anything from a handful to very many tiny wireless devices with sensors. One very popular type of nodes are the motes developed primarily at U.C. Berkeley and Intel, Figure 1. Motes have very constrained resources. An example of a sensor mote is the mica2dot. A typical configuration may have a 4MHz, 8-bit processor, with 128KB of instruction memory, 4KB of RAM, and 512KB of external flash memory. The radio runs at 433 MHz and 38.4 Kbps. Given the limited resources of these sensor nodes, it is a key technical challenge to design secure services, such as time-synchronization.

1.1 Time Synchronization in Sensor Networks

Time synchronization protocols provide a mechanism for synchronizing the local clocks of the nodes in a sensor network. There are several time synchronization protocols for the Internet, such as Network Time Protocol (NTP). However, given the non-determinism in transmissions in sensor networks, NTP cannot be directly used in wireless sensor networks.

Fig. 1. Mica mote family [14]

Time synchronization implementations have been developed specifically for sensor networks. Three of the most prominent are Reference Broadcast Synchronization (RBS) [3] Timing-sync Protocol for Sensor Networks (TPSN) [5], and Flooding Time Synchronization Protocol (FTSP) [10]. However, none of these protocols were designed with security as one the goals. Security is an important issue in sensor networks given their diverse and usually very sensitive applications. For example, it is crucial to protect people's privacy when sensors are used for elderly health care monitoring. Sensor networks are usually unattended after deployment, and their deployment location is un-trusted. In addition, nodes communicate using a radio channel, which makes all communications susceptible to eavesdropping. Therefore, sensor-network security can easily be breached either by passive attack, such as eavesdropping, or active attacks, such as denial of service attacks, which can be launched at, for example, the routing or the physical layer.

To provide more secure wireless communications in sensor networks, Karlof et. al. proposed and implemented TinySec [7], which uses symmetric private key encryption to authenticate and encrypt messages. If an adversary physically captures a node, however, he will gain access to the network-wide key and can participate in the authenticated communication without being recognized as an attacker. In this work we focus on attacks of this type, where the adversary compromises a node and injects erroneous time synchronization information in the network. To the best of our knowledge, there has been no previous published work on time synchronization attacks in sensor networks, their effect on different sensor network applications, or secure time synchronization protocols.

2 System Model

In this section, we define the problem of secure time synchronization and discuss the clock model, communication model, trust assumptions, threat and attacker models, and security goals we wish to accomplish.

2.1 Clock Model

Every sensor node has a notion of time that is based on the oscillation of a crystal quartz. The sensor clock has a counter that is incremented at rate f where f is the frequency of the oscillation. The counter counts time steps, and the length of these time steps is prefixed. The clock estimates the real time $T(t)$, where,

$$T(t) = k \int_{t_0}^{t} \omega(\tau)d\tau + T(t_0) \qquad (1)$$

$\omega(\tau)$ is the frequency of the crystal oscillation and k is a constant [17]. Ideally this frequency should be 1, i.e. $dC/dt = 1$. However, in reality the frequency of a clock fluctuates over time due to changes in temperature, pressure, and voltage. This will result in a frequency different than 1. This difference is termed *clock drift*. There are a number of ways to model the clock drift. In addition to frequency fluctuation in one clock, the crystals of different clocks oscillate at different rates. This difference causes what is called the *offset* between two clocks [17].

2.2 Communication Model

In order to perform time synchronization, the network relies on transmitting the time synchronization messages. The messages are sent through wireless channel. We assume that the wireless links are symmetric, meaning if node A hears node B, then node B also can communicate with node A. However, it is worth mentioning that in reality the wireless links are not always symmetric; in fact, it has been shown through experiments that the wireless channel is asymmetric.

2.3 Adversary Model

As stated above, wireless sensor networks use a wireless channel for communications. This gives an adversary a large window of opportunity, from passive eavesdropping to more serious attacks such as message injection.

We mentioned that in sensor networks there are one or more base stations, which are sinks and aggregation points for the information gathered by the nodes. Since base stations are often connected to a larger and less resource-constrained network, we assume a base station is trustworthy as long as it is available. Beside the base stations, we do not place any trust requirements on the sensor nodes because they are vulnerable to physical capture.

While it is possible that the adversary has access to sensor nodes very identical to the ones deployed or has more powerful nodes such as laptops, in this paper we only consider a mote-class adversary. We assume that the nodes are not tamper resistant.

An outsider attacker has no special access to the sensor network, such as passive eavesdropping, but an insider attacker has access to the encryption keys or other code used by the network. We consider only an insider attacker. We assume that the adversary has been able to capture and corrupt a fraction of the total nodes in the network. The adversary therefore also has access to the secret keys for authorized communication with other nodes.

The goal of the adversary in our setting is to inject false time synchronization information in the network, without being detected by the honest nodes.

3 Time Synchronization Protocols for Sensor Networks

As mentioned in section 1.1, sensor network is used for monitoring different real world phenomena. Since the existing time synchronization protocol do not fit the special needs of sensor network, a number of clock synchronization protocols have been developed to meet the memory and energy constraint of these networks. There are three main ways to synchronize nodes together. In the first approach an intermediate node is used to synchronize the clocks of two nodes together, such as Reference Broadcast Synchronization [3]. The second approach assumes that the clock drift and offsets are linear, and nodes perform pair-wise synchronization to find their respective drift and offset, such as TPSN [5]. In the third approach, one node declares itself the leader, and all the other nodes in the network synchronize their clocks to the leader, such as Flooding Time Synchronization Protocol [9]. In the following sections, we review these three approaches to the problem of time synchronization in sensor networks.

3.1 Reference Broadcast Synchronization

In RBS a reference message is broadcast to two receivers and the receivers synchronize their respective local clocks to each other, Figure 2. Each receiver records its local time when it gets the reference message. Then the two receivers exchange their local times. It is possible to increase the precision of the estimated time by broadcasting m reference messages instead of one, as follows [3]:

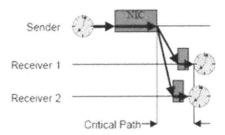

Fig. 2. Example of RBS [3]. RBS does not synchronize the receivers to the sender. Instead it synchronizes the receivers to each other

- A transmitter broadcasts m reference messages.
- Each of the n receivers record the time they received the reference messages based on their respective local times.
- The receivers exchange their local times.
- Each receiver can calculate its phase offset as the LS linear regression of the phase offsets implied by each reference message observed by the receivers as follows: Each node finds the time difference between itself and a neighbor for each reference broadcast message. Then the node finds the least-squares (LS) error fit for these points. The node can then find the offset and skew of its clock from the slope and intercepts point of the line.

The advantage of RBS is that it eliminates the non-determinism on the transmitter side. RBS does not use a multi-hop scheme to synchronize all the nodes to the same global clock. For the multihop case, the nodes that fall within the overlapping region of two reference transmitters perform clock conversion, to go from one clock estimate in one region to the other region. By doing the clock conversion, it is possible to find the time difference between events that happen at different parts of the network.

3.2 Time-Sync Protocol for Sensor Networks

TPSN initially creates a spanning tree of the sensor network. The tree starts at the root of the network, which is generally a base station, and each node establishes its level based on the 'level-discovery' message that it receives. While the tree is being built, the nodes perform pairwise synchronization along the edges of the tree. Each node exchanges synchronization messages with its parent in the spanning tree. By comparing the reception time of the packets with the time of transmission, which is placed on the packet by the parent, the node can find and correct for its own phase offset. We give an example to clarify this scheme. Let us assume there are two nodes, N1 and N2, where N1 is the parent of N2, Figure 3. When N2 wants to synchronize its clock, it sends a *synchronization pulse* to N1 along with the value T1, the time the packet is transmitted from N2.

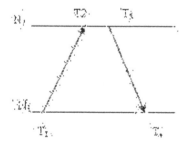

Fig. 3. An example of TPSN

When N1 receives this packet, it sends back an *acknowledgment* packet with times T2, reception time, and T3, retransmission time. N2 receives this ACK packet at T4, and using the four time values it can find the clocks drift and propagation delay, using the equations [5]:

$$d = ((T_2 - T_1) + (T_4 - T_3))/2 \qquad (2)$$

TPSN is a sender-initiated time-synchronization protocol, and the sender synchronizes to the receiver's clock. TPSN has a better performance than RBS by time stamping the messages at the MAC layer of the radio stack. It also uses a two way message exchange instead of a one way exchange as in RBS [5]. However, TPSN does not account for clock drift. It also does not efficiently handle dynamic topology changes because it has to compute the spanning tree of the network every time a change happens.

3.3 Flooding Time Synchronization Protocol

In FTSP, a root node broadcasts its local time and any nodes that receive that time synchronize their clocks to that time. The broadcasted synchronization messages consist of three relevant fields: *rootID*, *seqNum*, and *sendingTime* (the global time of the sender at the transmission time). Upon receiving a message, a node calculates the offset of its global time from the global time of the sender embedded in the message [9]. The receiving node calculates its clock skew using linear regression on a set of these offsets versus the time of reception of the messages.

Given the limited computational and memory resources of a sensor node, it can only keep a small number of reference points. Therefore, the linear regression is performed only on a small subset of the received nodes. Since this regression requires that set of updates, however, a node cannot calculate its clock skew until it receives a full of reference messages. Therefore, there is a non-negligible initiation period for the network.

FTSP also provides multi-hop time synchronization in the following manner: Whenever a node receives a message from the root node, it updates its global time. In addition, it broadcasts its own global time to its neighbors. All nodes act in a similar manner, receiving updates and broadcasting their own global time to their neighbors. To avoid using redundant messages in the linear regression described above, each node retains the highest sequence number it has received and the *rootID* of the last received message used. A synchronization message is only used in the regression if the *seqNum* field of the message (the sequence number of the flood associated with that message) is greater than the highest sequence number received thus far and the *rootID* of the new message (the origin of the flood associated with that message) is less than or equal to the last received *rootID*. FTSP is more robust against node failures and topology changes than TPSN since no topology is maintained and the algorithm can adapt to the failure of a root node. If a node does not hear a time synchronization message for a ROOT_TIMEOUT period, it declares itself to be the new root. To make sure there is only one root in the network, if a root hears a time synchronization message from another root with lower ID than itself, it gives up its root status.

4 Attacks on Time Synchronization Protocols

In this section, we discuss different attacks on the time-synchronization protocols explained above. We discuss possible attacks on RBS, TSPN, and FTSP in each subsection.

All the attacks on time synchronization protocols have one main goal, to somehow convince some nodes that their neighbors' clocks are at a different time than they actually are. Since global time synchronization is build upon synchronization at the neighborhood level, this will disrupt the mechanisms by which the protocols above maintain global time in the network or allow events at distant points in the network to be given to be give time-stamps which reflect the actual difference between their times of occurrence.

4.1 Attacks on RBS

As mentioned above, in RBS a reference message is broadcast by the base station. Two nodes that receive this message exchange their local time for clock synchronization. However, if a node is compromised, it can send a falsified synchronization message to its neighbor during this exchange period. The effect is that the honest node will calculate an incorrect phase offset and skew. In the field of Robust Estimation, the *breakdown point* of an estimator is defined as the smallest fraction of contamination in data that can cause the estimator to take on values arbitrarily away from the real value [16]. It is known that the average and LS estimators have a very low breakdown point. For example, the breakdown point for the LS estimator is $1/n$. Therefore, as n gets larger, the breakdown point gets closer to 0, meaning that a very small fraction of contaminated data can cause the estimation to get far from the real value. If the number of nodes is large in a sensor network, a small fraction of the compromised nodes can cause the time synchronization estimates to get far from true global time.

It has been proposed [16] to use Least Median of Squares (LMS) instead of LS to fit a more robust model to the data sets. LMS has a high breakpoint of 50%, and therefore is resilient to a high fraction of outliers. LMS is explained in more detail in the Appendix.

It is also possible to attack the multi-hop version of RBS. As mentioned before, RBS does not keep a global time; instead, it tries to find the time difference between the occurrences of two events at different parts of the network. Since the nodes on the boundary of the overlapping regions perform clock conversion, a compromised node placed in any of the overlapping regions can affect multiple regions by giving an erroneous value in the clock conversion. Once the adversary sends a miscalculated clock con-version, this error will be propagated as it is sent throughout the network.

4.2 Attacks on TPSN

TPSN creates a spanning tree of the network at the initial phase of level-discovery and then performs pair-wise clock synchronization. As mentioned above, TPSN is a sender-based synchronization protocol, so the adversary can only affect the protocol through manipulating its children in the spanning tree. That is, the adversary will not be able to cause any disruption by initiating a time-synchronization request since only the parent node will reply. However, the compromised node can effect its children by sending incorrect time stamps for the reception time and the transmit time of the time-synchronization request. The adversary can propagate this desynchronization down the spanning tree toward the leaves on its own branch. Therefore, the closer the compromised node is to the root, the more effect it has on time synchronization in the network, and the larger a region it can contaminate. Alternatively, the adversary can place more compromised nodes at different branches of the tree, and affect a wider region of the network.

The compromised node can also lie about its level in the tree, meaning it can claim a lower level than its real level. By doing so, it can convince the other nodes at its level to request time-synchronization updates from the compromised node. In addition, the compromised node can avoid participating in the tree building phase, and by doing so cut off a number of nodes that would have been its children from the root. If those nodes can-not find any other node as their parent, they will not be able to synchronize to the rest of the network. The effects of this attack would be greatest in a sparse network topology.

4.3 Attacks on FTSP

The major innovations of FTSP, in terms of multi-hop synchronization, over RBS are that the root is chosen dynamically and any node may claim to be the root if it has not heard time updates for a preset interval. One possible attack on this protocol is for the compromised node to claim to be the root node with ID 0 and begin at a higher sequence number than the actual root node so all the updates originating at the actual root node will be ignored. This can be easily done since the protocol allows any node to elect itself root to handle the situation where nodes have not heard from the root node for a long period. Once a compromised node becomes the root, it can give false updates to its neighbors, which will in turn propagate that false time to their neighbors and so on. Every node that accepts the false updates will calculate a false offset and skew for its clock.

5 Effects of Time Synchronization Attacks on Sensor Network Applications

To motivate our discussion of time-synchronization attacks, we describe here in detail the effects of time synchronization attacks on a set of sensor network applications and services that are dependant on time synchronization.

In many application areas, time synchronization allows engineers to design simpler and more elegant algorithms. If security is a high priority, however, the simplest countermeasure against an attack on time synchronization is to build the algorithm such that it does not rely on a time-synchronization service whenever possible. That said, for certain classes of applications under certain conditions, algorithms cannot provide correct results without an accurate and reliable time-synchronization service. Rather than try to describe those conditions, we have tried to select a representative set of applications that rely on a time synchronization service and demonstrate the effects on them of a time synchronization attack. While we believe these algorithms to be representative of the set of algorithms relying on time synchronization, the set is not exhaustive.

5.1 Shooter Localization

The shooter localization algorithm designed by Vanderbilt University [9,12] has three key stages: Detection: To find the location of the sound of the muzzle blast, the nodes report the time of arrival (TOA) of the sound on three audio frequencies. The sound is processed to determine whether it will be reported to the base station. The report contains an age field. Routing-Integrated Time Synchronization: In this step, the Directed Flooding Framework is used. As the message is routed toward the base station, the routing nodes update the age field by adding the elapsed time between reception and retransmission of the packet. MAC layer time stamping is used, as in FTSP, to update the age field precisely. This age field is used at the base station to find the exact time of the shot [13]. Sensor Fusion: Once the muzzle blast TOA

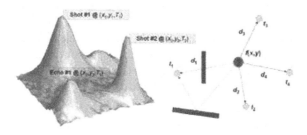

Fig. 4. Consistency function to determine the time and location of the sound from a gunshot [9]

data is submitted to the base station, a sensor fusion algorithm is used to estimate the shooter location and the trajectory of the projectile. The fusion algorithm searches for the maximum of a consistency function, which corresponds to the location of the muzzle blast. The consistency function is defined on four-dimensional space-time space as follows: let (x,y,z) be the shooter position and let t be the shot time. The time of arrival of the ith measurement is [12]:

$$t_i(x, y, z, t) = \frac{\sqrt{(x - x_i)^2 + (y - y_i)^2 + (z - z_i)^2}}{\nu} \tag{3}$$

Where v is the speed of sound, and (x_i, y_i, z_i) is the coordinates of the sensor which is making the ith measurement. If the sensor were in a line with the shot, then the real shot time and the estimated time should be equal. However, in reality this does not happen. As a result, the considered measurements are those that satisfy:

$$|t_i(x, y, z, t) - t_i| \leq \tau \tag{4}$$

where $\tau = \frac{\delta_1}{\nu + \tau_2 + \tau_3}$ is the uncertainty value, δ_1 is the maximum localization error, τ_2 is the maximum time synchronization error, and τ_3 is the maximum allowed signal detection uncertainty. It is usually assumed that the uncertainty value is dominated by the localization error and not the other two terms. Therefore, from the above definition the value of the consistency function is defined as the number of observations that support the claim that the shot was taken from location (x, y, z) at time t with uncertainty τ, where it is assumed that an upper bound is known for τ based on the localization, time synchronization and the signal detection algorithms :

$$C_\tau(x, y, z, t) = count_{i=1,...,N}(|t_i(x, y, z, t) - t_i| \leq \tau) \tag{5}$$

The maximum of this consistency function gives the location and time of the shot, as in Figure 4.

However, if the time synchronization protocol has been attacked by compromised nodes, the consistency function can be affected. The maximum time synchronization error might become comparable to the localization error. This might change the shape of the consistency function and, consequently, change the value and location of the local maximum. This would yield an incorrect estimate of the location and time of the shot. For example, if the consistency function is as in Figure 4, there are three local maxima in the consistency function. If, due to a time synchronization attack, some of the nodes that reported time t_1 for the first shot report a shot at $t_1 + \Delta$, the first maximum will be pushed down and the location and time of the shot will incorrectly be estimated at another peak in the consistency function.

5.2 TDMA-based Channel Sharing

In TDMA-based channel sharing protocols, time is divided into intervals and each node is assigned a schedule in which to transmit and receive messages. In this section, we are going to discuss two different protocols that use TDMA-based approaches. The first is the flexible power scheduling [6] and the second is *PEDAMACS* [2].

Flexible Power Scheduling

The goal of this protocol is to reserve time slots for transmission and reception of data messages on each node so that the nodes can go to sleep during the idle slots.

This saves power because nodes do not have to be listening transmissions the entire time. We give an outline of the protocol below:

Time is broken into cycles and each cycle consists of m time slots. During each slot the node has one of the following six states [6]:

- *Transmit (T)*: The node may transmit a message to its parent.
- *Receive (R)*: The node may receive a message from its child.
- *Advertisement (A)*: The parent node broadcasts an advertisement to find an available reservation slot from the child
- *Transmit Pending (TP)*: The node sends a reservation request to its parent.
- *Receive Pending (RP)*: The node receives a reservation request from its child.
- *Idle (I)*: The node powers down during this period if this slot in its schedule is unoccupied.

The algorithm proceeds as follows: When a node is in an A slot, its parent advertises for a reservation slot and marks the corresponding slot in its own schedule as RP. If a child node hears this advertisement, but still has not met its demand, it marks the corresponding slot in its schedule as TP, and sends a reservation request when the time slot arrives. To initialize the algorithm, the base station picks a reservation slot at random and sends out an advertisement for it. It then waits for a reply during the following cycle. When a new node joins the network, it sets all slots in its schedule to I and then listens for one cycle for an advertisement. In order to keep the time synchronized, once a node selects a parent, it periodically synchronizes its time to the parent. This is similar to TPSN in the sense that the child synchronizes its clock to the parent in a spanning tree of the network.

We can summarize the FPS algorithm's main steps during each cycle in the following [6]: After initialization, each node picks an advertisement slot A randomly from the idle slots. Then it picks a reservation slot (RP) at random from the idle slots. If supply $\geq$ demand

- Turn off the radio during the idle slots
- Schedule an advertisement during an A slot

Otherwise,

- Leave the radio on and listen for advertisement Schedule a reservation request during a TP slot

For each time slot, the node checks its power schedule, and performs an action according to its schedule. For example, if the slot is marked T, it transmits a message. At the end of the cycle, the node clears the current slot and the previous RP from the schedule.

It is worth mentioning that the reservations are not renegotiated unless the child node cancels the reservation or the parent node has a time-out due to having no transmissions for a given period of time. In addition, if the parent node does not receive a message from a child for a large number of cycles where it has R in its schedule, it assumes there has been a change in the topology of the network, and the R slot is recycled in the parent schedule [6].

If the compromised node is a parent, it can send wrong time synchronization messages, and that will cause the schedule of the child to be in-consistent with the parent's schedule, i.e. the time slots corresponding to the same action in the two schedules will not be aligned in time. Since the parent can send wrong messages to different children, the time slots in multiple nodes will conflict with each other, and as a result, there will be more collisions due to transmission at overlapping time slots. Even though the protocol does not need a fine-grained time synchronization protocol to work, when the error in local clocks gets accumulated over time due to an attack, it can affect and degrade the performance of FPS protocol.

PEDAMACS

PEDAMACS is a TDMA-based MAC protocol used in sensor networks. It assumes that the base station is attached to a reliable and infinite power supply and is able to reach all the nodes in the network. The protocol consists of three phases [2]:

- Topology discovery
- Topology collection
- Scheduling

The network uses three different ranges for communication, where the longest range is used by the base station to reach all the nodes in one hop. The second range is used to determine the interference with other nodes, and the last level is for a node to reach its one hop neighbors. Topology discovery is done using Carrier Sense Multiple Access (CSMA). In the topology collection phase, the base station gathers information about the nodes in the network and builds a tree of the shortest paths. Next, the base station creates a transmission schedule based on its knowledge of the network topology and broadcasts it to all the nodes. This schedule is structured such that nodes which do not interfere with each other during transmission may transmit in the same time slot. The base station periodically broadcasts the current time to all the nodes in the network to facilitate this scheduling. A guard interval on the time slots is used to compensate for clock drift between updates to the current time [2].

A compromised node might spoof the base station and broadcast an alternate time to its neighbors or it might attempt to jam the network to prevent the time updates from being received. If the induced difference in time between 2 nodes that are trying to transmit on adjacent slots is greater than the guard interval, packet loss may occur due to contention in the channel.

Since the base station can, by definition, communicate directly with every other node without routing packets through intermediary nodes, the base station can simple use an authenticated broadcast protocol to prevent a compromised node from injecting false packets. We see no counter-attack to a jamming attack.

Estimation

To illustrate the effects of corrupted time synchronization on estimating state based on sensor readings from a sensor network, we give a simple example using the

Kalman filter. The Kalman filter estimates the state of a discrete-time controlled process that governed by a linear stochastic difference equation [13]:

$$x_k = Ax_{k-1} + Bu_{k-1} + w_{k-1} \qquad x \in \Re^n \qquad (6)$$

given the measurement $z_k \in \Re^m$, where

$$z_k = Hx_k + v_k \qquad (7)$$

The random variables w and v represent process and measurement noise and are assumed to be independent random variables with normal distribution,

$$p(w) \sim N(0, Q) \qquad p(v) \sim N(0, R)$$

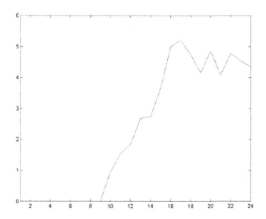

Fig. 5. The y axis shows the norm of the difference between the results from the Kalman filter before and after de-synchronization. The x axis is the time of the corresponding observation.

The Kalman filter estimates the state at every time step. We simulated the movement of an object using 7, where the state is position and velocity of the object in two dimensions. We then used the Kalman filter to estimate the position and velocity of the object before and after modifying the time of some of the position observations, as might occur in an attack on the time synchronization in the sensor network. The norm of the error is shown in Figure 5. As seen in the Figure 5, we began the de-synchronization at time 10.

Authenticated Broadcast(μTesla)

μTESLA is an authenticated broadcast protocol for sensor networks that forms the basis for some of countermeasures proposed in this paper. It relies on an asymmetric mechanism to allow nodes receiving a broadcasted message to verify the authenticity of the message. That mechanism is based on the arrival time of the messages, so one of the requirements of μTESLA is that the base station and the nodes have loose time synchronization.

Before beginning a broadcast that the nodes in the network may authenticate, the base station must bootstrap the protocol. Each message is transmitted in a time slot, although more than one message may be transmitted in one slot, and each time slot requires exactly one key. So the base station must transmit to each of the other nodes, authenticated with a private key shared only by the base station and that node, the length and starting time of the time slots. It also transmits a private key K_0. To generate K_0, the base station first generates a single key K_n and keeps it private. It then generates enough additional keys from K_n to transmit the packets in a sequence of time slots.

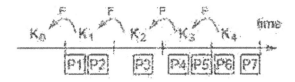

Fig. 6. Example of $\mu TESLA$ [11]

For example, if only one packet can be transmitted for each time slot, the base station would need $|S|$ additional keys. To generate a key from K_n, the base station applies a public one-way function F to K_n once for each of the required keys. That is, $K_i = F(K_i + 1)$. The sequence of keys constitutes a one way key chain. K_0 is last key in this sequence and is used to authenticate the messages in the first time slot, with K_i used for the ith interval and so on [11].

The nodes retain the packets they receive from the base station because they cannot authenticate them immediately. K_i is kept secret until after the *ith* interval. That is, at the $(i + k)th$ interval, for some k transmitted during the bootstrap process, the base station reveals K_i. With that key, the nodes can authenticate all the messages received at interval i (and all the messages in the previous time slots) [11].

To illustrate how the protocol works, consider Figure 6. P_1 through P_7 are the data packets, and K_0 through K_4 are the keys of the one way key chain. K_1 is used to decrypt P_1 and P_2, and K_3 is used to decrypt P_4 and P_5.

It is easy to see that μTESLA is vulnerable to time synchronization attacks. Although the protocol only requires loose clock synchronization, it does require that the network have a synchronized time. That is, the nodes must be able to unambiguously determine in which slot the messages occur. If a compromised node injects corrupted time updates that exceed the time synchronization error bound, the honest nodes would not be able to decide to which slot the messages or the keys belong. As a consequence, they would be unable to authenticate messages from the base station. It is interesting to note the symbiotic relationship between our countermeasures and μTESLA. On the one hand, our countermeasures rely on μTESLA or some other authenticated broadcast scheme. But without our countermeasures or some other scheme for securing time synchronization protocols, μTESLA is not secure. This is not to say that neither can function since they cannot function until the other has been established. Rather, if our countermeasures maintain time synchronization, they can continue to use μTESLA to do so.

6 Countermeasures for Time Synchronization Attacks

We propose countermeasures for single hop networks and multi-hop networks separately. Each time synchronization protocol facilitates single hop synchronization by having a node periodically broadcast its local version of the global time and allowing the other nodes to synchronize their clocks to that time. Since multihop time synchronization algorithms are all extensions to that basic building block, our proposals for multihop networks are a superset of our proposals for single hop networks.

6.1 Countermeasures for Single Hop Networks

RBS is intended to be used in single-hop networks, although there is a multi-hop extension of the protocol. TPSN and FTSP can also be run in a single-hop network, however, and reportedly perform better than RBS. Regardless of which of these protocols is used in a single-hop network, however, the scheme we present is applicable since in the single-hop scenario all three protocols behave similarly.

In single-hop networks, every node is within radio range of every other node. We consider separately single-hop networks in which all the nodes are within range of a base station, a locus of trust, and those networks in which there is no such base station. The former is the case in many small deployments or deployments that include a second tier of supernodes. Such supernodes may maintain their time using GPS or by communication with another network that has a reliable time synchronization service. For these networks, the central challenge for time synchronization security is to prevent any node from spoofing the root node and sending erroneous updates to the global time.

Such a network might employ a network-wide symmetric private key to encrypt and authenticate messages from the root node, including time synchronization updates, to prevent spoofing of the root node and falsification of the time updates. There exist implementations of such a scheme for sensor networks, as mentioned above. An insider attack might occur, however, if a node were physically compromised. An adversary would gain access to the network-wide key and could falsify time synchronization updates. To detect the presence of falsified time updates, the nodes could easily look for redundant sequence numbers in the packets. Or, if the updates were expected to be sent at a regular, predefined period, the nodes could detect injected packets by a change greater in the frequency of the time-synchronization updates greater than would reasonably be expected shifting network conditions. If a node detects the presence of falsified packets, it could rely on a private key shared only with the base station to request and authenticate packets from the base station. However, such a scheme would require each node in the network to have a private key with the base station, a highly inefficient proposition. Moreover, the radio traffic would increase linearly with the number of nodes requesting authenticated time updates. For these reasons, we recommend an authenticated broadcast scheme such as μTESLA as an effective means of maintaining time synchronization in single-hop networks with a base station. The base station should encrypt all time synchronization updates.

In single-hop networks that lack a base station or in which the base station fails, there is no trustable source of global time. In that situation, the nodes can elect a root node against which the other nodes can synchronize their clocks. FTSP provides one mechanism for electing a root node, as discussed above. Upon election, the new root node can bootstrap a new key chain for μTESLA using pair wise keys with the other nodes. How-ever, we see no way to prevent a corrupted node from being elected as the root node under the FTSP scheme. Is fact, under the FTSP scheme, a corrupted node could become the root with certainty, as discussed in the attack section. A corrupted root node could send erroneous reference broadcasts that would cause the nodes to calculate an erroneous skew and offset. To prevent this, we propose that, instead of allowing a single node to be the source of time synchronization updates, a subset of the nodes act as the root on a rotating basis. In addition, to prevent a corrupted node from impersonating other nodes, we recommend that all the nodes share a private key with that subset of the nodes in the single-hop network that may become a root. That is, if there are N nodes in the network and $M \ll N$ nodes may become a root node, there will be $N * M$ private keys in the entire network. Those keys could be used to bootstrap an authenticated broadcast scheme for each of the nodes in the rotation of root nodes. Any corrupted nodes might continue to send erroneous updates under this scheme, but the effects on the nodes' calculations of the skew and offset would be reduced. For instance, if there are C corrupted nodes in the network and every node has equal probability of being elected as a root node (a generous assumption, based on the previous discussion of the FTSP root-election scheme), then the probability that a corrupted node is elected a root node is C/N. Suppose the nodes use linear regression with L data points to calculate the clock skew. Then without a root rotation scheme, with probability C/N, all the data points would come from a corrupted node.

If we let M nodes be the root in a round robin fashion, then the probability of having at least one corrupted node being the root is:

$$p = \frac{\binom{C}{1}\binom{N-C}{M-1}}{\binom{N}{M}} \tag{8}$$

If we use LS to find the slope and intersection of the regression line using k data points, we have:

$$b = \frac{\sum_k (x_i - \overline{x})(y_i - \overline{y})}{\sum_k (x_i - \overline{x})^2} \qquad (9)$$

$$a = \overline{y} - b\overline{x}$$

where $\overline{x}, \overline{y}$ are averages. Now if one data point is corrupted, i.e. the value of one of the x_i (and as a result y_i) is shifted by Δ. That will cause the averages to shift by Δ/k. We call this shift in slope and y-axis intercept Δb and Δa. Therefore, if we have one compromised node introducing corrupted data in the network, we have a change of Δb in the clock skew. A compromised node is selected with probability p, so the expected value of the change in clock skew is:

$$E(b) = b * \Delta b$$

6.2 Countermeasures for Multi-Hop Networks

In multi-hop networks, at least one pair of nodes communicates with each via a third node that routes messages between them. As in single-hop networks, there may or may not be a base station, a source of trustable time synchronization updates, although most deployments do have such a base station. However, all the nodes in the network may not be in contact with the base station at all times. While there are various multi-hop schemes for time synchronization, as discussed above, they are all essentially schemes for repeating the synchronization scheme for single hop networks, estimation of a node's clock skew and offset by comparison with neighboring nodes, across larger networks. The difficulty in keeping each of the multi-hop schemes secure is in preventing corrupted nodes from increasing the synchronization error between the node with the reference clock and distant nodes. That is, the difficulty is in enabling distant nodes to verify the time synchronization updates they receive despite relying on nodes between the node with the reference clock and themselves to actively maintain-not simply route those updates.

For accurate time synchronization, the nodes must receive the global time of their single-hop neighbors, not nodes that are more distant in the network. If a node computed the offset of its global time from the global time it received from a distant neighbor, that offset would include nondeterministic delays due to the time required to route the update message by intermediate nodes. Therefore, an authenticated broadcast scheme as provided by $\mu TESLA$ does not suffice to securely and accurately synchronize the time of the network to the broadcasting node. The root node could not maintain time synchronization by periodically flooding the network with authenticated clock updates. Clock skew and offset could be calculated in the same way as for updates from a neighboring node. Such updates would, however, provide a useful approximation on the clock skew and offset for all the nodes in the network. Such updates might be sent an order of magnitude less frequently than the updates from a node's immediate neighbors. For each node, the error in both the skew and offset derived from these updates would be primarily determined by the number of hops between the node and the base station, since there would be nondeterministic delays for each hop. In the absence of a base station, the long term trends in the skew calculated by a node might also serve as a useful approximation. So, our first proposal for multi-hop networks is to use such an approximation to get an upper bound on the error that can be induced by an adversary. If the node receives updates from its neighbors that yields a skew and offset sufficiently far from the approximated skew and offset of the root node or the long-term trends, it could ignore its neighbors and use that approximated skew and offset instead. We are currently conducting tests to show the accuracy of such an approximation experimentally.

As discussed above, FTSP and TPSN rely on updates from a single neighbor node to calculate the offset and skew of its clock. One obvious means of increasing the reliability of these synchronization schemes, then, is to introduce redundancy into the system. This is our second proposal for multi-hop time synchronization protocols. In FTSP, it is especially easy to introduce redundancy. Rather than relying on a single update from a single node for each wave of updates from the nearest root node (i.e. for each seqNum), the nodes should record a subset S of the updates from their neighbors. This would increase the storage space required for the linear regression data points by a factor of S. In the current implementation of FTSP, the regression table holds 8 data points of 8 bytes each, 4 for the offset and 4 for the arrival time of that offset. If S were 5, for instance, this scheme would require accommodating 32 additional data points or $32 * (4 + 4) = 256$ bytes. Even on a mote class node, as described above, this is a reasonable additional memory requirement. Given this additional data, the nodes could take the median of the updates for any sequence number instead of whichever update is received first, which is the current scheme in FTSP. The nodes could initially have a set of private keys for their neighbors to authenticate these updates and prevent a compromised node from inserting false updates and corrupting that median. In addition, if a node A saw a trend in which the updates from a neighboring node B tended to deviate from the other neighbors A by more than some threshold, node A could cease considering node B in its calculation of its clock skew and offset.

In that situation, where a node has become skeptical of the updates it is receiving from its neighbors, it may cease sending updates to its neighbors (FTSP) or children (TPSN). This is our third proposal for increasing the security of multi-hop time synchronization protocols, a policy of containment. Under this policy, corrupted nodes would be unable to de-synchronize nodes beyond their immediate neighborhood. However, this policy requires that nodes have multiple sources of time updates in case their source of updates ceases sending updates. In tree-based schemes, such as TPSN, that means having a mechanism whereby the children of nodes in the neighborhood of a corrupted node can find another parent outside the neighborhood of any corrupted node. If the tree for routing time synchronization updates is distinct from the tree for routing any other rout-ing tree used in the network, the tree already satisfies this requirement. However, maintaining multiple trees has the cost of redundant state at each node. In flooding-based schemes such as FTSP, this requirement is satisfied for free since each node receives updates from multiple neighbors and the loss of updates from a single neighbor can be ignored.

In multi-hop networks, if the base station fails or a node becomes disconnected from it, a new root must be elected against which nodes can synchronize their clocks. The same problems arise as discussed above in single hop networks-there is a risk that the newly elected root node will be a corrupted node. To avoid that situation, we recommend the same root rotation scheme as discussed above for single-hop networks. Our final proposal for improving the security of multi-hop time synchronization is to make the LS linear regression used by each node to calculate the skew of its clock more robust. We propose using an algorithm similar to RANSAC [4], which would fit a model to a set of points that contain outliers in the following steps:

- Randomly select a subset of the data points of size m and build the initial model from these points
- Determine the set of data points that are within ϵ of the model and call this set M. This set defines the inliers of the original data set.
- If $|M|$ is greater than a threshold T, we need to re-estimate the model using all the points in M, and the algorithm terminates.
- If $|M|$ is less than T, select a new subset and repeat from the second step on.
- After N trials the largest M is selected, and the model is re-estimated using all the data points in M.

In order to determine how many sample points m we need in each subset, we can use the following formula, where p is the probability that at least one of the subsets does not contain an outlier (usually taken to be 0.99), ϵ is the acceptable proportion of outliers, and N is the number of sub-sets:

$$N = \frac{\log(1 - p)}{\log(1 - (1 - \epsilon)^m)} \qquad (10)$$

In our case, the outliers are generated by an adversary (or a node with an especially erratic clock). As discussed in the appendix, an alternative to both RANSAC and LS for linear regression that is more robust to outliers is LMS. However, LMS is a complex method and it may not be possible to implement it on a mote-class node. We have no such concerns about RANSAC, since it involves only a moderate modification of the basic LS scheme.

7 Conclusion

In this paper we showed that designing a secure time synchronization protocol is a crucial task in sensor networks by showing the adverse effects of time synchronization attacks on some important sensor network applications.

We described three major time synchronization protocols for sensor networks, the set of attacks on each protocol, and proposed countermeasures for these attacks. We are currently implementing and testing those countermeasures on sensor network testbeds. Based on our results, we can evaluate effectiveness of our proposals, including a comparison of the performance of the RANSAC algorithm to using LMS regression.

References

1. Balogh, G., Ledeczi, A., Maroti, M., Simon, G. Time of Arrival Data Fusion for Source Localization.
2. Coleri, S. PEDAMACS: Power Efficient and Delay Aware Medium Access Protocol for Sensor Networks. M.S. Thesis, UC. Berkeley, December 2002.
3. Elson, J., Estrin, D. Fine-Grained Network Time Synchronization using Reference Broadcast. The fifth symposium on Operating Systems Design and Implementation (OSDI), p. 147-163, December 2002.
4. Fischler, M. A., Bolles, R. C.. Random Sample Consensus: A Paradigm for Model Fitting with Applications to Image Analysis and Automated Cartography. Comm. of the ACM, Vol 24, pp 381-395, 1981.
5. Ganeriwawal, S., Kumar, R., Srivastava, M. Timing-Sync Protocol for Sensor Networks. The first ACM Conference on Embedded Networked Sensor Systems (SenSys), p. 138-149, November 2003.
6. Hohlt, B., Doherty, L., Brewer, E. Flexible Power Scheduling for Sensor Networks. Information Processing in Sensor Networks (IPSN), April 2004, Berkeley, CA.
7. Karlof, C., Sastry, N., Wagner, D. TinySec: A Link Layer Security Architecture for Wireless Sensor Networks. Proceedings of the Second ACM Conference on Embedded Networked Sensor Systems (SenSys), pages 162-175, November 2004.
8. Ledeczi A., Volgyesi P., Martoi M., et al. Multiple Simultaneous Acoustic Source Localization in Urban Terrain.
9. Maroti, M., Kusy, B., Simon, G., Ledeczi, A. The Flooding Synchronization Protocol. Proc. Of the Second ACM Conference on Embedded Networked Sensor Systems (SenSys), November 2004.
10. Oh, S., Russell, S., Sastry, S. Markov Chain Monte Carlo Data Association for General Multiple-Target Tracking Problems.
11. Perrig, A., Szewczyk, R., Wen, V., Culler, D., Tygar, J. D. SPINS: Security Protocols for Sensor Networks. Mobile Computing and Networking. Rome, Italy, 2001.
12. Simon, G., Maroti, M., Ledeczi, A.. Sensor Network-Based Countersniper System. Proc. Of the Second ACM Conference on Embedded Networked Sensor Systems (SenSys), November 2004.
13. Welch, G., Bishop, G. An Introduction to the Kalman Filter. University of North Carolina at Chapel Hill, Department of Computer Science, Chapel Hill, NC, USA. TR95-041. Available online at: http://www.cs.unc.edu/ welch/publications.html
14. Available on the web: www.xbow.com/Products/ productsdetails.aspx?sid=62
15. Available on the web: http://www.cs.unc.edu/ tracker/media/pdf/ SIGGRAPH2001_CoursePack_08.pdf
16. Available on the web: www.wabash.edu/econexcel/LMSOrigin/LMSIntro.doc
17. Rmer,Kay. Time Synchronization in Ad Hoc Networks. Proceedings of MobiHoc 2001, ACM, October 2001.

8 Appendix

In the LMS method, we still find the residuals, $r_i = Y_i - \hat{Y}_i$, as in the LS method, but instead of minimizing the sum of the squared of the residues we do the following:

$$\min med_i r_i^2$$

It is well known that LMS is more robust in the presence of outliers, so it is especially appropriate for this application of regression where data points from corrupted nodes appear as outliers.

Finding the least median squares is a challenging optimization problem, however, since we have to optimize the following equation [16]:

$$\min_{b_0, b_1} med_i SR_i = median\{(Y_1 - (b_0 + b_1 X_1))^2, ..., (Y_n - (b_0 + b_1 X_n))^2\}$$

There are software packages that can perform this optimization, but it may not be possible to port the algorithms to a mote-class node.

Secure and Resilient Time Synchronization in Wireless Sensor Networks

Kun Sun[1], Peng Ning[2], and Cliff Wang[3]

[1] Department of Computer Science, North Carolina State University ksun3@ncsu.edu
[2] Department of Computer Science, North Carolina State University pning@ncsu.edu
[3] U.S. Army Research Office, cliff.wang@us.army.mil

Summary. An accurate and synchronized clock time is crucial in many sensor network applications. A number of time synchronization schemes have been proposed for wireless sensor networks recently to address the resource constraints in such networks. However, most of these techniques assume benign environments, thus cannot survive malicious attacks in hostile environments, especially when there are compromised nodes. In this chapter, we present the clock model for time synchronization, and briefly introduce the insecure time synchronization techniques in wireless sensor networks. We then present the challenges and requirements to secure time synchronization and summarize the recent progress on secure time synchronization in wireless sensor networks.

1 Introduction

Wireless sensor networks have received a lot of attention recently due to their wide applications such as target tracking, environment monitoring, and scientific exploration in dangerous environments. As in all distributed systems, time synchronization is an important component of sensor networks to provide a common clock time in sensor nodes.

Many sensor network applications require a synchronized clocks in sensor nodes. The following lists some of these applications.

- *Data Fusion* The ability of the sensor network to aggregate the data collected can greatly reduce the number of messages that need to be transmitted across the network. Many data fusion algorithms (e.g., [31, 57]) have to process the sensor readings ordered by the time of occurrence (e.g., the time when a forest fire was sensed).
- *Target Tracking* In target tracking applications (e.g., [5, 55]), sensor nodes need both the location and the time when the target is sensed to correctly determine the target moving direction and speed.
- *Power Saving* Several approaches intend to improve the energy efficiency by frequently switching sensor nodes into power-saving sleep mode (e.g., [56]). A

group of nodes needs a common synchronized clock to synchronize their behaviors on switching between waking-up and sleep modes at the same time.

- *Slotted MAC Protocols* The time slotted MAC protocols (e.g., [36]) achieve the multiple access to the shared communication medium by assigning time slots to a group of nodes. Thus, sensor nodes need to have a synchronized clock to access their time slots without colliding with other nodes.

A number of time synchronization protocols (e.g., [11, 16, 21, 26, 33, 37, 40, 50]) have been proposed for sensor networks to achieve *pair-wise* and/or *global* time synchronization. Pair-wise time synchronization aims to obtain a high-precision clock synchronization between pairs of sensor nodes, while global clock synchronization aims to provide network-wide time synchronization in a sensor network.

As many other techniques, security is not the top priority when designing time synchronization in wireless sensor networks. However, without addressing security, all the time synchronization techniques in sensor networks cannot survive malicious attacks in hostile environments. Several secure time synchronization techniques have been proposed in sensor networks [14, 32, 53, 54]. Ganeriwal et al. [14] analyzed the attacks from the external attacker, especially the pulse-delay attack. Then they proposed a suite of secure pairwise and group time synchronization protocols for sensor networks. Manzo et al. [32] outlined the potential attacks on time synchronization in sensor networks, and proposed some countermeasures for these attacks. The details of these work will be covered in another two chapters of this editorial volume.

In this chapter, we will discuss the secure and resilient time synchronization technique [54], which mitigates the impacts of attacks from compromised nodes by providing redundant clock distribution. We also present the work on fault-tolerant cluster-wise time synchronization technique [53], which can guarantees an upper bound on the clock difference between normal nodes in a cluster when no more than $1/3$ of the nodes are compromised and collude with each other.

This chapter is organized as follows. We describe a clock model of sensor nodes in Section 2. Section 3 presents the existing time synchronization techniques in sensor networks. We present the attacks model in Section 4. Section 5 shows the requirements for secure time synchronization techniques. We discuss the secure and resilient time synchronization technique in Section 6. Section 7 describes the fault-tolerant cluster-wise time synchronization technique. Section 8 concludes this chapter.

2 Clock Model

A clock is an instrument for measuring time. We need to make a distinction between *real time* and *clock time*. Real time is an assumed Newtonian time frame that may not be directly observable, and clock time is the time that can be observed on the clocks. We usually use lowercase letters to denote the variables and constants about real time, and uppercase letters to denote those about clock time.

A *clock* C can be considered a mapping from real time to clock time, i.e., $T = C(t)$ is the clock time at the real time t. There are some important terms related to clock.

- *clock offset:* It is the difference between the time reported by a clock and the real time. The offset of the clock C_i at real time $t \geq 0$ is given by $C_i(t) - t$. The offset of clock C_i relative to clock C_j at time $t \geq 0$ is given by $C_i(t) - C_j(t)$.
- *frequency:* It is the rate at which a clock progresses. That is, the first derivative of the clock value with respective to real time. An ideal clock's frequency would be 1 all the times. The frequency of clock C at time t is $C'(t)$.
- *frequency offset:* It is the difference in the frequencies of the clock and the real time, also known as *skew*. The frequency offset of a clock C_i relative to clock C_j at time t is $C_i'(t) - C_j'(t)$.
- *drift:* It is the fluctuation of frequency of a clock. That is, the second derivative of the clock value with respective to real time. Drift may be caused by aging, changes in the environment, and other factors external to the oscillator.

A clock C is considered *well-behaved* if its frequency offset (i.e., skew) is bounded by a constant $\rho \rangle 0$ for all the real time points t_1 and t_2, where $t_1 \langle t_2$ [9]:

$$\frac{t_2 - t_1}{1 + \rho} \langle C(t_2) - C(t_1) \langle (1 + \rho)(t_2 - t_1). \tag{1}$$

The frequency offset between any two well-behaved clocks is bounded by $\lambda = \rho(2 + \rho)/(1 + \rho)$, which is less than 2ρ. Sensor nodes usually contain inexpensive crystal oscillators, and the typical clock frequency offset is tens of microseconds per second [46]. This is equivalent to a few seconds per day.

3 Overview of Time Synchronization

Time synchronization has been studied for a long time. We first introduce the traditional time synchronization techniques in Internet and distributed computer systems, and point out the reasons why these techniques cannot be directly applied in sensor networks. We then discuss the time synchronization techniques proposed in sensor networks.

3.1 Traditional Time Synchronization

Global Positioning System (GPS) GPS [1] is a constellation of satellites operated by the U.S. Department of Defense. GPS is originally intended to be used for precise positioning through the determination of pseudo-ranges from the satellites to the ground based receiver. In addition, the time difference of the receiver from the GPS clock time may be calculated. GPS can provide a time accuracy of several nanoseconds.

The GPS receiver is too large, expensive, and power-hungry for small, cheap, and battery powered sensor nodes. Moreover, GPS requires a clear sky view, which is not always available in some areas, such as inside of buildings or underwater.

NTP In Internet, computers can obtain a synchronized Internet time by using Network Time Protocol (NTP) [35]. NTP organizes all the computers in a client/server structure. The reliability of NTP is assured by redundant servers and diverse network paths. Several engineered algorithms have been proposed to reduce jitter, select from multiple sources and avoid improperly operating servers. NTP provides accuracies of low tens of milliseconds on WANs, sub-milliseconds on LANs, and sub-microseconds using a precision time source such as a cesium oscillator or GPS receiver.

Because NTP does not consider the energy and computation limitations of sensor nodes [12], it cannot be directly applied in sensor networks. NTP uses several engineered algorithms (i.e., data-filtering algorithm, peer-selection algorithm and combining algorithm) to reduce jitter, increase the robustness and avoid improperly functioning servers. Such algorithms are computationally intensive and assumes the CPU is always available to frequently discipline the oscillator. However, the CPU cycles in sensor nodes are also a scarce resource, and the sensor nodes cannot spend all the CPU cycles on time synchronization.

Probabilistic-based Synchronization   Cristian [7] proposed a probabilistic method to read remote clocks in distributed systems that are subject to unbounded random communication delays. When a process wants to synchronize to a remote process, it sends a time request to the remote process, and calculates the request round-trip time as the difference between the time when it initiates the request and the time when it receives the reply from the remote process. The reply contains the time when the remote process sends the reply. Then, the process adjusts its clock time to the sum of the time contained in the reply and half the round-trip time. Due to the non-deterministic message delay, to reduce the synchronization error, a process needs to perform multiple such trails and chooses the trial with the minimum round-trip time to synchronize its clock.

Cristian's method is probabilistic because it does not guarantee a process can always synchronize to a remote process with an a priori specified precision. To increase the probability of success to achieve a given precision, a process needs to increase the number of trials on estimating the remote process's clock time. Therefore, it requires a large number of message exchanges, which introduce a high communication overhead for resource constrained sensor nodes.

TEMPO   Gusella and Zatti [19] proposed a centralized clock synchronization service for the Ethernet local area network. A master node first measures the time differences between its local clock and those of other slave nodes. The master node computes the network time as the average of the times provided by normal clocks, and then sends to each slave node the correction that should be performed on its clock. This process is repeated periodically. It assumes the master node is always trusted. The similar idea can be used in sensor networks to achieve time synchronization in a group of nodes that can directly communicate with a sink node or a normal sensor node. However, when TEMPO is deployed to achieve a network-wide time synchronization, the bounds on the synchronization accuracy will increase a lot due to the delay uncertainty in multi-hop message communication.

Fault-tolerant Clock Synchronization   In distributed systems, fault-tolerant clock synchronization has undergone substantial research (e.g. [4,9,20,23–25,29,30, 39,44,47–49,51]). These techniques take either a *software* or a *hardware* approach

[44]. Hardware-based techniques require a synchronization circuitry continuously monitor all the clocks, and thus cannot be used in sensor networks. Software-based techniques can be further classified into convergence-averaging (e.g., CNV [25], LL [29, 30]), convergence-non-averaging (e.g., HSSD [9], ST [51]), or consistency algorithms (e.g., COM, CSM [25]). Some software-based (or hardware-assisted, hybrid) techniques [43] use hardware to generate timestamps, and thus can reduce the uncertainty involved in time synchronization. A common theme of these techniques is to use redundant messages to deal with malicious participants that may behave arbitrarily.

The traditional fault-tolerant time synchronization schemes usually assume there is unlimited computing resource and network bandwidth. All of these techniques involve heavy communication among the nodes, and sometimes heavy computation (e.g., digital signature generation/verification) as well. Thus, they are not suitable for sensor networks.

3.2 Time Synchronization in Sensor Networks

A number of time synchronization protocols have been proposed for sensor networks to achieve pair-wise and/or global clock synchronization. We briefly summarize these protocols.

Reference Broadcasting Elson et al. [11] developed the Reference Broadcast Synchronization (RBS) scheme for pair-wise as well as multi-domain time synchronization, which eliminates the uncertainty of send time and access time from the clock synchronization error by using a reference broadcast node. In RBS, one sender broadcasts a single pulse, two receivers can calculate their relative clock difference by exchanging the receiving time of the pulse from the sender. In addition, the sender can broadcast a number of pulses to improve the clock synchronization precision between the receivers. They also propose to use the least square linear regression technique to estimate the clock frequency offsets. Based on RBS, Palchaudhuri et al. [40] proposed a probabilistic time synchronization to reduce the communication overhead.

Timing-sync Protocol Generiwal et al. [16] proposed a hierarchical time synchronization scheme named TPSN for sensor networks, assuming time synchronization messages are timestamped at medium access control (MAC) layer. In the pairwise time synchronization, one sender synchronizes itself to a receiver by exchanging one pair of messages. If the two exchange messages can be timestamped at MAC layer right before being sent out, TPSN can provide a higher time precision than RBS. TPSN can provide a global time synchronization by building a spanning tree topology in the sensor network. Generiwal et al. [15] also proposed a synchronization scheme to estimate the long-term clock changes and to reduce the energy consumption in duty-cycling MAC layer in sensor networks.

Flooding Time Synchronization Maróti et al. [33] proposed a flooding time-synchronization protocol to synchronize a whole network. The node with the lowest node ID is elected as the leader that serves as the reference node. The leader periodically floods the network with a synchronization message that contains the leader's current time. Nodes that have not received this message record the time stamp in the message and the receiving time of the message, and broadcast the message to their neighbors. Each node collects eight messages and uses the linear regression to estimate the clock and the frequency differences to the leader.

Global Time Synchronization Li and Rus proposed a global time synchronization technique based on local diffusion of clock information [26]. Nodes achieve global synchronization by flooding their neighbors with information about its local clock value. After each node have received the clock values of all its neighbors, the node can use a derived consensus value to adjust its clock. They presented two protocols for both synchronous and asynchronous situations. Both protocols can converges to the average value of the clock readings in the network, within a certain error range. Their protocols have better performance in dense networks than in sparse networks.

Time Diffusion Synchronization Su and Akyildiz proposed a time diffusion synchronization (TDP) protocol to support a network-wide time synchronization [52]. Initially, a set of master nodes are elected based on its remaining power energy. Then, each master node establishes a tree hierarchy. After obtaining the round-trip time to each neighbor node, a master node calculates and broadcasts the average and standard deviation of the message delay to all neighbors. A neighbor node becomes a diffused node based on its remaining power energy and its clock property, and repeat the procedure as the master nodes. The diffusion procedure stops at a given number of hops from the master nodes. This protocol contains a number of algorithms on electing master nodes and synchronizing sensor nodes.

Tiny-sync and Mini-sync Sichitiu et al. [50] developed two lightweight pairwise synchronization schemes, Tiny-Sync and Mini-Sync, to deterministically estimate the bounds on both the relative clock offset and frequency offset between two sensor nodes. Both schemes use multiple round-trip measurements and a line-fitting technique to obtain the clock offset and frequency offset of two nodes. Tiny-Sync uses a heuristic to keep only two measurements in storage, but only achieves a suboptimal solution. Mini-Sync can provide an optimal solution with increased computation and storage overhead.

Lightweight Tree-Based Synchronization Greunen and Rabaey claimed that the maximum time accuracy required in sensor networks is relatively low (within a fraction of one second), and proposed two lightweight global time synchronization schemes in sensor networks [17]. The first scheme is centralized and needs to construct a spanning tree rooted at some reference node, which is responsible for initializing the synchronization process with a time interval decided by the depth of the tree and the required precision. The second scheme performs in a distributed fashion. When a node needs to synchronize its clock, it sends a synchronization request to the reference node by using any available routing protocol. Then all the nodes along the path from the reference node to the requesting node must be synchronized before the requesting node.

TSync Dai and Han [8] proposed two synchronization protocols by using an independent radio channel for synchronization messages to avoid the delay uncertainty due to message collisions. In the proactive synchronization protocol, a spanning tree rooted at a reference node is constructed. Then, the reference node use the reference broadcasting techniques [11] to synchronize the network. The reactive synchronization protocol is initiated by any node as opposed to a designated reference node.

Time Synchronization for Ad-hoc Networks Römer proposed a time synchronization scheme for wireless ad-hoc networks [45]. The basic idea is to generate time stamps using unsynchronized local clocks, instead of synchronizing the local clocks of nodes. When a message containing a time stamp is transmitted between two nodes, the time stamp is first transformed from the sender's local time to the standard Coordinated Universal Time (UTC), and then to the receiver's local time. The final timestamp is expressed as a time interval with a lower bound and an upper bound. Meier et al. [34] improved Römer's protocol by providing a tight bound on the transformed time interval on the receiving node.

Time Synchronization in IEEE 802.11 The IEEE 802.11 standard [3] requires time synchronization in wireless networks for keeping hopping synchronized and other functions like power saving. In an infrastructure basic service set, an access point (AP) transmits periodic beacon frames to the stations, which adjust their clocks to the clock value in the frames from the AP. In an independent basic service set, time is divided into beacon intervals. At the beginning of each interval, each station calculates a random delay and is scheduled to transmit a beacon when the delay timer expires. If a beacon arrives before the random delay timer has expired, the station cancels the pending beacon transmission and the remaining random delay. Upon receiving a beacon, a station sets its clock to the timestamp of the beacon if the value of the timestamp is later than the stations clock time. It guarantees that clocks only move forward and never backward.

4 Attack Model

A sensor node is *normal* if it correctly executes a given time synchronization algorithm and its clock is well-behaved; otherwise, it is a *malicious* node. In an *arbitrary attack model* [13], malicious nodes can arbitrarily deviate from the protocol and collude with each other.

It is possible to use authentication to prevent external attackers from impersonating normal insider nodes when every two nodes can share a unique pair-wise key. Such pair-wise key can be provided by several key predistribution schemes proposed recently [6, 10, 27]. Furthermore, local broadcast authentication can be achieved by using the techniques proposed in [18, 42, 58]. However, because sensor nodes are usually not tamper resistant, an attacker may attack time synchronization services through compromising nodes and obtaining keying materials.

4.1 Attacks in Single-Hop Pair-Wise Time Synchronization

The single-hop pair-wise time synchronization is to discover the clock difference between two neighbor nodes that can communicate with each other directly. It can be classified into two categories: *receiver-receiver synchronization* (e.g., RBS [11]), in which a reference node broadcasts a reference packet to help pairs of receivers to identify the clock differences, or *sender-receiver synchronization* (e.g., TPSN [16]), where a sender communicates with a receiver to estimate the clock difference. Malicious nodes may launch following potential attacks on single-hop pair-wise time synchronization.

- *Sybil attack:* A malicious node may attempt to forge multiple identities by launching Sybil attacks [38]. If colluding malicious nodes can exchange their keying materials, one malicious node may impersonate other remote malicious nodes in its local network [41]. The Sybil attack may have a severe consequence on the time synchronization protocol. For majority voting based resilient algorithm, the Sybil attack may increase the number of colluding malicious nodes in a neighborhood and thus may potentially defeat the time synchronization protocol.

- *wormhole attack:* In sensor networks, remote malicious nodes may pretend to be in normal nodes' local area through wormholes [22]. Thus, time synchronization messages at a remote area may be tunneled to local area to interrupt the local time synchronization process.

- *replay attack:* A malicious node may launch replay attacks by recording the current synchronization messages from other nodes and impersonating these nodes to send the buffered messages later. This type of attack can usually be defeated through the use of freshness token such as a sequence number.

- *pulse-delay attack:* A malicious node may distort the pair-wise time synchronization by jamming the signal between two normal nodes and then replaying the delayed signal to introduce synchronization error [14]. Even an external attacker can launch this attack.

- *silent attack:* Several time synchronization protocols (e.g., [16,17]) need to build a level or tree hierarchy in advance to achieve the network-wide time synchronization. However, a malicious node may perform well in the phase that builds up the hierarchy, but keep silent in the synchronization phase. For the nodes that synchronize to the source node through the malicious node, they may not receive the synchronization messages and thus cannot maintain a synchronized clock time.

- *malicious reference:* This attack is specific to receiver-receiver synchronization (e.g., RBS [11]). A compromised reference node may provide different non-compromised nodes different time values about the receipt of the reference packet.

An attacker can also launch signal jamming attacks, which can block normal nodes for receiving any synchronization messages. Because no scheme that requires inter-node communication can survive such attacks, we did not consider such attacks in our model.

4.2 Attacks in Multiple-Hops Pair-Wise Time Synchronization

Multi-hop pair-wise time synchronization protocols and most of the global time synchronization protocols (e.g., [11, 16, 50]) establish multi-hop paths in a sensor network, so that all the nodes in the network can synchronize their clocks to the source based on these paths and the single-hop pair-wise clock differences between adjacent nodes in these paths. Alternatively, diffusion based global synchronization protocols [26] achieve global synchronization by spreading local synchronization information to the entire network.

When a pair of nodes are synchronized through a multi-hop path (e.g., [11, 16, 50]), a compromised node in the path can introduce arbitrary errors. Moreover, a compromised node may severely disrupt the synchronization procedure. This implies multi-hop pair-wise and global time synchronization using multi-hop paths are vulnerable to compromised nodes. Even when the diffusion based global clock synchronization techniques [26] are used, compromised nodes may fluctuate their clock information periodically to prevent the convergence of the clocks.

5 Evaluation Metrics for Secure Time Synchronization

Given a secure time synchronization protocol, which targets at preventing or mitigating the attacks in time synchronization, we can use the following metrics to evaluate its performance in both benign and hostile environments.

- *synchronization precision:* It is the maximum clock difference between any two sensor nodes in a whole network. It is a metric closely related to the *synchronization error*, which is about the clock offset of a single node [46]. In hostile environments, we only care about the synchronization precision between normal nodes.
- *synchronization rate:* It is the percentage of sensor nodes in a sensor network that can correctly obtain a synchronized clock time.
- *memory overhead:* It is the size of memory allocated for storing the messages related to time synchronization in each sensor node. Currently, memory is still critical for resource constrained sensor nodes.
- *convergence time:* It has different meanings for external synchronization and internal synchronization.
  - *external synchronization* When all the clocks in the network are synchronized to an external clock source, the convergence time is the time interval between the start point of the synchronization process and the time point when the last sensor node that can be synchronized synchronizes its clock.
  - *internal synchronization* When all the nodes in a network need to agree on a consistent clock time without the help from an external clock source, the convergence time is the time interval between the start point of the synchronization process and the time point when the predetermined synchronization precision is achieved.
- *communication overhead:* It is the number of messages sent for time synchronization. The synchronization information may be piggy-backed in the messages for other applications, or sent by dedicated synchronization messages. The "piggy-back" method can avoid additional messages for synchronization, but it

may not provide a synchronized clock on demand, since it depends on the other applications. The communication overhead is related to the synchronization precision achieved.

6 Secure and Resilient Time Synchronization in Sensor Networks

When a pair of nodes are synchronized through a multi-hop path (e.g., [11, 16, 50]), a compromised node in the path can introduce arbitrary errors. The secure and resilient time synchronization technique [54] can mitigate such attacks by providing redundant ways for one node to synchronize its clock with another node, so that it can tolerate partially missing or false synchronization information from compromised nodes. A secure and resilient global time synchronization can be achieved by adopting a synchronization model where all sensor nodes synchronize their clocks to a common source, which is assumed to be well synchronized to the external clock.

Two schemes are developed using this general technique: *level-based time synchronization* and *diffusion-based time synchronization*. Targeted at static sensor networks, the level-based time synchronization constructs a level hierarchy initially, and uses (or reuses) this level hierarchy for multiple rounds of time synchronization. The diffusion-based time synchronization attempts to synchronize all clocks without relying on any structure assumptions, and thus can be used for dynamic sensor networks.

6.1 Synchronization Model

We assume there is a *source node* S that is well synchronized to the external clock, for example, through a GPS receiver. We would like to synchronize the clocks of all the sensor nodes in the network to that of the source node. We assume the source node is trusted, and all the other nodes know the identity of the source node.

We adopt the following model for secure and resilient clock synchronization:

1. Each node maintains a local clock. The local clock of the source node is the desired global clock.
2. For each of its neighbor nodes, each node maintains a single-hop pair-wise clock difference with a secure single-hop pair-wise time synchronization (e.g., [14]).
3. Each node also maintains a source clock difference between its local clock and the clock of the source node. Each node can directly obtain the source clock difference if it is a neighbor node of the source node. Otherwise, to tolerate up to t malicious neighbor nodes, each node needs to compute at least $2t+1$ *candidate* source clock differences through different neighbor nodes. It then chooses the median of the candidate source clock differences as its source clock difference.
4. Each node can estimate the global clock by using its local clock and its source clock difference.

We assume the sensor network of concern is dense so that each node has enough number of neighbor nodes to obtain $2t + 1$ candidate source clock differences. Moreover, we assume all the communication messages are authenticated with a pair-wise key shared between any two communicating nodes.

6.2 Level-Based Time Synchronization

Level-based time synchronization aims at static sensor networks, where the network topology does not change frequently. It consists of two phases: *level discovery phase* and *synchronization phase*.

Level Discovery Phase

The level discovery phase is to organize sensor nodes into a hierarchy rooted at the source node. Each node except for the source node has a set of parent nodes in the hierarchy, and each non-leaf node has a set of children nodes. Each neighbor of the source node sets the source node as its unique parent node; the nodes that are more than one hop away from the source node need to record $3t + 1$ parent nodes to tolerate up to t malicious parent nodes in the synchronization phase. This requirement of $3t + 1$ parent nodes is to defend against the *silent attack*, which is discussed in Section 6.4.

Synchronization Phase

In the synchronization phase, the source node periodically initiates the synchronization phase by unicasting synchronization messages to its neighbor nodes. All the sensor nodes obtain the source clock differences through their parent nodes, estimate their own source clock differences, and then help their children nodes to synchronize their clocks. Each neighbor of the source node can obtain the source clock difference from the source node directly. For a node that is more than one hop away from the source node, to tolerate up to t malicious nodes in its parent nodes, it has to collect at least $2t + 1$ candidate source clock differences through its parent nodes, and then sets its source clock difference as the median of the $2t + 1$ candidate source clock differences.

6.3 Diffusion-Based Time Synchronization

In the diffusion-based scheme, the source node initiates the synchronization process periodically by unicasting synchronization messages to its neighbor nodes. After obtaining a source clock difference from the source node, the neighbor nodes of the source node update their local clocks, and then unicast synchronization messages to help their neighbors except for the source node to synchronize their clocks. To tolerate up to t colluding malicious nodes among its neighbor node, a node more than one hop away from the source node needs to receive at least $2t + 1$ candidate source clock differences through different neighbor nodes, and updates its source clock difference as the median of the $2t + 1$ source clock differences. The node then sends synchronization messages to its neighbors from which it has not received synchronization messages.

6.4 Security Analysis

We have the following lemmas on the effectiveness of the two schemes.

Lemma 1. *The level-based time synchronization can synchronize all the normal nodes correctly, if each normal node at level l ($l\rangle 1$) receives at least $2t + 1$ source clock differences from distinct parent nodes and at most t out of these parent nodes are colluding malicious nodes.*

Lemma 2. *The diffusion-based time synchronization scheme can synchronize all the normal nodes correctly, if each normal node that is more than one hop away from the source node receives the source clock differences (of the neighbor nodes) from at least $2t + 1$ distinct neighbor nodes among which at most t nodes are colluding malicious nodes.*

The proof on both lemmas can be found in [54]. In summary, a normal node can correctly synchronize its clock to the source nodes when the following two conditions are satisfied:

- *Condition 1:* Each normal node can receive $2t + 1$ source clock differences;
- *Condition 2:* Among the $2t + 1$ source clock differences, there exist at most t malicious source clock differences.

Now consider the first condition. Our schemes are suitable for dense sensor networks in which a normal node can receive at least $2t + 1$ source clock differences. In the level-based scheme, a malicious node may refuse to provide its source clock difference to its children nodes. However, since a normal node records $3t + 1$ parent nodes in its parent set, even if up to t malicious nodes keep silent, the normal node can still receive $2t + 1$ source clock differences. Therefore, we can prevent the silent attack.

The second condition requests that a given normal node has at most t malicious nodes that appear to be its neighbors. Malicious nodes may launch various attacks to break this condition. We can prevent the replay attacks on the synchronization messages by including a per-node sequence number in the synchronization messages. The remote malicious nodes and normal nodes tunneled through wormholes can be detected with their locations and/or the message transmission delays, as indicated in [22, 28]. By using unique pair-wise keys to authenticate messages, our schemes can prevent malicious nodes from impersonating uncompromised normal nodes. One malicious node may impersonate other remote malicious nodes in its local network; however, such node may be detected and removed by using the techniques proposed in [41].

6.5 Time Synchronization with Multiple Source Nodes

To reduce the convergence time, increase the synchronization rate, and improve the synchronization precision, we can distribute multiple source nodes into the network and make sensor nodes synchronize to the nearest source nodes.

In hostile environments, it is possible for malicious attackers to compromise a small portion of the source nodes, though the source nodes are typically better protected from attacks than the normal ones. To tolerate up to s malicious source nodes, we extend the synchronization model in Section 6.1 by requiring each normal node obtain at least $2s+1$ source clock differences from distinct source nodes and chooses the median of the source clock differences as its final source clock difference.

6.6 Experimental Results

We studied both schemes through simulation in ns2 [2]. By experimental results, the diffusion-based scheme can synchronize more sensor nodes than the level-based scheme in sparse networks. While in dense networks, both schemes can synchronize almost all the sensor nodes.

The diffusion-based scheme has a much higher communication overhead than the level-based scheme. One reason is that, in the diffusion-based scheme, each node needs to unicast synchronization messages to all the neighbors from which it has not received the synchronization messages, while in the level-based scheme, each node only send to its children nodes. Another reason is the message collision. The more messages are sent in a local area, the more message collisions happen. In our simulation, one message can be retransmitted at most 4 times when it collides with other messages. The level-based scheme can provide a much better synchronization precision than the diffusion-based scheme. The major reason is that the diffusion-based scheme has a longer convergence time, during which a clock may drift away.

By using multiple source nodes, we can increase the robustness of our schemes against source node compromises; however, performance of our schemes may become worse with the increase of s and t. This is due to the fact that the coverage areas of different source nodes have overlaps. The messages from different source nodes may collide frequently, and as a result, both the communication overhead and the convergence time may increase a lot.

7 Fault-Tolerant Cluster-Wise Time Synchronization for Wireless Sensor Networks

In wireless sensor networks, it is usually necessary to have a cluster of nodes share a common view of a local clock time, so that the nodes can coordinate their actions in some applications, such as time slotted MAC protocols (e.g., [36]) and sleep/listen scheduling for power saving (e.g., [56]).

In benign environments, a local common clock can be easily achieved by having all the nodes synchronize to a given node. However, in hostile environments where some nodes may be compromised, it is quite challenging to synchronize the clocks among a cluster of nodes, since a compromised node may disrupt the time synchronization by sending different time to non-compromised nodes.

A novel fault-tolerant cluster-wise time synchronization scheme [53] has been proposed for securely synchronizing clusters of sensor nodes, where the nodes in each cluster can communicate with each other directly through broadcast. This scheme guarantees an upper bound on the clock difference between normal nodes when no more than $1/3$ of the nodes are compromised and collude with each other.

7.1 Fault-Tolerant Time Synchronization Model

The model is adapted from [9]. The clocks in sensor nodes are synchronized in rounds, each of which consists of R time units. Suppose these clocks are synchronized initially. We denote the real time point at which the first (or the last) normal node starts its f-th round as beg^f (or end^f). Over the time interval $[end^f, end^{f+1}]$ for any f, two well-behaved clocks may drift away at most $\delta = 2\rho(end^{f+1} - end^f)$, where ρ is the maximum frequency offset of all the normal nodes.

Suppose a node makes a clock adjustment at time t. We use $C(t)$ and $C^+(t)$ to represent the clock time before and after the clock adjustment, respectively. Suppose node i sends a message at $C_i(t_1)$, node j receives the message at $C_j(t_2)$ and adjusts its clock to $C_i(t_1)$. We have $C_i(t_2) - C_i(t_1)\langle\epsilon$, where ϵ is the upper bound for the synchronization error, which includes the maximum transmission delay and the clock drift during this delay.

Suppose there exist up to $m\langle\frac{n}{3}$ malicious nodes in a cluster of n nodes that can communicate with each other through broadcast. With $k = \frac{n-m\frac{\epsilon}{(\delta+\epsilon)}}{n-3m}$ and $\Delta = \delta + \epsilon(1 + 4\rho)$, the fault-tolerant cluster-wise time synchronization algorithm satisfies the following two conditions:

- *CS1:* For any two normal nodes i and j, there exists an upper bound on the clock difference between them for any real time point. That is, for all $f \geq 1$, and $t \in [beg^f, beg^{f+1}]$, $|C_i(t) - C_j(t)| \leq (2km + 1)\Delta + m\delta + 2\rho\epsilon$;
- *CS2:* If a node makes an adjustment to its clock at time t, there is an upper bound on the clock adjustment. That is, $|C^+(t) - C(t)| \leq k\Delta$.

7.2 Fault-Tolerant Cluster-wise Time Synchronization Algorithm

The algorithm executes one round of time synchronization every R time units. In each round of synchronization, only one node serves as the *synchronizer*, and only one synchronization message is broadcast by the synchronizer. To prevent malicious nodes from impersonating normal synchronizers, we adapt a recently proposed local broadcast authentication scheme for sensor networks [58] to authenticate the broadcast synchronization messages. We assume each node i has generated a one-way key chain, and exchanged the commitment $K_i^{(0)}$ with the other nodes using the pair-wise keys shared with them.

For simplicity, we assume the "starting time" is $beg^0 = end^0 = 0$. For any two normal nodes i and j, $|C_i(0) - C_j(0)| \langle \epsilon(1 + 4\rho)$. Each node maintains a counter f by increasing it by one in each round of time synchronization. Initially $f = 1$.

Algorithm 4 Synchronizer Ring

Initialization

$f \leftarrow 1; k \leftarrow \frac{n - m\frac{\epsilon}{\delta + \epsilon}}{n - 3m}; \Delta = \delta + \epsilon(1 + 4\rho); x \leftarrow (2km + 1)\Delta + m\delta;$

Task 1: Send

if $(C = f \times R)$ and $(Order(N_i) = f \bmod n)$ **then**

    Broadcast a message "$N_i | K_i^{(\lceil f/n \rceil)}$";

end if

Task 2: Receive

if (Receive a message "$N_i | K_i^{(\lceil f/n \rceil)}$" at T) **then**

    if $(f \times R - x \leq T \leq f \times R + x)$ and $(F(K_i^{(\lceil f/n \rceil)}) = K_i^{(\lceil f/n \rceil - 1)})$ and $(Order(N_i) = f \bmod n)$ **then**

        $\bar{\Delta} \leftarrow f \times R - T;$

        if $k\Delta \leq \bar{\Delta} \leq x$ **then**

            $\bar{\Delta} \leftarrow k\Delta;$

        else if $-x \leq \bar{\Delta} \leq -k\Delta$ **then**

            $\bar{\Delta} \leftarrow -k\Delta;$

        end if

        $C \leftarrow C + \bar{\Delta}; f \leftarrow f + 1;$

    else

        Drop the message;

    end if

end if

if Has not received a correct synchronization message by $f \times R + x$ (Note that this may be implemented as a timer.) **then**

    $f \leftarrow f + 1;$

end if

The algorithm consists of two tasks that run continuously on each normal sensor node. In the first task, if node i is the synchronizer for the f-th round of synchronization, when its clock time reaches $C = f \times R$, it immediately broadcasts a synchronization message "$N_i | K_i^{(\lceil f/n \rceil)}$" to all the other nodes, where N_i is node i's ID and $K_i^{(\lceil f/n \rceil)}$ is the key in node i's key chain that is used for authentication in the f-th round.

In the second task, when a node receives a synchronization message at its clock time T in the f-th round of time synchronization, if $T \langle f \times R - x$ or $T \rangle f \times R + x$, the node drops the message. In the algorithm, $x = (2km + 1)\Delta + m\delta$ is the maximum clock difference between any normal node and a normal synchronizer, where m is the number of malicious nodes, $k = \frac{n-m\frac{\epsilon}{\delta+\epsilon}}{n-3m}$, and $\Delta = \delta + \epsilon(1+4\rho)$ is the maximum clock difference between any two nodes if all the nodes are normal. Otherwise, it verifies that node N_i is the correct synchronizer and it is the first time to receive the $K_i^{(\lceil f/n \rceil)}$ and $F(K_i^{(\lceil f/n \rceil)}) = K_i^{(\lceil f/n \rceil - 1)}$, where F is the one-way function and $K_i^{(\lceil f/n \rceil - 1)}$ is the key received from node i in the $(f - n)$-th round or the commitment. If the message cannot pass these verifications, the node drops the message. Otherwise, the node calculates the clock difference $\bar{\Delta} = f \times R - T$ and performs the following clock adjustment: if $|\bar{\Delta}| \langle k\Delta$, the node adjusts its clock time by adding $\bar{\Delta}$; if $k\Delta \leq \bar{\Delta} \leq x$, it increases its clock time by $k\Delta$; if $-x \leq \bar{\Delta} \leq -k\Delta$, it decreases its clock time by $k\Delta$. The node also increments the counter f by 1. If the node does not receive an authenticated synchronization message for the current round by the time $f \times R + x$, it increments the counter f by 1 and enters the next round.

Algorithm 4 shows the pseudo code. Because all the nodes serve as the synchronizer in a round robin fashion, we refer to our scheme as *Synchronizer Ring (SR)* algorithm. To ensure that clocks are never set back, we may further adopt the technique proposed in [25], which spreads each synchronization adjustment over the next synchronization period.

We proved that the proposed algorithm satisfies the two conditions in Section 7.1, and have the following theorem. Refer to [53] for further detail.

Theorem 1. *When* $n \rangle 3m$, *Algorithm SR is a fault-tolerant clock synchronization algorithm with* $(2km + 1)\Delta + m\delta + 2\rho\epsilon$ *as the upper bound of the clock difference and* $k\Delta$ *as the upper bound of clock adjustment, where* $k = \frac{n-m\frac{\epsilon}{(\delta+\epsilon)}}{n-3m}$ *and* $\Delta = \delta + \epsilon(1+4\rho)$.

Theorem 1 gives an upper bound of the maximum clock difference between normal nodes when no more than $m \langle \frac{n}{3}$ nodes are compromise. Actually, the maximum clock difference is reached only when the m colluding malicious nodes serve as the synchronizer in a row, and the probability that this happens is only $P_m = \frac{(n-m)!m!}{(n-1)!}$.

7.3 Comparison with Previous Techniques

In each round of time synchronization of SR algorithm, only one node serves as the synchronizer, and no other nodes need to respond to the message from the synchronizer. As a result, there will be no collision between the messages involved in time synchronization (when there is no malicious attack). In contrast, almost all of the existing fault-tolerant time synchronization algorithms (e.g., CNV [25], HSSD [9]) require the participants send or forward synchronization messages around the same time. Thus, it is very likely to have message collisions in such schemes when they are applied in wireless sensor networks.

Moreover, SR takes advantage of the broadcast medium as well as a recently proposed authentication technique for sensor networks [58], and thus does not have

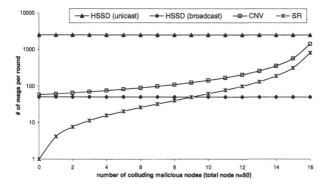

Fig. 1. Communication Overhead with the Same Maximum Clock Difference

to use costly digital signatures for broadcast authentication. In comparison, several traditional fault-tolerant techniques (e.g., CSM [25], HSSD [9]) require digital signatures, which make them undesirable for resource constrained sensor networks. Note that these algorithms cannot use the authentication technique in [58]. The major reason is that they require forwarding of received messages, while a malicious node may manipulate a message before forwarding it to other nodes.

Table 1 compares SR algorithm with traditional fault-tolerant time synchronization algorithms when they are used to synchronize a cluster of n fully connected nodes.

Table 1. Performance Comparison.

Algorithm	Degree of Fault-Tolerance	Comm. Overhead (# msgs/round)	Maximum Clock Difference
CNV [25]	$\frac{n-1}{3}$	n^2 (unicast)	$\frac{n}{n-3m}(2\epsilon + 2\rho R)$
COM [25]	$\frac{n-1}{3}$	n^{m+1} (unicast)	$(6m+4)\epsilon + 2\rho R$
CSM [25]	$\frac{n-1}{2}$	n^{m+1} (unicast)	$(m+6)\epsilon + 2\rho R$
$HSSD$ [9]	$n-1$	n^2 (unicast)	$\epsilon + 2\rho R$
SR	$\frac{n-1}{3}$	1 (broadcast)	$\left(\frac{2mn}{n-3m}+1\right)\epsilon + \frac{(\frac{2nm}{n-3m}+m+1)}{1-4\rho(\frac{2nm}{n-3m}+m+1)}2\rho R$

We refer to the maximum number of malicious nodes that one algorithm can tolerate as its *degree of fault-tolerance*. In a cluster of n nodes, SR's degree of fault-tolerance is $\frac{n-1}{3}$, which is the same as Algorithms CNV and COM. However, Algorithms CSM and $HSSD$ can provide better tolerance against colluding attacks.

We also compare the communication overhead in all these algorithms when setting the same bound for the maximum clock difference. We make a conservative assumption that the existing algorithms in Table 1 may use broadcast instead of unicast to send the synchronization messages. This reduces the number of messages per round from n^2 to n for CNV and HSSD, and from n^{m+1} to n^m for COM and CSM. Figure 1 indicates that SR always has less communication overhead than CNV. Compared with HSSD, SR has less communication overhead when the number of colluding malicious nodes is small, but more communication overhead when the number of colluding nodes grows. However, HSSD has to be modified to reach this performance, because using broadcast in HSSD will cause substantial message collisions. Moreover, the digital signatures required by HSSD make it undesirable for sensor networks.

8 Conclusions

Time synchronization is an important component for many applications in sensor networks. It is critical to provide secure time synchronization in hostile environments. This chapter summarized current time synchronization algorithms for traditional networks and sensor networks. Further we discussed the security and resilience properties of sensor network time synchronization. Finally, we presented two recent efforts on securing the time synchronization services in sensor networks. With more attentions to be attracted on protecting the time synchronization in sensor networks, we anticipate more secure time synchronization protocol to be developed.

References

1. GPS time transfer. http://tf.nist.gov/timefreq/time/gps.htm.
2. The network simulator – ns-2. http://www.isi.edu/nsnam/ns/.
3. IEEE standard 802.11. wireless lan medium access control (MAC) and physical layer (PHY) specification, 1999.
4. B. Barak, S. Halevi, A. Herzberg, and D. Naor. Clock synchronization with faults and recoveries. In *Proceedings of the 19th Annual ACM Symposium on Principles of Distributed Computing*, pages 133–142, 2000.
5. K. Chakrabarty, S. S. Iyengar, H. Qi, and E. Cho. Grid coverage for surveillance and target location in distributed sensor networks. *IEEE Transactions on Computers*, 51:1448–1453, 2002.
6. H. Chan, A. Perrig, and D. Song. Random key predistribution schemes for sensor networks. In *IEEE Symposium on Research in Security and Privacy*, pages 197–213, 2003.
7. F. Cristian. Probabilistic clock synchronization. *Distributed Computing*, 3(3):146–158, 1989.
8. H. Dai and R. Han. Tsync: a lightweight bidirectional time synchronization service for wireless sensor networks. *ACM SIGMOBILE Mobile Computing and Communications Review*, 8(1):125–139, 2004.
9. D. Dolev, J. Y. Halpern, B. Simons, and R. Strong. Dynamic fault-tolerant clock synchronization. *Journal of the ACM*, 42(1):143–185, 1995.
10. W. Du, J. Deng, Y. S. Han, and P. Varshney. A pairwise key pre-distribution scheme for wireless sensor networks. In *Proceedings of 10th ACM Conference on Computer and Communications Security (CCS'03)*, pages 42–51, October 2003.
11. J. Elson, L. Girod, and D. Estrin. Fine-grained network time synchronization using reference broadcasts. *ACM SIGOPS Operating Systems Review*, 36:147–163, 2002.

12. J. Elson and K. Römer. Wireless sensor networks: A new regime for time synchronization. In *Proceedings of the First Workshop on Hot Topics in Networks (HotNets-I)*, pages 149–154, October 2002.
13. A. Galleni and D. Powell. Consensus and membership in synchronous and asynchronous distributed systems. Technical Report 96104, LAAS, April 1996.
14. S. Ganeriwal, S. Capkun, C. Han, and M. B. Srivastava. Secure time synchronization service for sensor networks. In *Proceedings of 2005 ACM Workshop on Wireless Security (WiSe 2005)*, pages 97–106, September 2005.
15. S. Ganeriwal, D. Ganesan, H. Shim, V. Tsiatsis, and M. B. Srivastava. Estimating clock uncertainty for efficient duty-cycling in sensor networks. In *Proceedings of the 3rd international conference on Embedded networked sensor systems (SenSys)*, pages 130–141, 2005.
16. S. Ganeriwal, R. Kumar, and M. B. Srivastava. Timing-sync protocol for sensor networks. In *Proceedings of the First International Conference on Embedded Networked Sensor Systems (SenSys)*, pages 138–149, 2003.
17. J. Greunen and J. Rabaey. Lightweight time synchronization for sensor networks. In *Proceedings of the Second ACM International Workshop on Wireless Sensor Networks and Applications (WSNA)*, pages 11–19, September 2003.
18. N. Gura, A. Patel, A. Wander, H. Eberle, and S.C. Shantz. Comparing elliptic curve cryptography and RSA on 8-bit CPUs. In *Proceedings of Workshop on Cryptographic Hardware and Embedded Systems (CHES 2004)*, August 2004.
19. R. Gusella and S. Zatti. The accuracy of the clock synchronization achieved by tempo in berkeley unix 4.3bsd. *IEEE Transactions on Software Engineering*, 15(7):847–853, 1989.
20. J.Y. Halpern, B.B. Simons, H.R. Strong, and D. Dolev. Fault-tolerant clock synchronization. In *Proceedings of Third Annual ACM Symposium on Principles of Distributed Computing*, pages 89–102, 1984.
21. A. Hu and S. D. Servetto. Asymptotically optimal time synchronization in dense sensor networks. In *Proceedings of the Second ACM International Workshop on Wireless Sensor Networks and Applications (WSNA)*, pages 1–10, September 2003.
22. Y.C. Hu, A. Perrig, and D.B. Johnson. Packet leashes: A defense against wormhole attacks in wireless ad hoc networks. In *Proceedings of INFOCOM 2003*, April 2003.
23. C.M. Krishna, K.G. Shin, and R.W. Butler. Ensuring fault tolerance of phase-locked clocks. *IEEE Transactions on Computers*, 34(8):752–756, 1985.
24. L. Lamport. Time, clocks, and the ordering of events in a distributed system. *Communications of the ACM*, 21(7):558–565, 1978.
25. L. Lamport and P.M. Melliar-Smith. Synchronizing clocks in the presence of faults. *Journal of the ACM*, 32(1):52–78, 1985.
26. Q. Li and D. Rus. Global clock synchronization in sensor networks. In *Proceedings of IEEE INFOCOM 2004*, pages 214–226, March 2004.
27. D. Liu and P. Ning. Establishing pairwise keys in distributed sensor networks. In *Proceedings of 10th ACM Conference on Computer and Communications Security (CCS'03)*, pages 52–61, October 2003.
28. D. Liu, P. Ning, and W.K. Du. Detecting malicious beacon nodes for secure location discovery in wireless sensor networks. In *Proceedings of the 25th International Conference on Distributed Computing Systems (ICDCS '05)*, pages 609–619, June 2005.
29. J. Lundelius and N. Lynch. A new fault-tolerant algorithm for clock synchronization. In *Proceedings of the Third Annual ACM Symposium on Principles of Distributed Computing*, pages 75–88, 1984.
30. J. Lundelius-Welch and N. Lynch. A new fault-tolerant algorithm for clock synchronization. *Information and Computation*, 77(1):1–36, 1988.

31. S. Madden, M. J. Franklin, J. M. Hellerstein, and W. Hong. TAG:a tiny aggregation service for ad-hoc sensor networks. In *Proceedings of the 5th Annual Symposium on Operating Systems Design and Implementation (OSDI)*, December 2002.
32. M. Manzo, T. Roosta, and S. Sastry. Time synchronization attacks in sensor networks. In *Proceedings of the 3rd ACM workshop on Security of ad hoc and sensor networks*, pages 107–116, 2005.
33. M. Maroti, B. Kusy, G. Simon, and A. Ledeczi. The flooding time synchronization protocol. In *Proceedings of the Second ACM Conference on Embedded Networked Sensor Systems (SenSys'04)*, pages 39–49, Nov 2004.
34. L. Meier, P. Blum, and L. Thiele. Internal synchronization of drift-constraint clocks in ad-hoc sensor networks. In *Proceedings of the 5th ACM international symposium on Mobile ad hoc networking and computing MobiHoc '04*, 2004.
35. D.L. Mills. Internet time synchronization: The network time protocol. *IEEE Transactions on Communications*, 39(10):1482–1493, 1991.
36. S. Mishra and A. Nasipuri. An adaptive low power reservation based mac protocol for wireless sensor. In *Proceedings of the IEEE International Conference on Performance Computing and Communications*, pages 713–736, 2004.
37. M. Mock, R. Frings, E. Nett, and S. Trikaliotis. Clock synchronization for wireless local area networks. In *Proceedings of the 12th Euromicro Conference on Real-Time Systems (Euromicro-RTS 2000)*, June 2000.
38. J. Newsome, R. Shi, D. Song, and A. Perrig. The sybil attack in sensor networks: Analysis and defenses. In *Proceedings of IEEE International Conference on Information Processing in Sensor Networks (IPSN 2004)*, April 2004.
39. A. Olson and K.G. Shin. Fault-tolerant clock synchronization in large multicomputer systems. *IEEE Transactions on Parallel and Distributed Systems*, 5(9):912–923, 1994.
40. S. PalChaudhuri, A.K. Saha, and D.B. Johnson. Adaptive clock synchronization in sensor networks. In *Information Processing in Sensor Networks (IPSN)*, pages 340–348, April 2004.
41. B. Parno, A. Perrig, and V. Gligor. Distributed detection of node replication attacks in sensor networks. In *IEEE Symposium on Security and Privacy*, May 2005.
42. A. Perrig, R. Szewczyk, V. Wen, D. Culler, and D. Tygar. SPINS: Security protocols for sensor networks. In *Proceedings of Seventh Annual International Conference on Mobile Computing and Networks*, pages 521–534, July 2001.
43. P. Ramanathan, D.D. Kandlur, and K.G. Shin. Hardware-assisted software clock synchronization for homogeneous distributed systems. *IEEE Transactions on Computers*, 39(4):514–524, 1990.
44. P. Ramanathan, K.G. Shin, and R.W. Butler. Fault-tolerant clock synchronization in distributed systems. *IEEE Computer*, 23(10):33–42, 1990.
45. K. Römer. Time synchronization in ad hoc networks. In *Proceedings of the 2nd ACM international symposium on Mobile ad hoc networking & computing*, pages 173–182, 2001.
46. K. Römer, P. Blum, and L. Meier. Time synchronization and calibration in wireless sensor networks. In Ivan Stojmenovic, editor, *Wireless Sensor Networks*. John Wiley Sons, 2005. To appear.
47. F.B. Schneider. A paradigm for reliable clock synchronization. Technical Report TR 86–735, Cornell University, Department of Computer Science, 1986.
48. F.B. Schneider. Understanding protocols for Byzantine clock synchronization. Technical Report TR 87–859, Cornell University, Department of Computer Science, 1987.
49. K.G. Shin and P. Ramanathan. Clock synchronization of a large multiprocessor system in the presence of malicious faults. *IEEE Transactions on Computers*, 36(1):2–12, 1987.

50. M.L. Sichitiu and C. Veerarittiphan. Simple, accurate time synchronization for wireless sensor networks. In *IEEE Wireless Communications and Networking Conference WCNC03*, 2003.
51. T. K. Srikanth and S. Toueg. Optimal clock synchronization. *Journal of the ACM*, 34(3):626–645, 1987.
52. W. Su and I. F. Akyildiz. Time-diffusion synchronization protocol for wireless sensor networks. *IEEE/ACM Transactions on Networking (TON)*, 13(2), 2005.
53. K. Sun, P. Ning, and C. Wang. Fault-tolerant cluster-wise clock synchronization for wireless sensor networks. *IEEE Transactions on Dependable and Secure Computing (TDSC)*, 2(3):177–189, July–September 2005.
54. K. Sun, P. Ning, and C. Wang. Secure and resilient clock synchronization in wireless sensor networks. *IEEE Journal on Selected Areas in Communications*, 24(2), February 2006.
55. D. Tian and N. D. Georganas. A coverage-preserving node scheduling scheme for large wireless sensor networks. In *First ACM International Workshop on Wireless Sensor Networks and Applications WSNA02*, pages 32–41, September 2002.
56. W. Ye, J. Heidemann, and D. Estrin. An energy-efficient mac protocol for wireless sensor networks. In *Proceedings of IEEE INFOCOM 2002*, June 2002.
57. J. Zhao, R. Govindan, and D. Estrin. Computing aggregates for monitoring wireless sensor networks. In *Proceedings of the 1st International Workshop on Sensor Network Protocols and Applications*, May 2003.
58. S. Zhu, S. Setia, and S. Jajodia. LEAP: Efficient security mechanisms for large-scale distributed sensor networks. In *Proceedings of 10th ACM Conference on Computer and Communications Security (CCS'03)*, pages 62–72, October 2003.

Securing Timing Synchronization in Sensor Networks

Srdjan Čapkun[1], Saurabh Ganeriwal[2], Simon Han[2], and Mani Srivastava[2]

[1] Informatics and Mathematical Modelling Department, Technical University of Denmark (DTU), DK-2800 Lyngby, Denmark; sca@imm.dtu.dk
[2] Networked and Embedded Systems lab, 56-125B, EE-IV, University of California Los Angeles,{saurabh,simonhan,mbs}@ee.ucla.edu

1 Introduction

Time synchronization is critical in sensor networks at many layers of its design. It enables better duty-cycling of the radio, accurate and secure localization, beamforming and other collaborative signal processing. Examples of existing sensor network applications where precise time is needed include: measuring the time-of-flight of sound; distributing an acoustic beam forming array; forming a low-power TDMA radio schedule; integrating a time-series of proximity detections into a velocity estimate; suppressing redundant messages by recognizing duplicate detections of the same event by different sensors; ordered logging of events during system debugging; integrating multi sensor data; or coordinating on future action. Imagine the detrimental affect on the functionality of all these applications, if a malicious adversary is able to abuse the underlying time synchronization protocol. Nodes will have faulty estimates about the location of other nodes. Packets will be lost if the sleep-wakeup schedule of nodes do not intersect. This can further trigger unnecessary packet retransmissions if MAC layer acknowledgements are enabled. It will be trivial for adversaries to perform replay attacks. Collaborative data processing and signal processing techniques will be adversely affected.

Although time synchronization problem has been thoroughly studied in sensor networks [13] and there are several prototype implementations, RBS [4], TPSN [6], FTSP [10] that achieve microseconds accuracy, none of these protocols were designed to operate in adversarial settings. Realizing the inadequacy of existing time synchronization solutions, we develop several schemes for achieving secure time synchronization in sensor networks. Our approach involves integrating security mechanisms into existing protocols as well as developing new protocols from scratch. Our contributions are multi-fold.

First, we perform an in-depth security analysis of the sender-receiver synchronization protocols [4, 6]. We show that as sensor networks are deeply coupled with the physical world that they monitor, a malicious adversary can subvert the time synchronization protocol by exploiting weaknesses at the interface between the sensor network and the physical world. Examples of time synchronization protocols vulnerable to these attacks include TPSN [6], FTSP [10], LTS [15], Mini/Tiny Sync [12]. We note that the same attacks are feasible on receiver-receiver synchronization [6].

Second, we integrate security mechanisms into the basic approach of sender-receiver synchronization; we propose a protocol for secure pairwise time synchronization in sensor networks. We show that for a nominal overhead, our protocol can counter attacks from external attackers. We extend this basic scheme and we propose and analyze three protocols for secure pairwise synchronization over multiple hops: *opportunistic*, *direct* and *transitive*. Each of these protocols offers a different point of operation in the energy - security - accuracy subspace and their use depends on application needs.

Third, we propose protocols for secure group time synchronization. We show that these protocols can be resilient to attacks from external attacker as well as to attacks from a subset of compromised group nodes. Typical applications of secure group synchronization include object tracking, beamforming, intruder detection and fire monitoring. These applications will function accurately only if the synchronization error between any two nodes in a monitoring group is bounded within some pre-specified limits. In our protocol, this is achieved, in part, through data fusion, the process of transforming and merging individual sensor readings into a high-level sensing result.

Finally, we discuss how protocols for secure pairwise and group synchronization can be used to secure network-wide synchronization.

2 Synchronization in sensor networks

Approaches for synchronizing a pair of nodes can be broadly classified as sender-receiver [6] or receiver-receiver [4] synchronization. In the sender-receiver protocol, a single node synchronizes its clock to the clock of the reference node through bidirectional communication. Receiver-receiver synchronization enables simultaneous synchronization of clocks of a group of nodes to the clock of a reference node; this technique relies on the measurements of the differences in reception times of the signals sent by the reference node. Protocols for network-wide clock synchronization [4, 6] rely on these pairwise synchronization techniques to enable all network nodes to the establish the same time reference. This is typically achieved by the formation of multi-hop paths in the sensor network, through which all network sensors synchronize to the reference nodes (base stations). Another approach involves using diffusion techniques.

Synchronization through multi-hop paths is typically performed such that the first sensor on the path synchronizes to the base station, and all other nodes synchronize to their preceding neighbors on the path in an ascending order.

In this work, we focus on the sender-receiver protocol; this protocol is the basis for several synchronization schemes such as TPSN [4], LTS [15], Mini/Tiny Sync [12]. We consider protocols for instantaneous synchronization between sensor nodes; these protocols remove the offset error between a pair of nodes at the time of protocol execution.

2.1 Sensor node clock

Every sensor node maintains its own clock and this is the only notion of time that a node has. The clock is an ensemble of hardware and software components; it is essentially a timer that counts the oscillations of a quartz crystal running at a particular frequency. Let us represent this clock for node A by C_A. The difference in the clocks of two sensor nodes is referred as the offset error between them. There are three reasons for the nodes to be representing different times in their respective clocks - (1) The nodes might have been started at different times, (2) The quartz crystals at each of these nodes might be running at slightly different frequencies, causing the clock values to gradually diverge from each other (termed as the skew error), (3) The frequency of the clocks can change variably over time because of aging or ambient conditions such as temperature (termed as the drift error). These errors can be summarized as follows.

$$1 \text{ Offset } \delta = C_A(t) - C_B(t)$$
$$2 \text{ Skew } \eta = \frac{\partial C_A(t)}{\partial t} - \frac{\partial C_B(t)}{\partial t}$$
$$3 \text{ Drift } \lambda = \frac{\partial^2 C_A(t)}{\partial t^2} - \frac{\partial^2 C_B(t)}{\partial t^2}$$

2.2 Sender-receiver synchronization

Pairwise sender-receiver synchronization is performed by a handshake protocol between a pair of nodes. This protocol is executed in three steps as follows:

Pairwise Sender-receiver Synchronization
1 $A(T1) \rightarrow (T2)B : A, B, sync$
2 $B(T3) \rightarrow (T4)A : B, A, T2, T3, ack$
3 A calculates offset between the nodes

Here, T1, T4 represent the time measured by local clock of node A, C_A. Similarly T2, T3 represent the time measured by C_B. At time T1, A sends a synchronization pulse packet to B. Node B receives this packet at T2, where T2 is equal to T1+δ+d. Here, δ and d represents the offset between the two nodes and end-to-end delay respectively. At time T3, B sends back an acknowledgment packet. This packet contains the values of T2 and T3. Node A receives the packet at T4. Similarly, T4 is related to T3 as T4 = T3−δ+d. Node A can now calculate the clock offset and the end-to-end delay as:

$$\delta = \frac{(T2 - T1) - (T4 - T3)}{2}$$
$$d = \frac{(T2 - T1) + (T4 - T3)}{2} \tag{1}$$

Subsequently, A synchronizes its clock to B's clock.

3 Attacks on timing synchronization

In this section, we review attacks on sender-receiver synchronization protocols. First, we present our system and attacker models.

3.1 System model

Our system consists of a set of sensor nodes and a set of mutually synchronized (e.g., through GPS [7]) reference nodes (base stations). Sensor nodes and base stations communicate using radio transmissions. We assume that the radio link between neighboring devices is bidirectional. We further assume that all nodes have internal clocks and are able to measure time with certain precision (e.g. in the case of Mica2 motes, $\mu seconds$).

The network is operated by an authority. The authority controls the network membership and assigns a unique identity to each node. Each node is able to generate symmetric cryptographic keys and, more generally, to accomplish any task required to secure its communications. All network nodes can establish pairwise secret keys. This can be achieved by manually pre-loading all keys into the nodes in a network setup phase, by probabilistic key pre-distribution schemes [1, 5], or through an on-line key distribution center [8]. The node also share pairwise secret keys with base stations.

3.2 Attacker model

We adopt the following attacker model. We assume that the attacker controls the communication channel in a sense that it can eavesdrop messages, modify transmitted messages and schedule transmissions. The attacker can jam the transmission and in that way prevent the transmission of the information contained in the message. Finally, we assume M to be *computationally bounded*.

We distinguish two attacker models: internal and external. In the external attacker model, we assume that none of the nodes involved in the protocol are compromised. An external attacker thus cannot authenticate itself as an honest network node to other network nodes or to the central authority. An internal attacker controls one or more network nodes. We assume that when a node is compromised, its secret keys and other secrets that it shares with other nodes are known to the attacker; subsequently, compromised nodes can authenticate themselves to the authority and to other network nodes.

3.3 Attacks

We observe two types of attacks: internal and external. Internal attacks are those in which internal attackers report false clock references to their neighboring nodes. External attacks are those in which an (external) attacker manipulates the communication between pairs of trusted nodes and causes the nodes to desynchronize, or to remain unsynchronized even after a successful run of the synchronization protocol.

We now focus on external attacks. By a successful external attack we mean that, after the protocol execution by two mutually trusted nodes, these nodes calculate a faulty value of the offset δ but accept it as being correct.

There are several ways that the attacker can influence the calculation of the offset: (i) by modifying the values of T2 and T3 in transmission, (ii) by assuming the identity of one of the network nodes, and (iii) by delaying the transmission of the messages between the nodes and thus increasing the value of T2 (or T4). First two attacks can be easily prevented by traditional security primitives that protect the integrity and authenticity of the transmitted messages.

The third attack, which we call the *pulse-delay attack*, is more challenging to detect. Notably, T2 is measured as the time at which the initial synchronization pulse packet sent by A is received at B. If an attacker delays the time at which B receives the synchronization pulse, he will modify the computation of the offset at A. To delay the synchronization pulse, an attacker can simply jam the initial pulse, and replay it at some arbitrary time in the future (Figure 1). This attack cannot be prevented by the use of conventional cryptographic primitives. Here, we assume that the attacker can jam the communication between two nodes by transmitting signals which will disrupt packet reception at the receiver. By jamming, we consider stealth, disruptive jamming that cannot be detected at the receiver. Currently available sensor network platforms use 433MHz Chipcon1000, 2.4 GHz IEEE 802.15.4 compliant (Direct Sequence (DSSS)) or Bluetooth (Frequency Hopping (FHSS)). Even if DSSS and FHSS resist well to various types of jamming, because of their low transmitting RF power (1mW), these sensor platforms are vulnerable to broadband jamming (BBN). Recently, Xu et al. [17] showed that jamming attacks are indeed feasible against Mica2 motes, and that detecting these attacks requires significant resources. The attacker can perform a similar pulse-delay attack on the acknowledgment packet to modify T4.

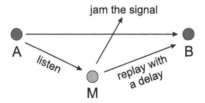

Fig. 1. Pulse Delay Attack.

Pulse-delay attack

Figure 1 shows the pulse-delay attack. We have already shown that we use two equations to derive the end-to-end delay and the clock offset between the sender and the receiver. These are: T2=T1+δ+d and T4=T3-δ+d. If an attacker performs a pulse-delay attack (e.g., on the initial sync packet), the equations will change to:

T2* =T1+δ+d+Δ and T4=T3-δ+d, where T2*=T2+Δ and Δ is the pulse-delay introduced by the attacker. The clock offset and the end-to-end delay then become:

$$\delta = \frac{(T2 - T1) - (T4 - T3) + \Delta}{2}$$
$$d = \frac{(T2 - T1) + (T4 - T3) + \Delta}{2} \tag{2}$$

This shows that the attacker can, by varying the pulse-delay, arbitrary change the computed clock offset. An important observation is that in the process of carrying out a pulse-delay attack, the attacker also changes the computed end-to-end delay. Note that the attacker has no control over the calculation of the end-to-end delay. As we will show in later sections, we use this observation to provide a mechanism for detecting pulse-delay attacks.

4 Secure Pairwise Time Synchronization

In this section, we present several schemes for secure sender-receiver synchronization, over single-hop and multi-hop communication links. We first describe the protocol for Secure Pairwise Synchronization (SPS) over a single-hop communication link.

4.1 Secure Pairwise Synchronization (SPS)

We propose the following protocol:

Secure Pairwise Synchronization (SPS)
1 $A(T1) \rightarrow (T2)B : A, B, N_A, sync$
2 $B(T3) \rightarrow (T4)A : B, A, N_A, T2, T3, ack, MAC_{K_{AB}}[B, A, N_A, T2, T3, ack]$
3 A calculates delay $d = \frac{(T2-T1)+(T4-T3)}{2}$
If $d \leq d^*$ then $\delta = \frac{(T2-T1)-(T4-T3)}{2}$
else abort

In this protocol, message integrity and authenticity are ensured through the use of Message Authentication Codes (MAC), and of a key K_{AB} shared between A and B. This prevents external attackers from successfully modifying any values in the synchronization pulse or the acknowledgment packet. Furthermore, the attacker cannot assume an identity of node B as it does not hold the secret key K_{AB}. Replay attacks are avoided by using a random nonce, N_A, during the handshake. Potentially more harmful attacks are pulse-delay attacks. In our protocol, these attacks are detected through a comparison of the computed message end-to-end delay d, with the maximal expected message delay d^*. Note that the calculation of the end-to-end delay d comes as an auxiliary benefit of the protocol. We have added no extra overhead on the functionality of sender-receiver synchronization. If the computed delay is greater than the maximal expected delay, we abort the offset calculation.

4.2 Performance evaluation

Clearly, how much the attacker can influence the synchronization relies on our estimate of the maximal delay d* and in order to make a judicious choice let us first analyze the end-to-end delay, d, in the absence of any external attackers. The three significant contributors to the end-to-end delay are:

1. Waiting time at the medium access control (MAC) layer to access the channel: This delay can be completely random varying from a few microseconds to a few minutes.
2. Time taken in transmitting the packet bit-by-bit at the radio of the sender node: This time will be in hundreds of microseconds but is deterministic in nature. It depends on the packet size and the radio speed.
3. Propagation time over the wireless link between the sender and receiver node: This is only a few nanoseconds.

Time stamping the packets below the MAC layer is feasible on typical sensor networking platforms [6], which removes the most significant variable factor, MAC access waiting time. The only other variable component is the propagation time over the wireless link. The transmission delay only depends on the size of the packet which can be fixed. Note that although the relevant contents in the sync and ack packets are of different lengths, we add some redundant bits to each packet to ensure that they are of the same length. This ensures that the transmission delay is symmetric for the two packet exchanges. Since the transmission delay is many orders of magnitude larger than the end-to-end delay, a stable value of the end-to-end delay can be calculated.

Measurements on Mica motes

Timing-sync protocol for sensor networks (TPSN), proposed in [6], is one of the most popular approaches for time synchronization in sensor networks that is based on sender-receiver synchronization. This protocol uses MAC layer time stamping to achieve an accuracy of around $10\mu s$ for Mica2 motes. We needed to incorporate the cryptographic functionality in this protocol. Specifically, as the packets are time stamped below the MAC layer, we needed a cryptographic library that can calculate the Message Authentication Code (MAC), on-the-fly, as the packets are being transmitted. TinySec [9], a symmetric cryptographic library on motes, enables this MAC calculation. Essentially, our prototype implementation of SPS integrates the time stamping library provided by TPSN with TinySec and performs a thresholding on the computed delay at the end of the two-way packet exchange.

We ran this prototype implementation of SPS on 5 different pair of motes. For every pair we ran multiple independent runs to calculate the value of the computed delay (d in equation (1)) from the protocol. In all, we computed the delay for 1000 independent runs. The first plot in Figure 2 shows the actual delay measured in every run and the second plot shows the corresponding histogram of the measured delay. Table 1 summarizes the statistics of the measurements.

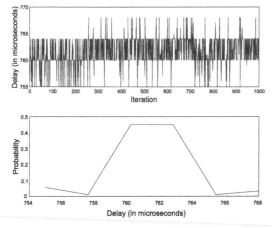

Fig. 2. End-to-end delay over a single link.

Maximum delay	Minimum delay	Average delay d_{avg}	Standard deviation σ
$768\mu s$	$755\mu s$	$762\mu s$	$2.82\mu s$

Table 1. Statistics of the end-to-end delay over a link.

The histogram of the computed delay closely resembles a Gaussian distribution. This is consistent with the results reported by authors in [4, 6]. Since the end-to-end delay follows a Gaussian distribution $d \sim N(d_{avg}, \sigma)$, the true delay will fall in the interval $[d_{avg} - 3\sigma, d_{avg} + 3\sigma]$, with 99.97% confidence. To account for this, the maximal expected delay is set to $d_avg + 3\sigma$, i.e., $d* = d_{avg} + 3\sigma = 762 + 3*2.82 \approx 771\mu s$.

Minimum synchronization precision

If the end-to-end delay was a constant, we should have been able to synchronize the nodes perfectly (zero error). The delay variation introduces synchronization error. Intuitively, the minimum synchronization precision (maximum error) will occur when the end-to-end delay variations in the two directions (from A to B and from B to A) are maximal, i.e., in one direction d is equal to $d_{avg} - 3\sigma$ and in the other direction d is equal to $d_{avg} + 3\sigma$. Using equations (1) and (2), the synchronization error for this worst-case scenario will be 3σ (around $10\mu s$).

Maximum attacker impact

We define the maximum attacker impact to be the maximum difference between the clocks of two nodes that the attacker can cause without getting detected. In the worst case, the actual end-to-end delay is equal to the minimum, $d = d_{avg} - 3\sigma$. In this case, the attacker can introduce a maximum pulse-delay factor of $\Delta = 12\sigma$. Node A will calculate the end-to-end delay, d, as:

$$d = d_{avg} - 3\sigma + (12\sigma/2) = d_{avg} + 3\sigma \le d^* \tag{3}$$

As the computed end-to-end delay is equal to the maximal delay, the thresholding verification will pass. Using equation (2), the calculated offset will be off by $\Delta/2$. Thus, the maximum attacker impact is 6σ (around $20\mu s$).

Discussion

In the absence of any malicious behavior, SPS is able to achieve the same accuracy as TPSN. The extra computation overhead of MAC calculation does not impact the accuracy of time synchronization as we are able to do on-the-fly MAC calculation. This might not be feasible on emerging new 802.15.4 compliant platforms such as Micaz, Telos etc., where the radio speeds are higher than Mica2 motes by an order of magnitude. However, 802.15.4 also mandates the implementation of the cryptographic library (equivalent of TinySec) in hardware, thereby providing a faster and different interface. This opens up an interesting design challenge. The implementation of SPS for this new class of platforms forms the part of our future work.

4.3 Multihop Synchronization

So far, we have assumed that the two nodes that need to synchronize are in each-other's power range and thus can directly communicate. In this section, we propose and analyze three protocols for secure sender-receiver synchronization over multiple hops: opportunistic, direct and transitive. We assume that two sensors can obtain sets of communication paths between them, either through a priori knowing network topology, through topology discovery [2], or through routing information [3]. We present the protocols through a representative example: A - C - D - B. In this example, node A wants to synchronize to node B which is not in its direct communication range. In fact the shortest path between these two nodes comprises of three hops, going through nodes C and D.

Secure Opportunistic Multi-hop synchronization (SOM)

The SOM protocol is executed as follows:

Secure Opportunistic Multi-hop synchronization (SOM)
1 $A(T1) \to C \to D \to (T2)B : A, B, N_A, sync$
2 $B(T3) \to D \to C \to (T4)A :$
$B, A, N_A, T2, T3, ack, MAC_{K_{AB}}[B, A, N_A, T2, T3, ack]$
3 A calculates delay $d = \frac{(T2-T1)+(T4-T3)}{2}$
$\quad$ If $d \le d_M^*$ then $\delta = \frac{(T2-T1)-(T4-T3)}{2}$
$\quad$ *else* abort

This protocol is essentially the same as the protocol used for one-hop synchronization, with a difference that in this protocol we assume that there are several forwarding nodes between the sender and the receiver. Note that this protocol assumes that there exist a secret key between A and B, K_{AB}, which are multiple hops apart. The expected end-to-end delay d, computed at A will be much longer in this protocol than in the one-hop case. We take this into account by estimating a maximum expected (and allowed) end-to-end delay by a larger value $d_M^* \rangle\rangle d^*$. Like in the one-hop time synchronization protocol, the variance of d_M^* will determine how fine-grained synchronization is, and will upper-bound attacker's possible impact on the synchronization precision. The end-to-end delay in SOM is equal to the cumulative sum of the transmission delays between each pair of nodes on the path and the MAC access delays of the forwarding nodes. Here, the processing delays at nodes can be neglected, as they are two-three orders of magnitude smaller than transmission and MAC access delays. Although the transmission delays can be accurately predicted from radio speed, MAC access times can be very unpredictable and can range from microseconds to a few minutes depending on the condition of the channel. Furthermore, channel access times also depend on the intensity of the communication between the nodes in the network, which is application-specific.

We performed an empirical study to find out the delay variations over multiple hops in a typical sensor network. The results were obtained in simulations using the Avrora simulator [14]. Avrora provides a cycle-accurate simulation of the AVR microcontroller, allowing real programs to be run with precise timing. We use the same TinyOS code - network stack, time stamping, etc. as in the previous section to obtain these results. Table 2 summarizes the results for varying number of hops between the sender and receiver. The statistics were obtained from 1000 independent runs.

Hop distance	Maximum delay (μs)	Minimum delay (μs)	Average delay d_{avg}^M (μs)	Standard deviation σ_M (μs)
2	32094	18761	25120	3861
3	62926	37510	49940	5450
4	92509	56260	74781	6738
5	120841	76259	99667	7827
6	149174	97092	124393	8841

Table 2. Statistics of the end-to-end delay over multiple hops.

It can be observed from Table 1 and 2 that the average delay and the standard deviation for the multihop case are significantly higher, by three orders of magnitude, than the corresponding numbers over a link. We observed that the end-to-end delay over multiple hops also follows a Gaussian distribution. Given this, the same analysis can be worked out for SOM as in SPS. The maximal delay should be set to $d_M^* = d_{avg}^M + 3\sigma_M$. The minimum synchronization precision and the maximum attacker impact will be given by $3\sigma_M$ and $6\sigma_M$ respectively.

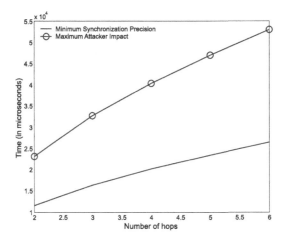

Fig. 3. Secure Opportunistic Multi-hop Synchronization.

Figure 3 plots these two quantities for varying number of hops. Note that the values are significantly higher, by three orders of magnitude, than the SPS protocol. We would like to point out that the values of d_{avg}^{M} and σ_M will change with changes in the network topology and the underlying communication traffic. The objective of this empirical study is to give representative numbers so that a meaningful relative comparison can be made with the other protocols. Even if SOM protocol provides very low synchronization accuracy, it can provide protection against external as well as internal attackers. This is because the intermediate nodes in SOM do not perform any processing on the packet. They simply receive the packet and forward it to the next hop node. Thereby, even if node D has been compromised and the attacker tries to modify the value of T2, T3 at node D, it will result in a MAC verification fail at node A. In the following sections, we propose two protocols - Secure Direct Multi-hop Synchronization (SDM) and Secure Transitive Multi-hop Synchronization (STM) that provide much better accuracy than SOM. However, both of them assume that intermediate nodes (C and D in the example) are trustworthy and hence, are not resilient to attacks from compromised nodes.

Secure Direct Multi-hop Synchronization (SDM)

The SDM protocol is executed as follows:

Secure Direct Multi-hop Synchronisation (SDM)

1 $A(T1) \rightarrow (T2)C(T3) \rightarrow (T4)D(T5) \rightarrow (T6)B : A, B, N_A, sync$

2 $B : m1 = (B, A, T6, T7, ack)$

   : $M_1 = MAC_{K_{BD}}[B, D, N_A, m1]$

  $B(T7) \rightarrow (T8)D : B, D, N_A, m1, M_1$

3 $D : m2 = B, D, A, T4, T9, (T6 - T5), (T8 - T7), ack$

   : $M_2 = MAC_{K_{DC}}[D, C, N_A, m2]$

  $D(T9) \rightarrow (T10)C : D, C, N_A, m2, M_2$

4 $C : m3 = B, D, C, A, T2, T11, (T4 - T3), (T10 - T9), (T6 - T5), (T8 - T7), ack$

   : $M_3 = MAC_{K_{CA}}[C, A, N_A, m3]$

  $C(T11) \rightarrow (T12)A : C, A, N_A, m3, M_3$

5 A calculates delay $d = \frac{[(T2-T1)+(T4-T3)+(T6-T5)]+[(T8-T7)+(T10-T9)+(T12-T11)]}{}$

  If $d \leq d_T^*$ then $\delta = \frac{[(T2-T1)+(T4-T3)+(T6-T5)]-[(T8-T7)+(T10-T9)+(T12-T11)]}{2}$

  else abort

Unlike SOM, both SDM and STM (proposed in the next section) require only neighboring nodes (A and C, C and D, D and B) to share pairwise secret keys. There is no need for A and B to share a secret key between them. To analyze this protocol, we first observe the communication between a pair of nodes, A and C, over the course of the protocol: $A(T1) \rightarrow (T2)C; C(T11) \rightarrow (T12)A$. These four timestamps are interrelated as follows:

$$T2 = T1 + \delta_{AC} + d_{AC}; \ T12 = T11 - \delta_{AC} + d_{AC} \tag{4}$$

Note that we have added subscripts to both the offset and the end-to-end delay for clarity. Similarly, if you observe the communication between the other two node-pairs, (C,D) and (D,B), the following relationships can be derived:

$$T4 = T3 + \delta_{CD} + d_{CD}; \ T10 = T9 - \delta_{CD} + d_{CD} \tag{5}$$

$$T6 = T5 + \delta_{DB} + d_{DB}; \ T8 = T7 - \delta_{DB} + d_{DB} \tag{6}$$

Adding the left hand and right hand side of equations (5) and (6), and introducing two new terms, δ_{AB} and d_{AB}, we can get the following two equations:

Here, $\delta_{AB} = \delta_{AC} + \delta_{CD} + \delta_{DB}$ and $d_{AB} = d_{AC} + d_{CD} + d_{DB}$. The clock offset and the end-to-end delay can be calculated from equations (7) and (8). Just like SOM, the thresholding technique in SDM also requires the estimation of a new maximal delay d_T^*. However, an important and crucial difference is that the end-to-end delay in SDM is not corrupted by the MAC access delays; all the time stamps from T1 to T12 are below the MAC layer. In fact, the end-to-end delay can be easily estimated knowing the number of hops. In the above example the end-to-end delay is equal to the cumulative sum of d_{AC}, d_{CD} and d_{DB}. All these three delays, d_{AC}, d_{CD} and d_{DB}, are equal to the end-to-end delay over a single link and hence, all of them follow the same Gaussian distribution calculated in Section 4.1, i.e., $d_{AC} = d_{CD} = d_{DB} \sim N(d_{avg}, \sigma)$. As a result, the end-to-end delay for SDM becomes a cumulative sum of independent Gaussian variables, i.e., $d_{AB} \sim N(nd_{avg}, \sigma\sqrt{n})$, where n is the number of hops. Given this, the maximal delay for an n-hop network should be set to $d_T^* = nd_{avg} + 3\sigma\sqrt{n}$. The minimum synchronization precision and the maximum attacker impact will be given by $3\sigma\sqrt{n}$ and $6\sigma\sqrt{n}$ respectively.

Note that σ is three orders of magnitude lower than M for any value of n and hence, the minimum synchronization precision as well as the maximum attacker impact in SDM is much lower, around three orders of magnitude, than SOM. However, the packet size of ack packets in SDM is larger than SOM resulting in an increased overhead. Every ack packet in SDM has to carry the state information (timestamps) about all the previous packets with it. For example, the acknowledgment packet from C to A contains 10 timestamps. In SOM, every acknowledgment packet has to just contain 2 timestamps, the same as SPS.

Secure Transitive Multi-hop Synchronization (STM)

Secure transitive synchronization scheme is essentially performed as SPS synchronization along all the links in the path from the source to the destination. The STM protocol is executed as follows:

Secure Transitive Multihop Synchronization (STM)

$1\ A \to C \to D \to B:\ A, B, NA, sync$

$2\ B: m1 = (B, D, notify)$

$\quad: M_1 = MAC_{K_{BD}}[B, D, N_A, m1]$

$\quad B \to D: B, D, N_A, m1, M_1$

$3\ D$ sync to B (SPS)

$\quad D: m2 = (B, D, C, notify)$

$\quad: M_2 = MAC_{K_{DC}}[D, C, N_A, m2]$

$\quad D \to C: D, C, N_A, m2, M_2$

$4\ C$ sync to D

$\quad C: m3 = (B, D, C, A, notify)$

$\quad: M_3 = MAC_{K_{CA}}[C, A, N_A, m3]$

$\quad C \to A: C, A, N_A, m3, M_3$

$5\ A$ sync to C

Here, the proper scheduling of node synchronization is achieved in this protocol through an explicit notification by the receiver node to its upstream neighbor (sender node). Authentication is achieved by attaching a MAC at the end of this notification packet. Unlike SOM or SDM, STM does not require the estimation of any new maximal delay parameter. It runs the SPS protocol on every link and hence, the threshold verification gets divided into stages. Every link is evaluated separately using the same maximal delay, d^*, as in Section 4.1. This local verification has both advantages and disadvantages. Imagine an external attacker which can carry out arbitrary pulse-delay attacks (e.g., on the link joining C and D). In both SOM and SDM, the thresholding verification is done only when the acknowledgment reaches back to node A. Thus, the system will detect the malicious attack only after running the complete protocol, which in our example pertains to transmission and reception of two packets per node. In STM, only nodes C and D will need to resynchronize when the thresholding verification at node C will fail. In fact, other nodes A and B won't even come to know about this malicious attack. Thereby, the overhead of countering a malicious attack in STM is lower than both SOM and SDM. However, local verification gives extra freedom to the external attacker. It can introduce multiple pulse-delay attacks on every link simultaneously, thereby; the cumulative sum result of these attacks can be huge. Note that the pulse-delay factor introduced at every link can be at most 12σ as we run SPS on every link (refer Section 4.1). Thereby, the maximum attacker impact for an n hop network on STM can be $6\sigma n$, which is higher than the corresponding number for SDM, $6\sigma\sqrt{n}$. We note that both STM and SDM will achieve the same minimum synchronization precision, $3\sigma\sqrt{n}$. Also, STM requires one extra packet transfer per node - the notification packet. In the absence of any malicious behavior, SOM and SDM require a total of 2n packet transmissions for synchronizing two nodes that are n hops away. STM requires 3n packet transmissions.

5 Group Synchronization

Several sensor network applications require all the nodes in a group to be time synchronized with each other. A few notable ones are - (1) Object tracking: The size, shape, direction, location, or velocity of objects is determined by fusing proximity detections, done at the same time, from sensors at different locations, (2) Consistent state updates: The current state of an object is most accurately determined by the node that has seen the object most recently. This requires all the nodes in the cluster to have the same notion of time, (3) Duplicate detection: The time of an event helps nodes in the cluster determine if they are seeing two distinct real-world events, or a single event seen from two vantage points. If they are indeed seeing the same event, they can further fuse their observations to get much meaningful information about the event. All these applications will function accurately only if the synchronization error between nodes in a group is bounded. In the following sections, we propose two group synchronization protocols - Lightweight Secure Group Synchronization (L-SGS) and Secure Group Synchronization (SGS). These protocols differ in their resilience to attacks and in the number of messages that the group nodes mutually exchange. Just like secure pairwise synchronization protocols, both L-SGS and SGS can detect the pulse-delay and packet modification attacks from external attackers. In addition, SGS can also detect the false timing reports by internal attackers but at the cost of a higher communication overhead.

5.1 System Model

In this model, we assume that group membership is known to all group nodes and that group members can authenticate each other using pairwise secret keys. We further assume that all group nodes reside in each other's power ranges; we do show that the proposed solutions can be easily extended to groups in which nodes are multiple hops apart. Given this assumption, we propose L-SGS and SGS exploiting the broadcast property of the wireless communication medium. Thereby, a message broadcast by a node in the group can be received by all the nodes in the cluster (here we assume that sensors have omnidirectional antennas). We represent the sending time of the packet at node i by T_i. T_{ij} represents the time at which the packet broadcasted by node i is received at j. Notice that, these times are measured by two different clocks. T_i is measured in the local clock of node i (C_i) whereas T_{ij} is measured by the local clock of node j (C_j). We represent the offset (or the difference between the local clocks) between the two nodes by δ_{ij}. The delay for the packet transfer from i to j is represented by d_{ij}. Although this delay is the same for every pair of nodes, we add the subscripts for clarity.

5.2 Lightweight Secure Group Synchronization (L-SGS)

L-SGS is executed as follows:

Lightweight Secure Group Synchronization (L-SGS)
1 $G_1 \rightarrow * : G_1, sync$
2 $G_i(T_i) \rightarrow (T_{i1})G_1 : G_i, N_i$
3 $G_1(T_1) : m = T_{i1}, N_i, G_i{}^{i=2,\dots,N}$
$\quad : M = MAC_{K_{1i}}[G_1, T_1, ack, T_{i1}, N_i, G_i]^{i=2,\dots,N}$
$\quad G_1(T_1) \rightarrow (T_{1i})* : G_1, T_1, ack, m, M$
4 G_i : compute $d = ((T_{i1} - T_i) + (T_{1i} - T_1))/2$
$\quad$ If $d \leq d^*$ then $\delta = ((T_{i1} - T_i) - (T_{1i} - T_1))/2$
$\quad$ *else* abort

In this protocol, one of the group member nodes (G_1) initiates time synchronization (step 1) to which the rest of the group members reply with messages containing their ids and challenge nonces (step 2). In step 3 of the protocol, G_1 replies with a single broadcast message to all other group nodes, containing MACs of the challenges and node ids. Note that the last protocol message (step 3) contains N-1 triples T_{i1}, N_i, G_i, one for each G_i, containing the receipt time of the challenge packet from G_i (T_{i1}), the nonce of G_i and the node-id of G_i respectively. It also contains N-1 MACs, one for each (G_1, G_i) pair, which enables each node G_i to authenticate the packet broadcasted by G_1. In the last protocol step $G_2, ..., G_N$ synchronize to G_1. Note that this protocol does not need to be preceded by any explicit leader election algorithm. The initiating node G1 does not need to be elected by other group members, but it can simply be the node that first broadcasts the sync packet after sensing a particular event. Therefore, L-SGS does not require any explicit scheduling in the group.

Performance Evaluation

We can observe L-SGS as an extension of the SPS protocol. In SPS, a single receiver synchronizes to a single sender; in L-SGS, multiple receivers synchronize to a single sender. L-SGS makes use of the broadcast property of the wireless channel. The total number of messages transmitted in an n-node cluster over the course of L-SGS is n+1. We note, however, that the last protocol message is significantly larger than other messages. As L-SGS relies on the same primitives as SPS, the resistance of L-SGS to external attacks is the same as with SPS, which means that L-SGS is resilient to pulse-delay and message modification attacks. The minimum synchronization precision and the maximum attacker impact of L-SGS is the same as SPS, equal to 10s (3σ) and 20s (6σ) respectively. However, L-SGS is not resilient to internal attacks. If node G_1 is malicious, it can produce a set of messages containing false times T_1 and T_{i1} for each node i and therefore have these nodes mutually desynchronized, while these nodes would believe to be synchronized.

Implementation

Step 3 of L-SGS requires G_1 to calculate N-1 MACs on-the-fly. Clearly, this won't be feasible for large values of N. Given this, we briefly mention some of the alternative design options that we plan to explore in detail in the future. First, the broadcast message in Step 3 can be replaced by multiple unicast messages. This will incur a significant communication overhead on G_1. Second, we can use public key signatures [16], which would make sure that a single signature (MAC) by node G_1 can be verified by all group nodes; this would, however, incur a significant computational cost at G_1. Instead we can achieve the same result by using a symmetric group key to generate a single MAC; albeit this will come at the cost of establishing the symmetric group key, shared by every node, in the cluster.

5.3 Secure Group Synchronization (SGS)

In this section, we propose a secure group synchronization protocol that also provides resiliency to internal adversaries; however, as we show, this protocol requires higher communication overhead than L-SGS. We start the section by proposing a simple consistency check that is the basic building block of the protocol.

Triangle consistency

Consider a set of 3 nodes, [i, j, k], that have successfully establish pairwise offsets with each other. Imagine a triangle connecting these three nodes such that the link weight on every edge is equal to the offset between them. We refer to the traversal of such a triangle starting and ending at the same node as making a cycle. As we traverse in a cycle, we keep on accumulating the link weight. For example, a valid cycle starting from node i will be $[i \rightarrow j, j \rightarrow k, k \rightarrow i]$. The final accumulated weight at the end of the cycle will be $\delta_{ij} + \delta_{jk} + \delta_{ki}$. It can be easily observed that this should be equal to 0. Corollary: If the result of any cycle is not 0, there exists an internal adversary in the set. We note that the result of a cycle might not be exactly zero because of the drift and skew error. In practice, we will use a simple threshold based policy to account for this.

Protocol

In this protocol, every group member (G_i) broadcasts a packet containing their ids and challenge nonces (step 1). In step 2 of the protocol, every member (G_i) broadcasts another packet containing the response to the challenges issued by all the other nodes in the first step. This packet contains N-1 triples T_{ji}, N_j, G_j, one for each G_j, containing the receipt time of the challenge packet from G_j (T_{ji}), the nonce of G_j and the node-id of G_j respectively. It also contains N-1 MACs, one for each (G_i, G_j) pair, which enables each receiver node G_j to authenticate the packet broadcasted by G_i in the second step. Note that the sending node Gi also includes the sending time (T_i') in the response packet. The first two steps are reminiscent of sender-receiver synchronization; we are now establishing pairwise relationships between multiple senders and multiple receivers simultaneously using the broadcast property of the wireless communication medium. In the next step, each node Gi performs threshold verification on all the computed delays, d_{ij}, corresponding to the challenge-response with node G_j. This provides the resiliency to pulse-delay modifications from external attackers. If this step is successful, G_i will be able to construct a set O_i, containing clock offsets with other nodes in the group. Step 4 and 5 provide the resiliency to SGS against internal attackers. As mentioned earlier, a simple triangle consistency check can be used to detect an internal attacker in a set of three nodes. Imagine a set of 3 nodes $[G_i, G_j, G_k]$. After step 3, G_i has the offsets with both G_j and G_k but it does not know the offset between G_j and G_k to perform the triangle consistency check. Step 4 of the protocol provides G_i with this information. In this step every node G_i broadcasts its offset set, O_i. This packet also contains N-1 MACs, one corresponding to every other node in the group, so that the contents of the packet can be authenticated at every receiver node. Following this, each node performs multiple triangle consistency checks. If this step is successful, G_i increments its clock by the largest offset, $\max 0, \max O_i$. As a result, G_i (and all the other nodes in the group) will get implicitly synchronized to the fastest clock in the group. SGS is executed as follows:

Secure Group Synchronisation (SGS)

1 $G_i(T_i) \rightarrow (T_{ij})* : G_i, N_i, sync; (i = 1, ..., N), j \neq i$

2 $G_i(T_i') : m = T_{ji}, N_j, G_j^{j=1,...,N;j\neq i}$

    $: M = MAC_{K_{ij}}[G_i, T_i', ack, T_{ji}, N_j, G_j]^{j=1,...,N;j\neq i}$

    $G_i(T_i') \rightarrow (T_{ij}')* : G_i, T_i', ack, m, M; (i = 1, ..., N), j \neq i$

3 $G_i :$ compute $d_{ij} = ((T_{ij} - T_i) + (T_{ji}' - T_j'))/2; j = 1, ..., N; j \neq i$

    if all $d_{ij} = d^*$

    then $O_i = \delta_{ij} = ((T_{ij} - T_i) - (T_{ji}' - T_j'))/2^{j=1,...,N;j\neq i}$

    else abort 4 $G_i : M = MAC_{K_{ij}}[G_i, O_i]^{j=1,...,N;j\neq i}$

    $G_i \rightarrow * : G_i, O_i, M; (i = 1, ..., N), j \neq i$

5 $G_i :$ check triangle consistencies

    if all triangles are consistent, synchronize to the fastest clock

Performance Evaluation

The number of messages transmitted in an n-node cluster over the course of SGS is 3n. SGS can also be viewed as an extension of the SPS protocol, whereby multiple nodes establish the pairwise clock relationships simultaneously. We note that the ack packet size in SGS (step 2) is significantly larger than other messages. Moreover, the computation of N-1 on-the-fly MACs opens up the same design challenges as in L-SGS. As SGS is based on the same primitives as SPS, the minimum synchronization precision and the maximum attacker impact of SGS is the same as L-SGS and SPS, equal to $10\mu s$ (3σ) and $20\mu s$ (6σ) respectively. Just like L-SGS and SPS, it can counter external pulse-delay and message modification attacks. However, in addition SGS is resilient to internal attackers.

Resiliency to internal adversaries

In SGS, there is no fixed node to which the nodes in the group will synchronize. Synchronization is implicitly done to the fastest clock in the cluster. The identity of the node with the fastest clock does not get revealed till the end of the protocol. This removes the vulnerability of the protocol to a single point of failure such as G_1 in L-SGS. Furthermore, if an internal attacker tries to create inconsistencies in the running of the protocol, the triangle consistency check will fail and the process will be aborted. Let us illustrate this by an example. Consider nodes 1-7 lying in the neighborhood of each other. Without any loss of generality, let us assume that node 1 is malicious and node 7 is having the fastest clock. Node 1 can easily project itself as the fastest clock by creating a fictitious clock. At the end of the protocol, all the nodes will be synchronized to node 1 rather than node 7. Note that this is not a valid attack. Although all the nodes are synchronized to some fictitious clock, they are consistently wrong. The relative error between them is still bounded. Having realized this, the aim of node 1 will be to introduce inconsistencies in the system by portraying itself as the fastest clock to only a few subsets of nodes. Node 1 has full control over the values of the receipt times, T_{j1} for j= 2, 3, .., 7, that it reports in the response packet (step 2). It can report arbitrary values creating inconsistent offset values at different nodes. We note that the thresholding verification in step 3 will provide some resiliency against arbitrary values reported by node 1 but it alone cannot guarantee safeguard against all possible malicious attacks. For example, imagine the scenario where node 1 just targets node 2. It reports a faulty value of the sending time T_1' from the fictitious clock. It also reports the value of T_{21} from the same fictitious clock but reports an arbitrary value of receipt times for rest of the nodes, T_{j1} for j=3, .., 7. As a result, the thresholding mechanism on d_{j1} will fail at nodes j=3,.., 7 and hence, they will abort SGS. The thresholding mechanism will pass at node 2 and it will update its clock so that it synchronizes to the fictitious clock by node 1. Thereby, in absence of steps 4 and 5, node 2 will be out of sync with rest of the nodes in the group. In SGS, when nodes exchange their offset sets, nodes 3-7 will not have valid pairwise offsets with node 1. As a result, the triangle consistency check will fail at node 2. Thereby, node 2 will also detect the presence of malicious behavior in the system and will also abort SGS.

Extension to multiple hops

SGS is extendible to the scenario when group nodes are multiple hops away from each other. In this case, the steps 1 to 3 of the protocol will have to be changed. Currently we use the broadcast property of the wireless communication medium to establish pairwise offsets between the nodes in the group. If group nodes are multiple hops away, we would have to use any of the three protocols proposed in Section 4 for establishing these pairwise offsets. Steps 4 and 5 of the protocol will remain unchanged.

6 Discussion: Network-wide Synchronization

Network-wide synchronization is achieved in two stages - (1) Hierarchical tree is established in the network with a reference node as the root, (2) Pairwise synchronization is performed along the edges of this tree using SPS. Every node synchronizes its clock to its parent in the tree. As a result, eventually all nodes in the network get synchronized to the reference node, which is the root of the hierarchical tree. Using SPS to perform pairwise synchronization provides resiliency against external attackers. However, the more challenging problem is to achieve secure network-wide synchronization when a few nodes in the network have been compromised. In this scenario, a compromised node can mislead all the nodes in its sub-tree to a different notion of time than the rest of the network. In general, the problem of network-wide synchronization can be viewed as a composition of several multihop synchronizations between the reference node and rest of the nodes in the network. However the solutions for secure multihop synchronization rely on the assumption that all the intermediate nodes are trustworthy. Clearly, even one compromised node in the path can bring detrimental effects to the functionality of SDM or STM.

A possible solution for countering compromised nodes is to use redundancy by using disjoint multiple paths to synchronize nodes. To get an idea of how protocols can resist to internal attackers through independent routes, we refer the reader to a rich literature on security of routing protocols in multi-hop wireless networks [8, 11, 18]. We note that on a network-level this means that the reference node needs to maintain multiple trees simultaneously.

7 Conclusions

Existing solutions for time synchronization in sensor networks are not resilient to malicious behavior from external attackers or internally compromised nodes. We showcase the feasibility of a pulse-delay attack, whereby an attacker can introduce arbitrary long delays in the packet propagation time directly affecting the achieved synchronization precision. We then propose a suite of protocols for secure pairwise and group synchronization of nodes that lie in each others' power ranges and of nodes that are separated by multiple hops a part. Table 3 summarizes the key aspects

	Neighboring nodes	Multi-hop sync. (examples for $n = 5$)			Group sync. of n nodes	
Protocols	SPS	SOM	SDM	STM	L-SGS	SGS
Synchronization precision	3σ ($\approx 10\mu s$)	$3\sigma_M$ ($\approx 25ms$)	$3\sigma\sqrt{n}$ ($\approx 25\mu s$)	$3\sigma\sqrt{n}$ ($\approx 25\mu s$)	3σ ($\approx 10\mu s$)	3σ ($\approx 10\mu s$)
Max. attacker impact	6σ ($\approx 20\mu s$)	$6\sigma_M$ ($\approx 50ms$)	$6\sigma\sqrt{n}$ ($\approx 50\mu s$)	$6\sigma\sqrt{n}$ ($\approx 50\mu s$)	6σ ($\approx 20\mu s$)	6σ ($\approx 20\mu s$)
Resiliency to internal attackers	-	Yes	No	No	No	Yes
Transmitted messages	2	$2n$	$2n$	$3n$	$n + 1$	$3n$
Ack Packet size[3]	-	Same	Large	Same	Large	Large

Table 3. Summary of Secure Time Synchronization Protocols

of these protocols. These protocols offer different points of operation in the energy-accuracy subspace and the choice of the specific protocol should be made by the network designer depending on his application needs.

References

1. H. Chan, A. Perrig, and D. Song, "Random Key Predistribution Schemes for Sensor Networks," in *Proceedings of the IEEE Symposium on Research in Security and Privacy.* IEEE Computer Society, May 2003, p. 197.
2. B. Deb, S. Bhatnagar, and B. Nath, "A topology discovery algorithm for sensor networks with applications to network management," in *Proceedings of the IEEE CAS Workshop*, September 2002.
3. J. Deng, R. Han, and S. Mishra, "A performance evaluation of intrusion-tolerant routing in wireless sensor networks," in *Proceedings of the 2nd IEEE International Workshop on Information Processing in Sensor Networks*, April 2003.
4. J. Elson, L. Girod, and D. Estrin, "Fine-grained network time synchronization using reference broadcasts," *SIGOPS Operating System Review*, vol. 36, no. SI, pp. 147–163, 2002.
5. L. Eschenauer and V. D. Gligor, "A key-management scheme for distributed sensor networks," in *Proceedings of the ACM Conference on Computer and Communications Security.* ACM Press, 2002, pp. 41–47.
6. S. Ganeriwal, R. Kumar, and M. B. Srivastava, "Timing-sync protocol for sensor networks," in *Proceedings of the ACM Conference on Networked Sensor Systems (SenSys).* ACM Press, 2003, pp. 138–149.
7. I. Getting, "The Global Positioning System," *IEEE Spectrum*, December 1993.
8. Y.-C. Hu, A. Perrig, and D. B. Johnson, "Ariadne: a secure on-demand routing protocol for ad hoc networks," in *Proceedings of the ACM/IEEE International Conference on Mobile Computing and Networking (MobiCom).* ACM Press, September 2002, pp. 12–23.
9. C. Karlof, N. Sastry, and D. Wagner, "Tinysec: a link layer security architecture for wireless sensor networks," in *SenSys '04: Proceedings of the 2nd international conference on Embedded networked sensor systems.* New York, NY, USA: ACM Press, 2004, pp. 162–175.

10. M. Maroti, B. Kusy, G. Simon, and A. Ledeczi, "The flooding time synchronization protocol," in *SenSys '04: Proceedings of the 2nd international conference on Embedded networked sensor systems*. New York, NY, USA: ACM Press, 2004, pp. 39–49.

11. P. Papadimitratos and Z. J. Haas, "Secure Routing for Mobile Ad Hoc Networks," in *Proceedings of the Communication Networks and Distributed Systems Modeling and Simulation Conference (CNDS)*, January 2002.

12. M. Sichitiu and C. Veerarittiphan, "Simple, accurate time synchronization for wireless sensor networks," in *Proceedings of the IEEE Wireless Communications and Networking Conference (WCNC)*, 2003.

13. B. Sundararaman, U. Buy, and K. A, "Clock Synchronization for Wireless Sensor Networks: A Survey," Tech. Rep., March 2005.

14. B. Titzer, D. Lee, and J. Palsberg, "Avrora: Scalable sensor network simulation with precise timing," in *Proceedings of IPSN Track on Sensor Platform, Tools and Design Methods for Networked Embedded Systems*, 2005.

15. J. van Greunen and J. Rabaey, "Lightweight time synchronization for sensor networks," in *WSNA '03: Proceedings of the 2nd ACM international conference on Wireless sensor networks and applications*. New York, NY, USA: ACM Press, 2003, pp. 11–19.

16. R. Watro, D. Kong, S. Cuti, C. Gardiner, C. Lynn, and P. Kruus, "Tinypk: securing sensor networks with public key technology," in *SASN '04: Proceedings of the 2nd ACM workshop on Security of ad hoc and sensor networks*. New York, NY, USA: ACM Press, 2004, pp. 59–64.

17. W. Xu, W. Trappe, Y. Zhang, and T. Wood, "The Feasibility of Launching and Detecting Jamming Attacks in Wireless Networks," in *Proceedings of the ACM International Symposium on Mobile Ad Hoc Networking and Computing (MobiHoc)*, Urbana-Champaign, Illinois, USA, 2005.

18. A. Yaar, A. Perrig, and D. Song, "Pi: A path identification mechanism to defend against DDoS attacks," in *Proceedings of IEEE Symposium on Security and Privacy*, 2003.

Index